FOLIATIONS 2012

FOLIATIONS 2012

Proceedings of the International Conference
Łódź, Poland   25 – 30 June 2012

editors

Paweł Walczak
Uniwersytet Łódzki, Poland

Jesús Álvarez López
Universidade de Santiago de Compostela, Spain

Steven Hurder
University of Illinois at Chicago, USA

Rémi Langevin
Université de Bourgogne, France

Takashi Tsuboi
University of Tokyo, Japan

 **World Scientific**

NEW JERSEY · LONDON · SINGAPORE · BEIJING · SHANGHAI · HONG KONG · TAIPEI · CHENNAI

Published by

World Scientific Publishing Co. Pte. Ltd.

5 Toh Tuck Link, Singapore 596224

USA office: 27 Warren Street, Suite 401-402, Hackensack, NJ 07601

UK office: 57 Shelton Street, Covent Garden, London WC2H 9HE

British Library Cataloguing-in-Publication Data
A catalogue record for this book is available from the British Library.

FOLIATIONS 2012
Proceedings of the International Conference

ISBN 978-981-4556-85-9

Printed in Singapore

Preface

This volume contains a set of papers written by the participants (and their colleagues who wished but could not participate) of the international conference *Foliations 2012*, a satellite event of the 6th European Congress of Mathematics (Kraków, Poland, July 2–7, 2012) held at the Faculty of Mathematics and Computer Science of the University of Łódź (Poland) between June 25 and June 30, 2012. The conference was sponsored by the Faculty of Mathematics and Computer Science of the University of Łódź, the Japan Society for Promotion of Science, the National Science Center of Poland, Polish Academy of Sciences, the Stefan Banach International Mathematical Center in Warsaw and the City of Łódź. Participants numbered in approximately 60, about 40 of them from abroad (Brazil, Czech Republic, France, Germany, Israel, Japan, Romania, Russia, Spain and USA).

The organizing committee consisted of five people who became editors of this volume and Robert Wolak, a member of the Executive Organizing Committee of the 6ECM. They were supported by the staff of the Faculty of Mathematics and Computer Science and a group of young mathematicians and students from the University of Łódź.

The volume is organized similarly to that of *Foliations 2005* published by World Scientific Publ. Co. Pte. Ltd. in 2006 (ISBN 981-270-074-9). As therein, the papers contained in this volume are closely related to the lectures given at the conference, which was designed to cover various aspects of the theory of foliations, focusing on topology, geometry and dynamics of such objects.

All the papers contained in this volume were refereed by experts. Most of them are research papers containing original results, some contain surveys which bring light to the current state of some aspects of the foliation theory, some are of "mixed-type" (surveys with a bit of new results). The volume contains also a list of open problems presented during the conference at the problem session, then collected and prepared for publication by one of the editors of the volume. We hope that both, the conference itself and this volume of proceedings, should make a significant contribution to the progress of our field of science. We express our gratitude to the participants, the contributors of this volume, the sponsors and all the colleagues and students who helped us while organizing the conference and preparing the volume for publication. In particular, we would like to mention Marek Badura (who organized the www-page of the conference as well as the participants data base), Maciej Czarnecki and Szymon Walczak (the secretaries of the organizing committee), and Zofia Walczak (who produced the final TeX-file of the volume).

The Editors

Foliations 2012

Łódź, June 25–30, 2012, Poland

6th European Congress of Mathematics satellite conference

Organizing Committees

Scientific committee

Jesús A. Álvarez López	– Universidade de Santiago de Compostela, Spain
Steven Hurder	– University of Illinois at Chicago, United States
Rémi Langevin	– Université de Bourgogne, France
Takashi Tsuboi	– University of Tokyo, Japan
Paweł Walczak	– Uniwersytet Łódzki, Poland
Robert Wolak	– Uniwersytet Jagielloński

Local organizers

Marek Badura	– Uniwersytet Łódzki, Poland
Maciej Czarnecki	– Uniwersytet Łódzki, Poland
Paweł G. Walczak	– Uniwersytet Łódzki, Poland
Szymon M. Walczak	– Uniwersytet Łódzki, Poland
Zofia Walczak	– Uniwersytet Łódzki, Poland

Contents

FOLIATIONS 2012
ed. by Paweł WALCZAK *et al.*
World Scientific, Singapore, 2013
pp. 1–8

Characterization of the uniform perfectness of diffeomorphism groups preserving a submanifold*

KŌJUN ABE

Department of Mathematical Sciences
Shinshu University, Matsumoto, Japan
e-mail: kojnabe shinshu-u.ac.jp

KAZUHIKO FUKUI

Department of Mathematics
Kyoto Sangyo University, Kyoto, Japan
e-mail: fukui@cc.kyoto-su.ac.jp

1. Introduction and statement of results

In [2] we studied the conditions for the diffeomorphism groups of manifolds preserving a submanifold to be uniformly perfect. In this paper, applying the results we characterize the uniformly perfectness of the diffeomorphism groups of surfaces preserving a union of circles, and some extension to the cases of manifolds of higher dimensions.

Let M be a connected C^∞-manifold without boundary and let $D_c^\infty(M)$ denote the group of all C^∞-diffeomorphisms of M which are isotopic to the identity through C^∞-diffeomorphisms with compact support. It is known that M. Herman [6] and W. Thurston [11] proved $D_c^\infty(M)$ is perfect, which

---

*This research was partially supported by Grant-in-Aid for Scientific Research (No. 23540111), Japan Society for the Promotion of Science

means that every element of $D_c^\infty(M)$ is represented by a product of commutators.

Let (M, N) be a pair of manifolds such that N is a submanifold of M and let $D_c^\infty(M, N)$ be the group of C^∞-diffeomorphisms of M preserving N which are isotopic to the identity through compactly supported C^∞-diffeomorphisms preserving N. In [1], we proved that the group $D_c^\infty(M, N)$ is perfect if the dimension of N is positive. In [2], we studied the conditions for $D_c^\infty(M, N)$ to be uniformly perfect. Here a group G is said to be *uniformly perfect* if each element of G is represented as a product of a bounded number of commutators of elements in G.

Let $p : D^\infty(M, N) \to D^\infty(N)$ be the homomorphism given by the restriction. Then we proved the following.

Theorem A (Theorem 1.1 of [2]). *Let M be an m-dimensional compact manifold without boundary and N an n-dimensional C^∞-submanifold such that both groups $D_c^\infty(M - N)$ and $D^\infty(N)$ are uniformly perfect. If the connected components of $\ker p$ are finite, then $D^\infty(M, N)$ is a uniformly perfect group for $n \geq 1$.*

Theorem B (Theorem 1.2 of [2]). *Let M be an m-dimensional compact manifold without boundary and N be a disjoint uinion of circles in M. If the connected components of $\ker p$ are infinite, then $D^\infty(M, N)$ is not a uniformly perfect group.*

Applying Theorems A and B we shall determine the uniformly perfectness of the group $D^\infty(M, N)$.

First we consider the 2-dimensional case. Let M be an orientable closed surface and N a disjoint union of k circles in M. Then we have the following

Theorem 1. $D^\infty(M, N)$ *is uniformly perfect if and only if*

(1) $M = S^2$ and $k = 1$ and
(2) $M = T^2$, $k = 1$ and N represents a non-trivial element of $\pi_1(T^2)$.

Next we consider an extension of Theorem 1 to higher dimensions. Let (M_1, N) be a manifold pair as in Theorem 1. Let $M = M_1 \times M_2$ be the product manifold of M_1 and M_2. Take a base point y in M_2, $N \times \{y\}$ is identified with N. Then we have the following

Theorem 2. *If $D^\infty(M_1, N)$ and $D_c^\infty(M - N)$ are uniformly perfect, then $D^\infty(M, N)$ is uniformly perfect.*

Combining Theorem 1 with Theorem 2 we have the following

Corollary 3. *Let (M_1, N) be a manifold pair as in Theorem 1. Suppose that $M_1 = S^2$ and $k = 1$, or $M_1 = T^2$, $k = 1$ and N represents a non-trivial element of $\pi_1(T^2)$. If $D_c^\infty(M - N)$ is uniformly perfect, then $D^\infty(M, N)$ is uniformly perfect.*

Example of Corollary 3. Let $M_1 = S^2$, $k = 1$ and $M_2 = S^1$. Since $M - N$ is homotopy equivalent to $S^1 \vee S^1$, it follows from the result of Tsuboi [12] that $D_c^\infty(M - N)$ is uniformly perfect. Hence $D^\infty(M, N)$ is uniformly perfect.

Applying Theorem A, we have the following examples for product manifolds.

Examples. Let N be a closed manifold and $M = N \times [0, 1]$. Then $D_c^\infty(\text{int}M)$ is uniformly perfect. Applying Theorem A, $D^\infty(M, N)$ is uniformly perfect if $D^\infty(N)$ is uniformly perfect and the connected components of $\ker p$ are finite. The conditions that $D^\infty(N)$ is uniformly perfect and the connected components of $\ker p$ are finite are satisfied for the following manifolds:

(1) S^2, S^3 (S. Smale [28], A. Hatcher [18]),
(2) hyperbolic 3-manifolds (D. Gabai [17]) and
(3) elliptic 3-manifolds (refer to S. Hong-J. Kalliongis-D. McCullough-J. Rubinstein [8]).

2. Proof of Theorem 1

Let M be an orientable closed surface and N be a disjoint union of circles $S_1^1, \cdots, S_k^1$ in M. Let $p : D^\infty(M, N) \to D^\infty(N)$ be the restriction maps which is a fibration map (R. Palais [9]). Put $G = D^\infty(M, N)$.

Set $V = M - N$ which is a disjoint union of open surfaces $V_1, \cdots, V_\ell$. Let W_i be the compact surface with boundary adding the boundary circles to V_i. We call W_i a *block submanifold* if W_i is not homeomorphic to the disk.

Let $D^\infty(M \text{ rel} N)$ be the subgroup of $D^\infty(M, N)$ consisting of diffeomorphisms which are the identity on N. Note that this group coincides with $\ker p$.

(i) The case of $M = S^2$ and $k = 1$.

In this case $D^\infty(S^2 \text{ rel} S^1)$ is connected by the theorem of Smale [28]. Since any element of $D_c^\infty(S^2 - S^1)$ and $D^\infty(S^1)$ can be written by a product of at most two commutators of elements in $D_c^\infty(S^2 - S^1)$ and $D^\infty(S^1)$ respectively (T. Tsuboi [12], M. Herman [7]), any element of G can be written by a product of at most four commutators of elements in G from Theorem A. Hence G is uniformly perfect.

(ii) The case of $M = T^2$, $k = 1$ and N represents a non-trivial element of $\pi_1(T^2)$.

In this case W is a compact annulus since $V = T^2 - N$ is an open annulus. Thus $D^\infty(T^2 \operatorname{rel} N)$ is identified with $D^\infty(W \operatorname{rel} \partial W)$, which is the group of diffeomorphisms of W isotopic to the identity through diffeomorphisms fixing ∂W pointwisely. Hence $D^\infty(T^2 \operatorname{rel} N)$ is connected. Since any element of $D_c^\infty(V)$ and $D^\infty(S^1)$ can be written by a product of at most two commutators of elements in $D_c^\infty(V)$ and $D^\infty(S^1)$ respectively, any element of G can be written by a product of at most four commutators of elements in G from Theorem A. Hence G is uniformly perfect from Theorem A.

Then (1) follows from (i) and (ii).

We need the following lemma to prove (2). Let $r_i : D^\infty(W_i, \partial W_i) \to D^\infty(\partial W_i)$ be the restriction map which is a fibration map ([9]).

Lemma 4. *Let W_i be a block manifold.*

(1) If W_i is a compact annulus, then $\pi_0(\ker r_i) \cong \mathbf{Z}$.

(2) If W_i is not a compact annulus, then $\pi_0(\ker r_i) \cong \pi_1(D^\infty(\partial W_i)) \cong \overbrace{\mathbf{Z} \oplus \cdots \oplus \mathbf{Z}}^{s}$, where s is the number of components of ∂W_i.

Proof. If W_i is a compact annulus, then $\pi_0(\ker r_i)$ is isomorphic to $\mathbf{Z}$ which is generated by an element of $\pi_0(\ker r_i)$ obtained from the operation by Dehn twist along one of the boundary.

Assume that W_i is not a compact annulus. Then, from the result of C. Earle and A. Schatz [3], $D^\infty(W_i, \partial W_i)$ is contractible. It follows by the homotopy exact sequence that $\partial : \pi_1(D^\infty(\partial W_i)) \to \pi_0(\ker r_i)$ is isomorphic. Each element of $\pi_0(\ker r_i)$ is produced by an element of $\pi_0(\ker r_i)$ obtained from the operation by certain Dehn twists along each component of the boundary. Then Lemma 4 follows.

(iii) The case of $M = S^2$ and $k = 2$.

In this case, the two connected components S_1^1 and S_2^1 bound an annulus C in S^2. Then we construct a diffeomorphism obtained from the operation by Dehn twist along one boundary component of C and extend the diffeomorphism by defining to be the identity outside of C. We denote the extended diffeomorphism of S^2 by h. Then h is an element of G, and the iteration of h generates elements of infinite order in $\pi_0(D^\infty(S^2 \operatorname{rel} N))$ from Lemma 4 (1). Therefore G is not uniformly perfect from Theorem B.

(iv) The case of $M = S^2$ and $k \geq 3$.

$(iv - a)$ When all connected components of N are parallel, neighboring two components bound an annulus C. Then we can argue similarly as in (iii).

$(iv - b)$ We consider the case that there is a component of N not parallel to other components. Take a component L_1 bounding a disk. Then there are components L_i $(i = 1, 2, \cdots, \ell)$ and K_1, K_2 of N satisfying that L_i $(i = 1, 2, \cdots, \ell)$ are parallel to each other and K_1, K_2 are not parallel to each L_i. That is, L_ℓ, K_1 and K_2 bound a compact surface P of pants type. If $\ell > 1$, we have an annulus such that one of boundary bounds a disk. Then we can argue similarly as in (iii). Therefore we have only to consider a compact surface P of pants type with three boundaries L_1, K_1 and K_2, which bound disks respectively. Then we construct a diffeomorphism of P, say h, obtained from the operation by Dehn twist along L_1 and extend to S^2 by defining to be the identity on $S^2 - P$. We denote the extended diffeomorphism of S^2 by the same letter h. Then h is in $D^\infty(P \operatorname{rel} \partial P)$, and we see from Lemma 4 (2) that it does not belong to the identity connected component, and furthermore the iteration of h generates elements of infinite order in $\pi_0(D^\infty(P \operatorname{rel} \partial P))$. Therefore G is not uniformly perfect from Theorem B.

(v) The case of $M = T^2$, $k = 1$ and N represents a trivial element of $\pi_1(T^2)$.

In this case, we obtain a compact surface P of genus one with boundary N by removing the interior of the disk with boundary N. Then by the same way as that in the case (iv), we construct a diffeomorphism in G generating elements of infinite order in $\pi_0(D^\infty(T^2 \operatorname{rel} N))$. Therefore G is not uniformly perfect from Theorem B.

(vi) The case of $M = T^2$ and $k \geq 2$.

$(vi - a)$ First we consider the case that a connected component L of N represents a trivial element of $\pi_1(T^2)$. Then if the disk with boundary L contains a connected component of N in its interior, there is another component, say L_0, of N satisfying that the disk with L_0 as its boundary does not contain any components of N. If the disk with boundary L contains no connected components of N, put $L_0 = L$. We obtain a compact surface of genus one with L_0 as its boundary from T^2 by removing the interior of the disk. Thus by the same way as in that of $(iv - b)$, we have an element h of G, and the iteration of h generates elements of infinite order in $\pi_0(D^\infty(T^2 \operatorname{rel} N))$. Therefore G is not uniformly perfect from Theorem B.

$(vi - b)$ When every connected component of N represents a non-trivial element of $\pi_1(T^2)$, neighboring components bound an annulus. Then by the same way as that in $(iv - a)$, we have an element h of G, and the iteration of

h generates elements of infinite order in $\pi_0(D^\infty(T^2 \, \mathrm{rel} N))$. Therefore G is not uniformly perfect from Theorem B.

(vii) The case of the genus of $M \geq 2$.

($vii - a$) When a component of N represents a trivial element of $\pi_1(M)$, by the same way as that in ($vi - a$), we have an element h of G, and the iteration of h generates elements of infinite order in $\pi_0(D^\infty(M \, \mathrm{rel} N))$. Therefore G is not uniformly perfect from Theorem B.

($vii - b$) We consider the case of that every component of N represents a non-trivial element of $\pi_1(M)$.

($b-1$) We consider the case that there is a component L of N separating M. Put $M = W_1 \cup_L W_2$, where W_i ($i = 1, 2$) are compact submanifolds of M. Note that W_1 and W_2 are compact surfaces of genus ≥ 1. Let h be a diffeomorphism of M obtained by rotating one time along L satisfying that the support of h is contained in a small neighborhood U of L. Note that the restriction of h to L is the identity and h is in G. Then h is not in the identity component of $D^\infty(M \, \mathrm{rel} N)$ from Lemma 4 (2) since the restriction of h to W_i is a diffeomorphism obtained from the operation by Dehn twist along L. Furthermore the iteration of h generates elements of infinite order in $\pi_0(D^\infty(M \, \mathrm{rel} N))$. Therefore G is not uniformly perfect from Theorem B.

($b - 2$) Next we consider the case that every component of N does not separate M. In this case, let h be a diffeomorphism of M obtained by rotating one time along a component L satisfying that the support of h is contained in a small neighborhood U of L. Note that the restriction of h to L is the identity and h is in G. Then $V = M - L$ produces the block manifold W with two boundaries L_1 and L_2, and h produces the diffeomorphism $\tilde{h}$ of W obtained from the operation by Dehn twist along L_1 and L_2. Thus h is not in the identity component of $D^\infty(M \, \mathrm{rel} N)$ from Lemma 4 (2). Furthermore the iteration of h generates elements of infinite order in $\pi_0(D^\infty(M \, \mathrm{rel} N))$. Therefore G is not uniformly perfect from Theorem B.

Then (2) follows from (iii), (iv), (v), (vi) and (vii). This completes the proof.   □

3. Proof of Theorem 2

Let $M = M_1 \times M_2$ be the product manifold of a closed orientable surface M_1 and a manifold M_2, and N be a union of circles in M_1. Then for a fixed point y in M_2, $N \times \{y\}$ is identified with N. Let $p_1 : D^\infty(M_1, N) \to D^\infty(N)$

and $p : D^\infty(M, N) \to D^\infty(N)$ be the restriction maps which are fibration maps. Then we consider the commutative diagram of the homotopy exact sequences of the fibrations $p_1 : D^\infty(M_1, N) \to D^\infty(N)$ and $p : D^\infty(M, N) \to D^\infty(N)$;

$$
\begin{array}{ccccccc}
\pi_1(D^\infty(M_1, N)) & \xrightarrow{(p_1)_*} & \pi_1(D^\infty(N)) & \xrightarrow{\partial_0} & \pi_0(\ker p_1) & \longrightarrow & 1 \\
\downarrow{i_*} & & \downarrow{=} & & \downarrow{i_*} & & \\
\pi_1(D^\infty(M_1 \times M_2, N)) & \xrightarrow{p_*} & \pi_1(D^\infty(N)) & \xrightarrow{\partial_0} & \pi_0(\ker p) & \longrightarrow & 1,
\end{array}
$$

where $i : D^\infty(M_1, N) \to D^\infty(M_1 \times M_2, N)$ is the inclusion defined by $i(f) = f \times 1_{M_2}$ for $f \in D^\infty(M_1, N)$.

Then since ∂_0 is surjective, $i_* : \pi_0(\ker p_1) \to \pi_0(\ker p)$ is also surjective. Since $D^\infty(M_1, N)$ is uniformly perfect from the assumption, then the connected components of $\ker p_1$ are finite from Theorems A and B. Thus we have from the above fact that the connected components of $\ker p$ are finite. By using Theorem A again, $D^\infty(M, N)$ is uniformly perfect. This completes the proof.

References

[1] K. Abe and K. Fukui, *Commutators of C^∞-diffeomorphisms preserving a submanifold*, J. Math. Soc. Japan., **61**(2) (2009), 427–436.

[2] K. Abe and K. Fukui, *Erratum and addendum to "Commutators of C^∞-diffeomorphisms preserving a submanifold"*, to appear in J. Math. Soc. Japan.

[3] C.E. Earle and A. Schatz, *Teichmuller theory for surfaces with boundary*, J. Diff. Geom., **4** (1970), 169–185.

[4] D. Gabai, *The Smale conjecture for hyperbolic 3-manifolds: Isom$(M^3) \simeq$ Diff(M^3)*, J. Diff. Geom., **58**(1) (2001), 113–149.

[5] A. Hatcher, *A proof of the Smale conjecture*, Diff$(S^3) \simeq$ O(3), Ann. of Math., **117**(2) (1983), 553–607.

[6] M. Herman, *Simplicité du groupe des difféomorphismes de classe C^∞, isotopes à l'identité, du tore de dimension n*, C. R. Acad. Sci. Paris, **273** (1971), 232–234.

[7] M.R. Herman, *Sur la conjugaison differentiable des diffeomorphismes du cercle à des rotations*, Inst. Hautes Études Sci. Publ. Math., **49** (1979), 5–233.

[8] S. Hong, J. Kalliongis, D. McCullough and J. Rubinstein, Diffeomorphisms of Elliptic 3-Manifolds, Lecture Notes in Mathematics 2055, Springer.

[9] R.S. Palais, *Local triviality of the restriction map for embeddings*, Comment. Math. Helv., **34** (1960), 305–312.

[10] S. Smale, *Diffeomorphisms of the 2-sphere*, Proc. A.M.S., **10** (1959), 621–626.

[11] W. Thurston, *Foliations and groups of diffeomorphisms*, Bull. Amer. Math. Soc., **80** (1974), 304–307.

[12] T. Tsuboi, *On the uniform perfectness of diffeomorphism groups*, Advanced Studies in Pure Math., **52** 2008, Groups of diffeomorphisms, 505–524.

[13] T. Tsuboi, *On the uniform perfectness of the groups of diffeomorphisms of even-dimensional manifolds*, Comment. Math. Helv., **87** (2012), 141–185.

Received December 1, 2012.

FOLIATIONS 2012
ed. by Paweł WALCZAK *et al.*
World Scientific, Singapore, 2013
pp. 9–21

Recent progress in geometric foliation theory

MAREK BADURA

Katedra Geometrii, Wydział Matematyki i Informatyki
Uniwersytet Łódzki, Łódź, Poland
e-mail: marekbad@math.uni.lodz.pl

MACIEJ CZARNECKI

Katedra Geometrii, Wydział Matematyki i Informatyki
Uniwersytet Łódzki, Łódź, Poland
e-mail: maczar@math.uni.lodz.pl

1. Introduction

We shall try to give some short review of new results in this part of the theory of foliations which are more closely related to geometry than to analysis or topology.

We present some theorems from papers that have appeared not earlier than 2005. The survey [15] by Paweł Walczak and the second author covers a large part of this topics. Some of the papers mentioned there were finally published and problems studied more deeply, so the intersection is obviously nonempty.

In the first section we collect some facts on existence of foliations under assumptions on the carrying manifold, and some special conditions on leaves or their relationships (e.g. Riemannian, homogeneous). The next two sections are devoted to the most popular geometric properties coming from the second fundamental form. Next we turn to conformal methods applied to the theory of foliation. Finally some aspects of dynamics closely connected to geometry are presented.

2. Existence of foliations of prescribed properties

Generally, a sphere does not admit codimension 1 foliations by spheres. This condition was weakened by Caminha, Souza and Camargo. The theory of harmonic functions appplied to graphs of functions allowed them to prove that even sufficiently large scalar curvature is an obstruction for foliating $\mathbb{S}^n$.

Theorem 1 ([10]). *There is no smooth transversely orientable foliation of codimension 1 on $\mathbb{S}^n$, whose leaves are complete and have constant scalar curvature > 1.*

Only some homeomorphic types of 3–manifolds are predicted to carry foliations of nonnegatively curved leaves. Bolotov recognized them among those admitting Seifert foliations (i.e. by circles).

Theorem 2 ([8]). *Assume that M is a 3–dimensional Riemannian manifold and $\mathcal{F}$ is a smooth transversely oriented codimension 1 foliation with all the leaves of nonnegative curvature. Then M is homeomorphic to one of the following:*

(1) a toric bundle over a circle
(2) a lens space $L_{p/q}$
(3) $\mathbb{S}^2 \times \mathbb{S}^1$
(4) the connnected sum of two real projective 3–spaces $\mathbb{R}P^3$
(5) a toric semi–bundle i.e. gluing together two copies of oriented segment bundles over the Klein bottle
(6) a prism manifold i.e. quotient of the sphere by a free action of finite group of isometries.

A foliation $\mathcal{F}$ on a Riemannian manifold M is called *homogeneous* if the subgroup of $\mathrm{Isom}\,(M)$ consisting of all isometries preserving $\mathcal{F}$ acts transitively on each leaf of $\mathcal{F}$. In the paper [6] Berndt and Tamaru proved that there are exactly two isometric congruency classes of homogeneous codimension one foliations on hyperbolic spaces $\mathbb{F}H^n$ where $\mathbb{F}$ is the field of real, complex or quaternions for $n \geq 2$ or octonions (Cayley numbers) for $n = 2$. One of these classes contains a horosphere foliation.

These two authors together with Díaz Ramos classified in [7] hyperpolar homogeneous foliations on symmetric spaces of non-compact type. A foliation is *hyperpolar* if a connected closed flat submanifold of the carrying manifold intersects each leaf orthogonally.

Theorem 3 ([7]). *Every hyperpolar homogeneous foliation on a connected Riemannian symmetric space M of non-compact type is isometrically con-*

gruent to some product of foliations containing an affine product foliation of some Euclidean space and a foliation naturally coming from a subset of simple roots in the Lie algebra of the group G where $M = G/K$.

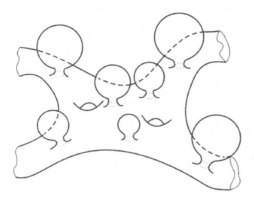

Figure 1.   A manifold which is not quasi–isometric to a leaf.

An old problem on (non)existence of foliations with a given leaf was elegantly and ultimately solved by Schweitzer in codimension 1 and class $C^{2,0}$ for all non-compact manifolds.

We say that a Riemannian manifold is of *bounded geometry* if there are lower and upper bounds for sectional curvature as well as an lower bound for injectivity radius.

Theorem 4 ([28]). *On any connected non-compact smooth manifold L of dimension ≥ 2 there is a Riemannian metric of bounded geometry such that L is not quasi–isometric to a leaf of a codimension 1 foliation of class $C^{2,0}$ on any compact manifold. Moreover, there are uncountably many quasi–isometric classes of Riemannian metrics on L which satisfy this negative condition.*

Figure 1 shows balloons of increasing volume which make a manifold being not quasi–isometric to a leaf.

There exist some results on nonexisting of Riemannian foliations on compact manifolds of negative curvature (see [15]). Under some additional assumption even singular Riemannian foliations do not exist in that case.

We say that a *singular* foliation is *Riemannian* if any geodesic starting orthogonally to a leaf is orthogonal to the leaves it meet. Such a foliation *admits sections* if for any point of any leaf of maximal dimension there is isometrically immersed complete totally geodesic submanifold meeting all the leaves orthogonally.

Theorem 5 ([29]). *There is no singular Riemannian foliation with sections on a compact manifold of negative curvature.*

For hyperbolic 3–manifolds Fenley obtained a property implying continuous extension of a codimension 1 foliation to the ideal boundary in the universal cover. In [16] he proved that almost transversality to a quasi-geodesic pseudo–Anosov flow is a sufficient condition for a foliation on an atoroidal closed 3–manifold to have the above extension.

3. Totally geodesic foliations

A submanifold is called *totally geodesic* if in any of its points the shape operator is zero. We say that a foliation is totally geodesic if all of its leaves are totally geodesic submanifolds.

We start from the simplest manifolds. In the real hyperbolic space $\mathbb{H}^n$ all the codimension 1 totally geodesic submanifolds are those isometric to $\mathbb{H}^{n-1}$. In the ball model D^n they are parts of spheres orthogonal to the ideal boundary $\mathbb{H}^n(\infty) \simeq S^{n-1}$.

Lee and Yi in [22] proposed the following boundary classification of totally geodesic codimension 1 foliations on $\mathbb{H}^n$. Fix a point $0 \in D^n$ and to any totally geodesic submanifold L assign a point on the ideal boundary which is the end of the geodesic ray starting at 0 perpendicularly to L (orientation of the ray is determined by the orientation of the leaf). Any continuous family of disjoint codimension 1 totally geodesic submanifolds of $\mathbb{H}^n$ could be parameterized on an open interval, say $(0, \pi)$. We define a function $z : (0, \pi) \to S^{n-1}$ called *arc–center* mapping a submanifold to the point on the ideal boundary as described above.

Theorem 6 ([22]). *The set of all codimension 1 totally geodesic foliations of class C^k on $\mathbb{H}^n$, with foliations orthogonal to horocycles excluded, is in one-to-one correspondence (up to isometry) with the set of all functions $z : [0, \pi] \to S^{n-1}$ of class C^{k-1} with $z(0) = e_1 = z(\pi)$ satisfying $|z'(r)| \leq 1$ for all r, modulo the action of $O(n-1) \times \mathbb{R} \times \mathbb{Z}_2$.*

On a Riemannian manifold (M, g) foliated by $\mathcal{F}$ the tensor g is called *bundle-like metric* with respect to $\mathcal{F}$ if any geodesic on M which is tangent to the normal distribution $\mathcal{F}^\perp$ at one point remains tangent to $\mathcal{F}^\perp$ at other points.

Computing connections, Bejancu and Farran proved that totally geodesic foliations on Riemannian manifolds with bundle-like metric under some curvature condition exist only in sufficiently large codimension

Theorem 7 ([5]). *Assume that $\mathcal{F}$ is a totally geodesic n–foliation of an $(n+p)$–dimensional Riemannian manifold (M, g), that the metric is bundle-like and that mixed sectional curvatures (i.e. in direction of a tangent and a normal vector to $\mathcal{F}$) do not vanish at some point. Then $n \leq p - 1$.*

A strong restriction on the nature of a manifold with a totally geodesic foliation and a bundle-like metric were found by Quiroga-Barranco. In some sense the only such manifolds are locally symmetric spaces of non-compact.

Theorem 8 ([27]). *Let $(M, \mathcal{F})$ be a foliated manifold with a finite volume complete bundle-like Riemannian metric h for which $\mathcal{F}$ is totally geodesic and the leaves are isometrically covered by a homogeneous space X modeled on a simple Lie group G of real rank ≥ 2. Assume that $\mathcal{F}$ has a dense leaf.*

Then there is a homogenenous Riemannian manifold Y covered by the maximal compact subgroup $H \subset G$ for which the universal cover $\widetilde{M}$ with the lifted metric $\tilde{h}$ is isometric to $X \times Y$ and an irreducible arithmetic lattice $\Gamma \subset G \times H$ such that $\Gamma \backslash \widetilde{M}$ is the finite covering of M while lifted foliation $\widetilde{\mathcal{F}}$ comes from natural projection $X \times Y \to Y$.

Thus up to fibrations with compact fibers, M is an irreducible locally symmetric space of non-compact type.

Another step towards classification of Riemannian submersion was made by Munteanu and Tapp. They used a special lift of geodesics to study Riemannian submersions on locally symmetric spaces.

Let (M, g) be a Riemannian manifold and $p \in M$. We say that a triple of vectors $X, V, W \in T_p(M)$ is *good* if given a Jacobi field $X(t)$ along a geodesic starting from p in the direction V such that $X(0) = X$ and $X'(0) = W$ and an analogous Jacobi field $V(s)$ along the geodesic starting from p in direction X such that $V(0) = V$ and $V'(0) = W$ the following condition holds

$$\exp(tV(s)) = \exp(sX(t)).$$

A *doubly ruled surface* is then the parameterized surface $(s, t) \mapsto \exp(tV(s))$.

Theorem 9 ([25]). *If (X, V, W) is a good triple in a locally symmetric space M, then $R(X, V)W = 0$, where R denotes the curvature tensor of M.*

For the Lorentzian 3–manifolds (i.e. semi–Riemannian with the metric of Lorentz type) Boubel, Mounoud and Tarquini in [9] gave a classification of time-like geodesically complete totally geodesic codimension 1 foliations showing that they are circle bundles over tori or foliations of 3–dimensional tori, or the 3-dimensional torus with some exclusions.

4. Umbilical and constant mean curvature foliations

A submanifold is called *totally umbilical* if in any of its points the shape operator is a homothety i.e. it is proportional to identity. If the mean curvature function H is constant over a submanifold we refer to it as *CMC submanifold*. We say that a foliation is umbilical (resp. CMC) if all of its leaves are totally umbilical (resp. CMC) submanifolds.

In [14] Lużyńczyk and the second author found a condition for the mean curvature of umbilical foliations orthogonal to a geodesics on $\mathbb{H}^n$. In $\mathbb{H}^n$ the only totally umbilical hypersurfaces are geodesic spheres, totally geodesic hypersurfaces, hyperspheres (equidistant from totally geodesic) and horospheres. Only the last three could serve as leaves of nonsingular foliations.

Theorem 10 ([14]). *Consider an umbilical codimension 1 foliation of $\mathbb{H}^n$ which is orthogonal to an arc–length parameterized geodesic line $\gamma : \mathbb{R} \to \mathbb{H}^n$. Denote by $h(t)$ the (constant) mean curvature of the leaf L_t through $\gamma(t)$ with sign of $h(t)$ depending on whether the point $\gamma(t)$ lies the positive side of the totally geodesic hypersurface for which L_t is equidistant or not, in the orientation given by the geodesic γ. Then*

(1) $-1 \leq h \leq 1$
(2) if $h(t_0) = -1$ (resp. 1) then h is constantly equal -1 (resp. 1) for all $t < t_0$ (resp. $t > t_0$)
(3) the function $2 (\tanh)^{-1} \circ h$ is non-expanding i.e. $(\tanh)^{-1} \circ h$ is $\frac{1}{2}$-Lipschitz.

Moreover, if a function $h : \mathbb{R} \to \mathbb{R}$ satisfies conditions (1)–(3) then the family of totally umbilical hypersurfaces in $\mathbb{H}^n$ orthogonal to an arc–length parameterized geodesic line of the mean curvature prescribed by h form an umbilical foliation of $\mathbb{H}^n$.

A kind of rigidity of umbilicity on manifolds of constant non-positive cuvature was proved by Gomes.

Theorem 11 ([17]). *Let $\mathcal{F}$ be a Riemannian codimension 1 foliation on* *a connected complete manifold of constant non-positive curvature. Then if $\mathcal{F}$ has a totally umbilical leaf then $\mathcal{F}$ is an umbilical foliation.*

In [15], a computational method by Chopp and Velling to obtain CMC foliations on $\mathbb{H}^3$ sharing common ideal boundary was announced. Coskunuzer proved lastly generalized this result to any dimension.

Theorem 12 ([11]). *Let Γ be the boundary of a star–shaped $C^{1,1}$ domain in S^n. For $H \in (-1, 1)$ we denote by L_H a hypersurface in $\mathbb{H}^{n+1}$ with the ideal boundary Γ and constant mean curvature H. Then the family $\{L_H : H \in (-1, 1)\}$ forms a CMC foliation of $\mathbb{H}^{n+1}$.*

CMC foliations on hyperbolic spacetimes modelled on the anti–de Sitter 3–space were described by Barbot, Béguin and Zeghib in [2] (cf. [15] for formulation).

For asymptotically flat 3–manifolds, Quing and Tian in [19] and Ma in [23] showed (under very technical assumptions) that, outside of some compact set, such manifolds could be foliated by constant mean curvature spheres. In another paper [26] by Neves and Tian, similar results for asymptotically flat 3–manifolds were obtained. Metzger showed that in case of flat asymptoticity prescribing of mean curvature is possible (see [24]).

The existence of some special vector field allowed Colares and Palmas in the paper [12] to foliate a $(n + 1)$–dimensional Lorentzian manifold by leaves which are almost umbilical hypersurfaces, i.e. having $n - 1$ constant and equal principal curvatures.

5. Conformal geometry of foliations

In the space $\mathbb{R}^{n+1}$ consider *the Lorentz form* given by

$$\langle x|y \rangle = -x_0 y_0 + x_1 y_1 + \ldots + x_n y_n,$$

the light cone $\mathcal{L}$ consisting of isotropic vectors, and *the de Sitter space* $\Lambda^n = \{x : \langle x|x \rangle = 1\}$.

Any oriented $(n - 2)$-sphere Σ in $\mathbb{S}^{n-1}$ is represented by a point σ of Λ^n. This one-to-one correspondence assigns to any σ an intersection of the vector hyperplane $\sigma^\perp$ with the sphere $\mathbb{S}^{n-1}$ which is a light cone's section at level $x_0 = 1$. Thus families of spheres are objects in Λ^n. Details could found in [20].

Observe that spheres $\Sigma_1, \Sigma_2 \subset \mathbb{S}^{n-1}$ are disjoint iff $|\langle \sigma_1 | \sigma_2 \rangle| > 1$.

An $(n-1)$-ball in $\mathbb{S}^{n-1}$ of boundary Σ_0 is a model of the hyperbolic space $\mathbb{H}^{n-1}$. Any totally geodesic hypersurface in $\mathbb{H}^{n-1}$ is a sphere orthogonal to

Σ_0 hence it is uniquely determined by its ideal boundary which is an $(n-3)$-sphere included in Σ_0. Thus the set of totally geodesic hypersurfaces of $\mathbb{H}^{n-1}$ in this model could be identified with the $(n-1)$-dimensional de Sitter space

$$\Lambda_{\Sigma_0}^{n-1} = \{\sigma \in \Lambda^n \; : \; \langle \sigma | \sigma_0 \rangle = 0\}.$$

Langevin and the second author proved that totally geodesic hypersurfaces foliate the hyperbolic space iff a curve representing them in de Sitter space is time-or-light–like.

Theorem 13 ([13]). *For an $(n-2)$-dimensional sphere $\Sigma_0 \subset S^{n-1}$, any unbounded curve $\Gamma : \mathbb{R} \to \Lambda_{\Sigma_0}^{n-1}$ such that*

$$|\langle \Gamma(t_1) | \Gamma(t_2) \rangle| \geq 1 \quad \text{for all } t_1, t_2 \in \mathbb{R}$$

represents a totally geodesic codimension 1 foliation of $\mathbb{H}^{n-1}$.

Moreover, any such foliation is transversely oriented and is represented by a curve Γ as above.

If the curve Γ is differentiable, the condition $|\langle \Gamma(t_1) | \Gamma(t_2) \rangle| \geq 1$ gives $\langle \Gamma' | \Gamma' \rangle \leq 0$, so at any point the curve Γ is locally contained in the closed time cone. For umbilical foliations of $\mathbb{H}^{n-1}$, the construction is more complicated — spheres representing leaves could intersect outside of the ball model which lead to the construction of local boosted time cones (cf. [13]).

Langevin and Walczak in [20] successfully developed and applied conformal methods obtaining interesting results. In particular, they proved that there are no umbilical codimension 1 foliations on a closed Riemannian manifold provided that its Ricci curvature is bounded from above by a negative constant.

Theorem 14 ([20]). *Let (M, g) be a closed Riemannian n-manifold with $\mathrm{Ric}_M \leq c < 0$ and let $\mathcal{F}$ be a codimension 1 transversely oriented foliation on M. Denote by $k_1, \ldots, k_{n-1}$ principal curvatures of leaves. Then*

$$\int_M \sum_{i<j} |k_i - k_j|^n \, d\mathrm{vol} \geq \binom{n}{2}^{\frac{2-n}{2}} (-2c)^{\frac{n}{2}} \mathrm{vol}\,(M).$$

Now we shall describe some conformally defined objects on $\mathbb{S}^3$. A *canal surface* is an envelope of 1-parameter family of 2-spheres. A *Dupin cyclide* is a surface which is a common envelope of two 1-parameter families of 2-spheres.

For a regular surface $S \in \mathbb{R}^3 \subset \mathbb{S}^3$, we define its principal curvatures k_1, k_2, *conformal principal curvatures* θ_1, θ_2 as conveniently normalized derivatives of the $k_i's$ in the corresponding principal directions, and *the Bryant invariant* Ψ (see [3] for details). In these terms, a surface is canal iff one of the conformal principal curvatures vanishes identically, while it is a Dupin cyclide iff both θ_i's vanish (cf. [4]).

Canal surfaces, Dupin cyclides and their relatives having constant conformal invariant θ_1, θ_2 and Ψ (*CCI-surfaces* for short) are the most natural objects which are conformally defined. So the question of (non)existence of foliations with such leaves is natural as well.

Theorem 15 ([20]). *There are no foliations with all leaves being Dupin cyclides either on $\mathbb{S}^3$ or on any compact quotient $\mathbb{H}^3/G$, where G is some group of Möbius transformations.*

Theorem 16 ([21]). *Any foliation of $\mathbb{S}^3$ by canal surfaces is either a Reeb foliation or is obtained from such a Reeb foliation inserting a zone $\mathbb{T}^2 \times [0, 1]$ filled by a union of toral and cylindrical leaves.*

Theorem 17 ([3]). *There are no foliations by CCI-surfaces on compact 3-manifolds of constant positive (resp. negative) curvature.*

6. Geometric dynamics of foliations

The first author in [1] has showed that a growth type satisfying simple conditions can be realized as a growth type of a leaf of a compact foliated manifold. It was shown how to obtain leaves of a given growth type on some compact foliated manifold.

Let us recall the notion of growth type. Let $\mathcal{I}$ be the set of nonnegative nondecreasing functions on $\mathbb{N}$. Let $f, h \in \mathcal{I}$. We say that h dominates f $(f \preceq h)$ if and only if for some $A \in \mathbb{R}_+$ and $B \in \mathbb{N}$, $f(n) \leq Ah(Bn)$ for any $n \in \mathbb{N}$. The preorder $\preceq$ induces an equivalence relation $\simeq$ in $\mathcal{I}$: $f \simeq h \iff f \preceq h \preceq f$. The equivalence class of $f \in \mathcal{I}$ is called the *growth type* of f and is denoted by $[f]$.

The set of all growth types has the partial order $\leq$ induced by the preorder $\preceq$. If $f \preceq h$, then $[f] \leq [h]$ and if $f \preceq h \preceq f$, then $[f] = [h]$.

We say that η is the *derived growth type* of the growth type ξ if $[\Sigma f] = \xi$ for any $f \in \eta$, where $\Sigma f(n) = \sum_{k=1}^{n} f(k)$.

For example, $[\Sigma \exp] = [\exp]$, $[\Sigma 1] = [n]$, $[\Sigma n^k] = [n^{k+1}]$ (the polynomial growth type of degree k is the derived growth type of the polynomial growth type of degree $k + 1$).

Theorem 18 ([1]). *If a growth type ξ, $[0] < \xi \leq [\exp]$, has a derived growth type, then there exists a C^1-foliation $\mathcal{F}$ of a suitable compact Riemannian manifold M containing a leaf L which has the growth type $gr(L) = \xi$.*

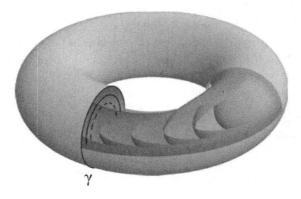

Figure 2.   Reeb component with a vanishing cycle γ.

Let $\mathcal{F}$ be a transversely orientable codimension one minimal foliation without vanishing cycles of a manifold M. In [30] Yokoyama has showed that if the fundamental group of each leaf of $\mathcal{F}$ has polynomial growth of degree k, then the foliation $\mathcal{F}$ is without holonomy.

A loop γ on the leaf $L \in \mathcal{F}$ is called *a vanishing cycle* if γ is not homotopically trivial in L and there is a mapping $F : S^1 \times [0,1] \to M$ such that arcs $F(x, [0,1])$ are transverse to $\mathcal{F}$ for every $x \in S^1$, each loop $F(S^1, t)$ is homotopically trivial in a leaf L_t for $t \in (0,1]$ and $F(S^1, 0)$ is homotopic to γ in L (see Figure 2).

Theorem 19 ([30]). *Let $\mathcal{F}$ a codimension one transversely orientable C^0 minimal foliation without vanishing cycles of a manifold M and k a non-negative integer. If all fundamental groups of the leaves of $\mathcal{F}$ have the same polynomial growth of degree k, then the foliation $\mathcal{F}$ is without holonomy.*

A foliation $\mathcal{F}$ is *quasi–isometric* if there is a constant $Q > 1$ such that for any two points $x, y \in M$ which lie on the same leaf of $\mathcal{F}$, $d_{\mathcal{F}}(x, y) \leq Q d_M(x, y) + Q$ where $d_{\mathcal{F}}$ is the distance measured along the leaf and d_M is the distance measured on the manifold.

A diffeomorphism $f : M \to M$ on a compact Riemannian manifold is *partially hyperbolic* (see [18]) if there are a Tf-invariant splitting, $TM =$

$E^s \oplus E^c \oplus E^u$, an integer $n \geq 1$, and constants $\lambda < 1 < \mu$ such that

$$||Tf^n v^s|| < \lambda < ||Tf^n v^c|| < \mu < ||Tf^n v^u||$$

for all $x \in M$, and unit vectors $v^s \in E^s(x), v^c \in E^c(x)$ and $v^u \in E^u(x)$.

For every partially hyperbolic system, there are unique foliations $\mathcal{F}^u$ and $\mathcal{F}^s$ tangent to the unstable E^u and stable E^s subbundles. In many but not all cases, there is also a foliation tangent to the centre subbundle E^c.

Hammerlindl in the paper [18] describes some properties of quasi-isometry for partially hyperbolic systems on a manifold.

Theorem 20 ([18]). *Suppose $f : M \to M$ is partially hyperbolic and the foliations $\mathcal{F}^u$, $\mathcal{F}^s$ and $\mathcal{F}^c$ are quasi-isometric when lifted to the universal cover, $\tilde{M}$. Then, the foliations $\mathcal{F}^{cs}$ and $\mathcal{F}^{cu}$ tangent to $E^c \oplus E^s$ and $E^c \oplus E^u$ are also quasi-isometric on $\tilde{M}$, and f has Global Product Structure, that is, for every $x, y \in \tilde{M}$:*

(1) $\mathcal{F}^u(x)$ and $\mathcal{F}^{cs}(y)$ intersect exactly once,

(2) $\mathcal{F}^s(x)$ and $\mathcal{F}^{cu}(y)$ intersect exactly once,

(3) if $x \in \mathcal{F}^{cs}(y)$, then $\mathcal{F}^c(x)$ and $\mathcal{F}^s(y)$ intersect exactly once, and

(4) if $x \in \mathcal{F}^{cu}(y)$, then $\mathcal{F}^c(x)$ and $\mathcal{F}^u(y)$ intersect exactly once.

Theorem 21 ([18]). *Suppose M is a compact connected manifold with an abelian fundamental group, $f : M \to M$ is partially hyperbolic with Global Product Structure.*

(1) If $\mathcal{F}^s$ and $\mathcal{F}^u$ are quasi-isometric on the universal cover, then any two leaves of $\mathcal{F}^c$ are homeomorphic.

(2) If $\mathcal{F}^u$ and $\mathcal{F}^s$ are quasi-isometric on the universal cover, then the foliations $\mathcal{F}^c$, $\mathcal{F}^{cs}$ and $\mathcal{F}^{cu}$ are without holonomy.

(3) If the centre foliation $\mathcal{F}^c$ exists and is without holonomy, then all leaves of $\mathcal{F}^c$ (on M) are homeomorphic.

(4) If $\dim E^s = \dim E^c = \dim E^u = 1$, and $\mathcal{F}^{cs}$ (or $\mathcal{F}^{cu}$) is without holonomy, then M is a circle bundle over $\mathbb{T}^2$.

References

[1] M. Badura, *Realizations of growth types*, Ergod. Th. & Dynam. Sys., **25**(2) (2005) 353–363. MR 2129100

[2] T. Barbot, F. Béguin and A. Zeghib, *Constant mean curvature foliations of globally hyperbolic spacetimes locally modelled on AdS₃*, Geom. Dedicata, **126** (2007), 71–129. MR 2328923

[3] A. Bartoszek, P.G. Walczak, *Foliations by surfaces of a peculiar class*, Ann.

Polon. Math. **94**(1) (2008), 89–95. MR2427507

[4] A. Bartoszek, R. Langevin and P.G. Walczak, *Special canal surfaces of* $\mathbb{S}^3$, Bull. Braz. Math. Soc. (N.S.), **42**(2) (2011), 301–320. MR 2833804

[5] A. Bejancu, H.R. Farran, *On totally geodesic foliations with bundle-like metric*, J. Geom., **85** (2006) 7–14. MR 2260127

[6] J. Berndt, H. Tamaru, *Homogeneous codimension one foliations on noncompact symmetric spaces*, J. Diff. Geom., **63**(1) (2003), 1–40. MR 2015258

[7] J. Berndt, J.C. Díaz-Ramos and H. Tamaru, *Hyperpolar homogeneous foliations on symmetric spaces of non-compact type*, J. Diff. Geom., **86**(2) (2010), 191–235. MR 2772550

[8] D.V. Bolotov, *Foliations of nonnegative curvature on closed 3-dimensional manifolds*, Mat. Sb., **200**(3) (2009), 3–16. MR 2529142

[9] C. Boubel, P. Mounoud and C. Tarquini, *Lorentzian foliations on 3-manifolds*, Ergod. Th. & Dynam. Sys., **26** (2006) 1339–1362. MR 2266364

[10] A. Caminha, P. Souza and F. Camargo, *Complete foliations of space forms by hypersurfaces*, Bull. Braz. Math. Soc. (N.S.), **41**(3) (2010), 339–353. MR 2718145

[11] B. Coskunuzer, *Foliations of hyperbolic space by constant mean curvature hypersurfaces*, Int. Math. Res. Not., **8** (2010), 1417–1431. MR 2628831

[12] A.G. Colares, O. Palmas, *Foliations by* $(n-1)$*–umbilical spacelike hypersurfaces*, Mat. Contemp., **34** (2008), 135–154. MR 2588610

[13] M. Czarnecki, R. Langevin, *Totally umbilical foliations on hyperbolic spaces*, in preparation.

[14] M. Czarnecki, M. Lużyńczyk, *Umbilical routes along gedesics in hyperbolic spaces*, in preparation.

[15] M. Czarnecki, P. Walczak, *Extrinsic geometry of foliations*, Foliations 2005, World Scientific 2006, 149–167. MR 2284780

[16] S.R. Fenley, *Geometry of foliations and flows. I. Almost transverse pseudo-Anosov flows and asymptotic behavior of foliations*, J. Diff. Geom., **81**(1) (2009), 1–89. MR 2477891

[17] A. Gomes, *Umbilical foliations on a Riemannian manifold*, Results Math., **52**(1–2) (2008), 51–53. MR 2430411

[18] A. Hammerlindl, *Dynamics of quasi-isometric foliations*, Nonlinearity, **25**(6) (2012), 1585–1599. MR 2924724

[19] J. Qing, G. Tian, *On the uniqueness of the foliation of spheres of constant mean curvature in asymptotically flat 3-manifolds*, J. Amer. Math. Soc., **20**(4) (2007), 1091–1110. MR 2328717

[20] R. Langevin, P.G. Walczak, *Conformal geometry of foliations*, Geom. Dedicata,**132** (2008), 135–178. MR 2396915

[21] R. Langevin, P.G. Walczak, *Canal foliations of* $\mathbb{S}^3$, J. Math. Soc. Japan, **64**(2) (2012), 659–682. MR 2916082

[22] K.B. Lee, S. Yi, *Metric foliations on hyperbolic spaces*, J. Korean Math. Soc., **48**(1) (2011), 63–82. MR 2778020

[23] S. Ma, *Uniqueness of the foliation of constant mean curvature spheres in asymptotically flat 3-manifolds*, Pacific J. Math., **252**(1) (2011), 145–179. MR 2862146

[24] J. Metzger, *Foliations of asymptotically flat 3-manifolds by 2-surfaces of prescribed mean curvature*, J. Diff. Geom. **77**(2) (2007), 201–236. MR 2355784

[25] M. Munteanu, K. Tapp, *On totally geodesic foliations and doubly ruled surfaces in a compact Lie group*, Proc. Amer. Math. Soc., **139**(11) (2011), 4121–4135. MR 2823057

[26] A. Neves, G. Tian, *Existence and uniqueness of constant mean curvature foliation of asymptotically hyperbolic 3-manifolds. II*, J. Reine Angew. Math., **641** (2010), 69–93. MR 2643925

[27] R. Quiroga-Barranco, *Totally geodesic Riemannian foliations with locally symmetric leaves*, C. R. Math. Acad. Sci. Paris, **342**(6) (2006), 421–426. MR 2209222

[28] P.A. Schweitzer, *Riemannian manifolds not quasi–isometric to leaves in codimension one foliations*, Ann. Inst. Fourier Grenoble, **61**(4) (2011), 1599–1631. MR 2951506

[29] D. Töben, *Singular Riemannian foliations on nonpositively curved manifolds*, Math. Z., **255**(2) (2007), 427–436. MR 2262739

[30] T. Yokoyama, *Codimension one minimal foliations whose leaves have fundamental groups with the same polynomial growth*, C. R. Math. Acad. Sci. Paris, **350**(5–6) (2012), 285–287. MR 2911940

Received February 1, 2013.

FOLIATIONS 2012
ed. by Paweł WALCZAK et al.
World Scientific, Singapore, 2013
pp. 23–39

A class of dimensional type estimations of topological entropy of groups and pseudogroups

ANDRZEJ BIŚ

Katedra Geometrii, Wydział Matematyki i Informatyki
Uniwersytet Łódzki, Łódź, Poland
e-mail: andbis@math.uni.lodz.pl

1. Introduction

In this article we address some problems concerning the upper and lower estimations of topological entropy of generalized dynamical systems, i.e. of finitely generated groups (and pseudogroups) of homeomorphisms of a compact metric space. These generalized dynamical systems appear naturally in foliation theory as holonomy groups or pseudogroups. They reflect both dynamical, geometrical and topological properties of considered foliations.

Ghys, Langevin and Walczak [5] introduced notions of a topological entropy of a pseudogroup and of a geometric entropy $h_{\mathrm{geom}}(F, g)$ of a compact foliated manifold (M, F) with Riemannian structure g, which is closely related to geometry of (M, F). It is known ([5, 25]) that $h_{\mathrm{geom}}(F, g)$ of codimension-one foliation F is positive if and only if the foliation F has some resilient leaf. Moreover, if $h_{\mathrm{geom}}(F, g) = 0$, then the Godbillon-Vey class of F is vanishing.

In classical dynamical systems there are relations between the topological entropy of a continuous map $f : X \to X$ and Hausdorff dimension. More than thirty years ago Bowen [3] provided a definition of topological entropy of a map which resembles the definition of Hausdorff dimension.

A dimensional type approach to topological entropy of a single continuous map one can find for example in [1, 6, 9, 12]. A cyclic group or semigroup $< f >$ generated by a single map $f : X \to X$ has linear growth. Therefore it is difficult to adopt ideas and techniques presented for groups of linear growth to finitely generated groups or pseudogroups which growth is rarely linear.

In the paper we present and apply the theory of Carathéodory structures (or C-structures), studied by Pesin ([10, 11]) and Pesin and Pitskel ([12]), which are the powerful generalization of the classical construction of Hausdorff measure (see Chapter 2 of [4]). Pesin introduced a C-structure axiomatically by describing its elements and relation between them.

It is known ([2]) that for a finitely generated pseudogroup (G, G_1) there exists a C-structure with upper capacity that coincides with the topological entropy of (G, G_1). We present those arguments to construct other entropy like invariants of dimensional types for finitely generated pseudogroups and to make the paper self contained.

The paper is organized as follows. In Section 2, we recall the definition of a finitely generated pseudogroup and its topological entropy. In Section 3, we present the main concepts of the theory of C-structures: a definition of a C-structure as well as dimension and capacity of a C-structure. In particular, for a finitely generated pseudogroup (G, G_1) we construct some Caratheodory structure with upper capacity that coincides with the topological entropy of (G, G_1) (see Corollary 2). In Section 4, we provide a general construction of a class of entropy like invariants for groups and pseudogroups. The main result of the paper are as follows.

In Theorem 1 we consider two classes of Caratheodory structures $\Gamma_{i,\delta} = (F_\delta, \xi_i, \eta_i, \psi_i)$ defined on a metric space X, $i = 1, 2$ and formulate a condition which provides that the upper capacity $\overline{CP^1}_Z$ of the first C-structure Γ_1 is smaller or equal to the upper capacity $\overline{CP^2}_Z$ of the second C-structure Γ_2 for any $Z \subset X$.

Theorem 1. *Assume that two given classes of C-structures $\Gamma_{i,\delta} = \{(F_\delta, \xi_i, \eta_i, \psi_i) : \delta > 0\}$ defined on a metric space X, where $i = 1, 2$, satisfy the following condition: there exists $K > 0$ such that for any $\delta > 0$ the inequalities*

$$\xi_1(B_N^H(x, \delta)) \leq K \cdot \xi_2(B_N^H(x, \delta)), \quad \eta_1(B_N^H(x, \delta)) \leq K \cdot \eta_2((B_N^H(x, \delta)))$$

and

$$\psi_1(B_N^H(x, \delta)) = \psi_2(B_N^H(x, \delta)) = \frac{1}{N}$$

hold for any $B_N^H(x, \delta) \in F_\delta$. *Then the inequality* $\overline{CP^1}_Z \leq \overline{CP^2}_Z$ *holds for any* $Z \subset X$.

We denote by E a class of continuous and decreasing functions f : $[0, \infty) \rightarrow [0, \infty)$ with $\lim_{x \to \infty} f(x) = 0$. Now, we fix a pseudogroup (H, H_1) acting on a metric space X. Any function $f \in E$ and the pseudogroup (H, H_1) determine a class of C-structures $\Gamma(f)_\delta = \{(F_\delta, \xi, \eta, \psi) : \delta > 0\}$ and the limit C-structure $\Gamma(f)$ on X. The upper capacity of a set $Z \subset X$, with respect the limit C-structure $\Gamma(f)$, is denoted by $\overline{CP(f)}_Z$. Theorem 2 describes two classes of lower and upper estimations of the topological entropy $h_{\text{top}}((H, H_1), Z)$ of a pseudogroup (H, H_1) restricted to a subset $Z \subset X$. More precisely

Theorem 2. *Given a finitely generated pseudogroup* (H, H_1) *acting on a compact metric space* X. *Assume that for* $f, g \in E$ *and for any* $x \in [0, \infty)$ *the inequalities* $f(x) \leq e^{-x} \leq g(x)$ *hold. Then, for any subset* $Z \subset X$ *we get*

$$\overline{CP(f)}_Z \leq h_{\text{top}}((H, H_1), Z) \leq \overline{CP(g)}_Z.$$

A pseudogroup is an extension of the group concept, therefore Theorem 1 and Theorem 2 are still true if we replace a finitely generated pseudogroup (H, H_1) of local homeomorphisms of a metric space X by a finitely generated group (H, H_1) of homeomorphisms of X.

In Section 5, we apply Theorem 2 to the estimate geometric entropy $h_{\text{geom}}(F, g)$ of a compact foliated manifold (M, F), which describes the global dynamics of (M, F). It is known that a compact foliated manifold (M, F) with fixed so called nice covering U determines a finitely generated holonomy pseudogroup $(H(U), H_1(U))$ acting on the transversal T_U. Here, the finite generating set $H_1(U)$ consists of elementary holonomy maps corresponding to overlapping charts of U.

In Corollary 2 we obtain two classes of lower and upper estimations of the geometric entropy $h_{\text{geom}}(F, g)$ of a compact foliated manifold (M, F) endowed with the Riemannian structure g.

Corollary 1. *Assume that for* $f_1, f_2 \in E$ *and for any* $x \in [0, \infty)$ *the inequalities* $f_1(x) \leq e^{-x} \leq f_2(x)$ *hold. For any nice covering* U *of a compact foliated manifold* (M, F) *endowed with a Riemannian structure* g, *denote by* $\text{diam}(U)$ *the maximum of the diameters of the plaques of* U *measured with respect to the Riemannian structures induced on the leaves. Then*

$$h_{\text{geom}}^{\text{lower}}(F, f_1) \leq h_{\text{geom}}(F, g) \leq h_{\text{geom}}^{\text{upper}}(F, f_2),$$

where

$$h_{\text{geom}}^{\text{lower}}(F, f_1) = \sup\left\{\frac{1}{\text{diam}(U)}\overline{CP(f_1)(H(U), H_1(U))}_{T_U} : U\text{-nice cover of } M\right\}$$

$$h_{\text{geom}}^{\text{upper}}(F, f_2) = \sup\left\{\frac{1}{\text{diam}(U)}\overline{CP(f_2)(H(U), H_1(U))}_{T_U} : U\text{-nice cover of } M\right\}.$$

2. Topological entropy of a finitely generated pseudogroup

Given a topological space X, denote by $\text{Homeo}(X)$ the family of all homeomorphisms between open subsets of X. For $g \in \text{Homeo}(X)$ denote by D_g its domain and by $R_g = g(D_g)$ its range.

Definition 1. A pseudogroup Γ on X is a collection of homeomorphisms $h : D_h \to R_h$ between open subsets D_h and R_h of X such that:

(1) If $g, f \in \Gamma$, then $g \circ f : f^{-1}(R_f \cap D_g) \to g(R_f \cap D_g)$ is in Γ.
(2) If $g \in \Gamma$, then $g^{-1} \in \Gamma$.
(3) $\text{id}_X \in \Gamma$.
(4) If $g \in \Gamma$ and $W \subset D_g$ is an open subset, then $g|_W \in \Gamma$.
(5) If $g : D_g \to R_g$ is a homeomorphism between open subsets of X and if, for each point $p \in D_g$, there exists a neighbourhood N of p in D_g such that $g|_N \in \Gamma$, then $g \in \Gamma$.

For any set $G \subset \text{Homeo}(X)$ which satisfies the condition

$$\bigcup_{g \in G}\{D_g \cup R_g : g \in G\} = X$$

there exists a unique smallest (in the sense of inclusion) pseudogroup $\Gamma(G)$ which contains G. Notice that $g \in \Gamma(G)$ if and only if $g \in \text{Homeo}(X)$ and for any $x \in D_g$ there exist maps $g_1, \ldots, g_k \in G$, exponents $e_1, \ldots, e_k \in \{-1, 1\}$ and an open neighborhood U of x in X such that

$$U \subset D_g \quad \text{and} \quad g|_U = g_1^{e_1} \circ \cdots \circ g_k^{e_k}|_U.$$

The pseudogroup $\Gamma(G)$ is said to be *generated by* G. If the set G is finite, then we say that $\Gamma(G)$ is *finitely generated*.

Now, consider a finitely generated pseudogroup (G, G_1) acting on a compact metric space (X, d). Let G_1 be a finite symmetric generating set of G and

$$G_n := \{g_{i_1} \circ \cdots \circ g_{i_n} : g_{i_j} \in G_1\}.$$

Usually, it is assumed that $\text{id}_X \in G_1$ which implies the inclusion $G_m \subset G_n$ for any $m \leq n$. We emphasize the generating set G_1 of the pseudogroup G

writing (G, G_1) instead of G. Following [5] we say that two points $x, y \in E \subset X$ are (n, ϵ, E)-separated by (G, G_1) if there exists $g \in G_n$ such that $x, y \in D_g$ and $d(g(x), g(y)) \geq \epsilon$. Let $s(n, \epsilon, E)$ denote the maximal number of (n, ϵ, E)-separated points of E. The quantity

$$h_{\text{top}}((G, G_1), E) := \lim_{\epsilon \to 0} \limsup_{n \to \infty} \frac{1}{n} \log s(n, \epsilon, E)$$

is called the *topological entropy of G restricted to E*, with respect to G_1. The topological entropy $h_{top}((G, G_1), E)$ can be defined not only in terms of (n, ϵ, E)−separated sets but also in terms of (n, ϵ, E)-spanning sets. We say that a set $F \subset E$ is (n, ϵ, E)-spanning whenever for any $x \in E$ there exists a point $y_0 \in F$ such that the inequality $d(g(x), g(y_0)) < \epsilon$ holds for any $g \in G_n$ such that $x, y_0 \in D_g$. The minimal cardinality of (n, ϵ, E)-spanning subset of E is denoted by $r(n, \epsilon, E)$. It is known (see [5, 25]) that

$$\lim_{\epsilon \to 0} \limsup_{n \to \infty} \frac{1}{n} \log s(n, \epsilon, E) = \lim_{\epsilon \to 0} \limsup_{n \to \infty} \frac{1}{n} \log r(n, \epsilon, E).$$

3. Hausdorff dimension and Carathéodory dimension structures

3.1. *Hausdorff dimension*

The notion of topological entropy can be introduced similarly to the definition of Hausdorff dimension. For a single map this approach was used by Bowen [3] and Misiurewicz [9].

A cyclic group or a cyclic semigroup $< f >$ generated by a single map $f : X \to X$ has linear growth. Therefore it is difficult to adopt ideas and techniques presented for groups of linear growth to finitely generated groups or pseudogroups which growth is rarely linear. Dimensional type approach to the entropy of finitely generated groups and pseudogroups was studied by the author in [2].

Now we briefly recall the notion of Hausdorff dimension, more detailed introduction and properties of Hausdorff dimension one can find in many monographs (for example: [4, 7, 8]).

A countable collection of subsets $U_i \subset \mathbb{R}^n$ is called a δ−cover of a set $E \subset \mathbb{R}^n$ if for any i the diameter $\text{diam}(U_i) \leq \delta$ and E is covered by the union of U_i. Let $\mathbb{I}$ denotes the family of all subsets of $\mathbb{N}$. For a subset $E \subset \mathbb{R}^n$, $s \geq 0$ and $\delta > 0$ we define

$$\mathcal{H}_\delta^s(E) = \inf \left\{ \sum_{i \in I} [\text{diam}(U_i)]^s : (U_i)_{i \in I} \text{ is a } \delta\text{-cover of } E, I \in \mathbb{I} \right\}. \quad (1)$$

As δ decreases, the collection of δ-covers of E is reduced, thus the infimum increases and approaches a limit with δ tending to 0.

Definition 2. The quantity

$$\mathcal{H}^s(E) = \lim_{\delta \to 0} \mathcal{H}^s_\delta(E)$$

is called the *s-dimensional Hausdorff measure* of E.

Definition 3. The real number $\dim_H(E)$, called the *Hausdorff dimension* of E, is such that $\mathcal{H}^s(E) = \infty$ if $s < \dim_H(E)$ and $\mathcal{H}^s(E) = 0$ if $s > \dim_H(E)$.

A direct conclusion is obtained from the above definition

$$\dim_H(E) = \inf\{s : \mathcal{H}^s(E) = 0\} = \sup\{s : \mathcal{H}^s(E) = \infty\}.$$

3.2. *Carathéodory dimension structure*

In this section, we describe a general approach to a construction of α-measures on a metric space, developed by Pesin ([10, 11]) which is a generalization of the Hausdorff measure and the classical Carathéodory construction. Pesin introduced axiomatically a structure, called the Carathéodory structure (or C-structure), by describing its elements and relation between them.

Let X be a set and F a collection of subsets of X. Following Pesin [10] we assume that there exist two set functions $\eta, \psi : F \to \mathbb{R}_+$ satisfying the following conditions:

A1. $\emptyset \in F$ and $\eta(\emptyset) = 0 = \psi(\emptyset)$; for any non-empty $U \in F$ we get $\eta(U) > 0$ and $\psi(U) > 0$.

A2. For any $\delta > 0$ there exists $\epsilon > 0$ such that $\eta(U) \leq \delta$ for any $U \in F$ with $\psi(U) \leq \epsilon$.

A3. For any $\epsilon > 0$ there exists a finite or countable subcolection $G \subset F$ which covers X and $\psi(G) := \sup\{\psi(U) : U \in G\} \leq \epsilon$.

Definition 4. Let $\xi : F \to \mathbb{R}_+$ be a set function. We say that the collection of subsets F and the set functions ξ, η, ψ satisfying conditions A1–A3, introduce a *Carathéodory dimension structure* or *C-structure* τ on X and we write $\tau = (F, \xi, \eta, \psi)$.

Now, consider a set X endowed with a C-structure $\tau = (F, \xi, \eta, \psi)$. For any subset $Z \subset X$, real numbers α and $\epsilon > 0$ we define

$$M_C(Z, \alpha, \epsilon) := \inf_G \{\sum_{U \in G} \xi(U) \cdot \eta(U)^\alpha\},$$

where the infimum is taken over all finite or countable subcollections $G \subset F$ which cover Z and satisfy the condition $\psi(G) \leq \epsilon$. Therefore, the limit $m_C(Z, \alpha) = \lim_{\epsilon \to 0} M_C(Z, \alpha, \epsilon)$ exists.

The set function $m_C(\cdot, \alpha)$ becomes an outer measure on X, according to the general measure theory it induces a σ-additive measure called the α-*Carathéodory measure*. Moreover

Lemma 1 (Proposition 1.2 in [10]). *There exists a critical value* α_C, $-\infty \leq \alpha_C \leq \infty$ *such that* $m_C(Z, \alpha) = \infty$ *for* $\alpha \leq \alpha_C$ *and* $m_C(Z, \alpha) = 0$ *for* $\alpha > \alpha_C$.

The **Carathéodory dimension of a set** $Z \subset X$ with respect to the C-structure τ, is defined as follows

$$\dim_{C,\tau} Z = \alpha_C = \inf\{\alpha : m_C(Z, \alpha) = 0\}.$$

3.3. *Carathéodory capacity of sets*

Assume that a C-structure $\tau = (F, \xi, \eta, \psi)$ satisfies condition A3. It is useful to require a slightly stronger condition. Pesin (p.16 in [10]) introduced another type of Carathéodory dimension characteristic of a set and defined A3' condition as follows:

A3'. There exists $\epsilon_0 > 0$ such that for any $\epsilon \in (0, \epsilon_0)$ one can find a finite or countable subcollection $G \subset F$ covering X such that $\psi(U) = \epsilon$ for any $U \in G$.

It is clear that condition A3' is stronger than condition A3. For any subset $Z \subset X$, real number α and $\epsilon > 0$ we define

$$R_C(Z, \alpha, \epsilon) := \inf_G \{\sum_{U \in G} \xi(U) \cdot \eta(U)^\alpha\},$$

where the infimum is taken over all finite or countable subcollections $G \subset F$ which cover Z and satisfy the condition $\psi(U) = \epsilon$ for all $U \in G$. Due to A3' the quantity $R_C(Z, \alpha, \epsilon)$ is well defined, it yields the existence of the limits

$$\underline{r}_C(Z, \alpha) = \underline{\lim}_{\epsilon \to 0} R_C(Z, \alpha, \epsilon) \quad \text{and} \quad \overline{r}_C(Z, \alpha) = \overline{\lim}_{\epsilon \to 0} R_C(Z, \alpha, \epsilon).$$

The behaviour of $\underline{r}_C(\cdot, \alpha)$ and $\overline{r}_C(\cdot, \alpha)$ is described by the following result.

Proposition 1 (Proposition 2.1 [10]). *For any* $Z \subset X$, *there exist* $\underline{\alpha}_C, \overline{\alpha}_C \in \mathbb{R}$ *such that*

(1) $\underline{r}_C(Z, \alpha) = \infty$ *for* $\alpha < \underline{\alpha}_C$ *and* $\underline{r}_C(Z, \alpha) = 0$ *for* $\alpha > \underline{\alpha}_C$,

(2) $\bar{r}_C(Z, \alpha) = \infty$ for $\alpha < \bar{\alpha}_C$ and $\bar{r}_C(Z, \alpha) = 0$ for $\alpha > \bar{\alpha}_C$.

Given $Z \subset X$, the *lower* and the *upper Carathéodory capacities* of a set Z are defined by

$$\underline{\text{Cap}}_C Z = \underline{\alpha}_C = \inf\{\alpha : \underline{r}_C(Z, \alpha) = 0\} = \sup\{\alpha : \underline{r}_C(Z, \alpha) = \infty\};$$
$$\overline{\text{Cap}}_C Z = \bar{\alpha}_C = \inf\{\alpha : \bar{r}_C(Z, \alpha) = 0\} = \sup\{\alpha : \bar{r}_C(Z, \alpha) = \infty\}.$$

The upper Carathéodory capacity of a set has the following property.

Lemma 2 (Theorem 2.1, [10]). *If $Z_1 \subset Z_2 \subset X$, then*

$$\overline{\text{Cap}}_C Z_1 \leq \overline{\text{Cap}}_C Z_2.$$

We will use the following properties of the lower and the upper Carathéodory capacities of a set. For $\epsilon > 0$ and any $Z \subset X$ we define

$$\Lambda(Z, \epsilon) := \inf_G \{\sum_{U \in G} \xi(U)\},$$

where the infimum is taken over all finite or countable subcollection $G \subset F$ covering Z for which the condition $\psi(U) = \epsilon$ holds for all $U \in G$.

Let us assume that the set function η satisfies the following condition

A4. $\eta(U_1) = \eta(U_2)$ for any $U_1, U_2 \in F$ for which $\psi(U_1) = \psi(U_2)$.

Then, the lower and upper Carathéodory capacities have the following properties.

Lemma 3 (Theorem 2.2, [10]). *If the set function η satisfies condition A4, then for any $Z \subset X$*

$$\overline{\text{Cap}}_C Z = \overline{\lim}_{\epsilon \to 0} \frac{\log \Lambda(Z, \epsilon)}{\log(\frac{1}{\eta(\epsilon)})} \quad \text{and} \quad \underline{\text{Cap}}_C Z = \underline{\lim}_{\epsilon \to 0} \frac{\log \Lambda(Z, \epsilon)}{\log(\frac{1}{\eta(\epsilon)})}.$$

Lemma 4 (Theorem 2.4, [10]). *Under condition A4 the equality*

$$\overline{\text{Cap}}_C(Z_1 \cup Z_2) = \max\{\overline{\text{Cap}}_C(Z_1), \overline{\text{Cap}}_C(Z_2)\}$$

holds for any subsets $Z_1, Z_2 \subset X$.

3.4. *Topological entropy coincides with a capacity of some C-structure*

Using the theory of C-structures we are able to illustrate the relationship between topological entropy and dimensional characteristic of a dynamical system. We apply Pesin's theory to a finitely generated pseudogroup

(H, H_1) acting on a compact metric space (X, d) to describe its topological entropy. To this end, we construct a C-structure determined by (H, H_1) acting on X.

Definition 5. An *n-ball* $B_n^H(x, r)$ of radius r and centered at $x \in X$, with respect to a finitely generated pseudogroup (H, H_1) acting on a compact metric space (X, d), is defined by

$$B_n^H(x, r) := \{y \in X : d(h(x), h(y)) < r$$
$$\text{for any } h \in H_{n-1} \text{ such that } x, y \in D_h\}.$$

Fix $\delta > 0$. Define a collection F_δ of subsets of X by

$$F_\delta := \{B_n^H(x, \delta) : x \in X, n \in \mathbb{N}\}$$

and three set functions $\xi, \eta, \psi : F_\delta \to \mathbb{R}$ as follows

$$\xi(B_n^H(x, \delta)) \equiv 1, \ \eta(B_n^H(x, \delta)) = \exp(-n), \ \psi(B_n^H(x, \delta)) = \frac{1}{n}. \quad (2)$$

It is easy to verify that F_δ and three set functions ξ, η, ψ satisfy conditions A1–A3 and A3', therefore they determine a C-structure $\Gamma_\delta = (F_\delta, \xi, \eta, \psi)$ on X.

The Carathéodory function $\overline{r}_C(Z, \alpha, \delta)$, where $Z \subset X$ and $\alpha \in \mathbb{R}$, depends on the covering F_δ and is given by

$$\overline{r}_C(Z, \alpha, \delta) = \limsup_{N \to \infty} \inf_G \left\{ \sum_{B_N^H(x, \delta) \in G} e^{-\eta(B_N^H(x, \delta)) \cdot \alpha} : Z \subset \bigcup_{B_N^H(x, \delta) \in G} B_N^H(x, \delta) \right\},$$

where the infimum is taken over all finite or countable subcollections $G \subset F_\delta$ consisted of n-balls which cover Z.

The C-structure Γ_δ generates an upper Carathéodory capacity of Z, denoted here by $\overline{CP}_Z(\delta)$, specified by the covers F_δ and the pseudogroup (H, H_1). Let

$$\overline{CP}_Z(\delta) := \inf\{\alpha : \overline{r}_C(Z, \alpha, \delta) = 0\}.$$

By Theorem 11.1 in [10] the limit $\overline{CP}_Z := \lim_{\delta \to 0} \overline{CP}_Z(\delta)$ exists. Notice that the functions η and ψ satisfy condition A4, therefore by Lemma 3 and Theorem 11.1 in [10] we obtain

Lemma 5. *For any* $Z \subset X$ *there exists a limit*

$$\overline{CP}_Z := \lim_{\delta \to 0} \limsup_{N \to \infty} \frac{1}{N} \log \Lambda(Z, \delta, N),$$

where $\Lambda(Z, \delta, N) = \inf_G \{\text{card}(G)\}$ *and the infimum is taken over all finite or countable collections* $G \subset F_\delta$ *of N-balls such that*

$$Z \subset \bigcup_{B_N^H(x,\delta) \in G} B_N^H(x, \delta).$$

Corollary 2 (Corollary 3.10 in [2]). *For a finitely generated pseudogroup* (H, H_1) *acting on a compact metric space* (X, d) *and any subset* $Z \subset X$ *we get*

$$\overline{CP}_Z = h_{\text{top}}((H, H_1), Z).$$

Proof. One can directly verify that $\Lambda(Z, \delta, N)$ coincides with $s(N, \delta, Z)$, the maximal cardinality of (N, δ, Z)-separated subset of Z, with respect to (H, H_1). The claim follows from the definition of the topological entropy of (H, H_1) restricted to Z. $\qquad\qquad\qquad\qquad\qquad\qquad\qquad\qquad\qquad\qquad$ $\square$

4. A new class of entropy-like invariants for pseudogroups

We fix a finitely generated pseudogroup (H, H_1) acting on a compact metric space (X, d) and choose $\delta > 0$. Define the collection F_δ of subsets of X by

$$F_\delta = \{B_n^H(x, \delta) : x \in X, n \in \mathbb{N}\}$$

and choose six set functions $\xi_i, \eta_i, \psi_i : F_\delta \to \mathbb{R}$, where $i = 1, 2$.

We assume that F_δ and three set functions ξ_i, η_i, ψ_i, where $i = 1, 2$, satisfy conditions A1–A3, A3' and A4, therefore they determine a C-structure $\Gamma_{i,\delta} = (F_\delta, \xi_i, \eta_i, \psi_i)$ on X. In the section we consider Carathéodory functions only for $\alpha \geq 0$.

The Carathéodory function $\overline{r}_{C,i}(Z, \alpha, \delta)$ determined by the C-structure $\Gamma_{i,\delta}$, where $Z \subset X$ and $\alpha \geq 0$, depends on the covering F_δ and is given by

$$\overline{r}_{C,i}(Z, \alpha, \delta) := \limsup_{N \to \infty} \inf_G \left\{ \sum_{B_N^H(x,\delta) \in G} \xi_i(B_N^H(x, \delta)) \cdot \eta_i(B_N^H(x, \delta))^\alpha : \right.$$

$$\left. Z \subset \bigcup_{B_N^H(x,\delta) \in G} B_N^H(x, \delta) \text{ and } \psi_i(B_N^H(x, \delta)) = \frac{1}{N} \right\},$$

where the infimum is taken over all finite or countable subcollections $G \subset F_\delta$ consisted of N-balls which cover Z.

The C-structure $\Gamma_{i,\delta}$ generates an upper Carathéodory capacity of Z, denoted here by $\overline{CP^i}_Z(\delta)$ which is specified by the covers F_δ, the group

(H, H_1) and $\delta > 0$. Let

$$\overline{CP^i}_Z(\delta) := \inf\{\alpha : \overline{r}_{C,i}(Z, \alpha, \delta) = 0\}.$$

By Theorem 11.1 in [10] the limit $\overline{CP^i}_Z := \lim_{\delta \to 0} \overline{CP^i}_Z(\delta)$ exists. Notice that the functions η_i and ψ_i satisfy condition A4, therefore by Lemma 3 and Theorem 11.1 in [10] we obtain.

Lemma 6. *For any $Z \subset X$ there exists a limit*

$$\overline{CP^i}_Z := \lim_{\delta \to 0} \limsup_{N \to \infty} \frac{1}{N} \log \Lambda_i(Z, \delta, N),$$

where $\Lambda_i(Z, \delta, N) = \inf_G\{\sum_{B_N^H(x,\delta) \in G} \xi_i(B_N^H(x, \delta))\}$ and the infimum is taken over all finite or countable collections $G \subset F_\delta$ of N-balls such that $Z \subset \bigcup_{B_N^H(x,\delta) \in G} B_N^H(x, \delta)$.

Now, we are able to prove Theorem 1 formulated in Section 1.

Proof of Theorem 1. Take a subset $Z \subset X$, $\alpha \geq 0$ and $\delta > 0$. Fix a positive integer N. Notice that for any $d > 0$ there exists a finite or countable subset $Z_1 \subset Z$ such that the family $G_2(N) := \{B_N^H(x, \delta) : x \in Z_1\}$ is a covering of Z and

$$\sum_{B_N^H(x,\delta) \in G_2(N)} \xi_2(B_N^H(x, \delta)) \cdot [\eta_2(B_N^H(x, \delta))]^\alpha$$

$$\leq \inf \left\{ \sum_{B_N^H(x,\delta) \in G} \xi_2(B_N^H(x, \delta)) \cdot [\eta_2(B_N^H(x, \delta))]^\alpha : \right.$$

$$\left. Z \subset \bigcup_{B_N^H(x,\delta) \in G} B_N^H(x, \delta) \right\} + d,$$

where the infimum is taken over all finite or countable subcollections $G \subset F_\delta$ consisted of N-balls which cover Z. On the other hand for the cover $G_2(N)$ of Z we obtain

$$\inf \left\{ \sum_{B_N^H(x,\delta)\in G} \xi_1(B_N^H(x,\delta)) \cdot [\eta_1(B_N^H(x,\delta))]^\alpha : Z \subset \bigcup_{B_N^H(x,\delta)\in G} B_N^H(x,\delta) \right\}$$

$$\leq \sum_{B_N^H(x,\delta)\in G_2(N)} \xi_1(B_N^H(x,\delta)) \cdot \left[\eta_1(B_N^H(x,\delta))\right]^\alpha$$

$$\leq \sum_{B_N^H(x,\delta)\in G_2(N)} K \cdot \xi_2(B_N^H(x,\delta)) \cdot \left[K \cdot \eta_2(B_N^H(x,\delta))\right]^\alpha$$

$$\leq K^{1+\alpha} \cdot \sum_{B_N^H(x,\delta)\in G_2(N)} \xi_2(B_N^H(x,\delta)) \cdot \left[\eta_2(B_N^H(x,\delta))\right]^\alpha$$

$$\leq K^{1+\alpha} \cdot \left[\inf \left\{ \sum_{B_N^H(x,\delta)\in G} \xi_2(B_N^H(x,\delta)) \cdot \left[\eta_2(B_N^H(x,\delta))\right]^\alpha : \right. \right.$$

$$\left. \left. Z \subset \bigcup_{B_N^H(x,\delta)\in G} B_N^H(x,\delta) \right\} + d \right],$$

where the infimum is taken over all finite or countable subcollections $G \subset F_\delta$ consisted of N-balls which cover Z.

As $N \to \infty$ we obtain

$$\overline{r}_{C,1}(Z,\alpha,\delta) \leq K^{1+\alpha} \cdot [\overline{r}_{C,2}(Z,\alpha,\delta) + d].$$

Since d can be chosen arbitrary small we get

$$\overline{r}_{C,1}(Z,\alpha,\delta) \leq K^{1+\alpha} \cdot \overline{r}_{C,2}(Z,\alpha,\delta).$$

Therefore

$$\overline{CP^1}_Z(\delta) = \inf\{\alpha : \overline{r}_{C,1}(Z,\alpha,\delta) = 0\}$$
$$\leq \inf\{\alpha : \overline{r}_{C,2}(Z,\alpha,\delta) = 0\} = \overline{CP^2}_Z(\delta).$$

Finally, passing with δ to zero due to Theorem 11.1 in [10] we get

$$\overline{CP^1}_Z \leq \overline{CP^2}_Z. \qquad \square$$

4.1. *Upper capacity of a C-structure* $\Gamma((H, H_1), f)$

Now, we fix a pseudogroup (H, H_1) acting on a compact metric space X and consider *a class* E of continuous and decreasing functions $f : [0, \infty) \to [0, \infty)$ with $\lim_{x\to\infty} f(x) = 0$. Any function $f \in E$ determines a C-structure $\Gamma(f)_\delta = (F_\delta, \xi, \eta, \psi)$ on X, where $\delta > 0$, given by

$$\xi(B_n^H(x,\delta)) \equiv 1, \quad \eta(B_n^H(x,\delta)) = f(n), \quad \psi(B_n^H(x,\delta)) = \frac{1}{n}.$$

For any subset $Z \subset X$ and $\alpha \geq 0$ we consider a function

$$\overline{r}_{C(f)}(Z, \alpha, \delta) := \limsup_{N \to \infty} \inf_G \left\{ \sum_{B_N^H(x,\delta) \in G} (f(N))^\alpha : \right.$$

$$\left. Z \subset \bigcup_{B_N^H(x,\delta) \in G} B_N^H(x, \delta) \ and \ \psi(B_N^H(x, \delta)) = \frac{1}{N} \right\},$$

where the infimum is taken over all finite or countable subcollections $G \subset F_\delta$ consisted of N-balls which cover Z.

The C-structure $\Gamma(f)_\delta$ determines an upper Carathéodory capacity of Z, denoted here by $\overline{CP(f)}_Z(\delta)$, specified by the covers F_δ, function f, and the pseudogroup (H, H_1). Let

$$\overline{CP(f)}_Z(\delta) := \inf \left\{ \alpha : \overline{r}_{C(f)}(Z, \alpha, \delta) = 0 \right\}.$$

By Theorem 11.1 in [10] the limit $\overline{CP(f)}_Z := \lim_{\delta \to 0} \overline{CP(f)}_Z(\delta)$ exists.

The limit C-structure, determined by the pseudogroup (H, H_1) and by the families of C-structures $\Gamma(f)_\delta = \{(F_\delta, \xi, \eta, \psi) : \delta > 0\}$, is denoted by $\Gamma((H, H_1), f)$. Denoting by $\overline{CP(f)}_Z$ the upper capacity of a set $Z \subset X$, with respect to $\Gamma((H, H_1), f)$ we can prove Theorem 2.

Proof of Theorem 2. Given a subset subset $Z \subset X$. Take $\alpha \geq 0$ and $\delta > 0$. Choose $f_1, f_2 \in E$ with $f_1(x) \leq f_2(x)$ for any $x \in [0, \infty)$. For any $d > 0$ there exists a finite or countable subset $Z_1 \subset Z$ such that the family $G_2(N) := \{B_N^H(x, \delta) : x \in X_1 \subset X\}$ is a cover of Z and

$$\sum_{B_N^H(x,\delta) \in G_2(N)} f_2(N)^\alpha \leq \inf \left\{ \sum_{B_N^H(x,\delta) \in G} (f_2(N))^\alpha : Z \subset \bigcup_{B_N^H(x,\delta) \in G} B_N^H(x, \delta) \right\} + d,$$

where the infimum is taken over all finite or countable subcollections $G \subset F_\delta$ consisted of N-balls which cover Z. Also, for the cover $G_2(N)$ of Z we obtain

$$\inf \left\{ \sum_{B_N^H(x,\delta) \in G} (f_1(N))^\alpha : Z \subset \bigcup_{B_N^H(x,\delta) \in G} B_N^H(x, \delta) \right\}$$

$$\leq \sum_{B_N^H(x,\delta) \in G_2(N)} (f_1(N))^\alpha \leq \sum_{B_N^H(x,\delta) \in G_2(N)} (f_2(N))^\alpha$$

$$\leq \inf \left\{ \sum_{B_N^H(x,\delta) \in G} (f_2(N))^\alpha : Z \subset \bigcup_{B_N^H(x,\delta) \in G} B_N^H(x, \delta) \right\} + d.$$

As $N \to \infty$ we obtain

$$\overline{r}_{C(f_1)}(Z, \alpha, \delta) \leq \overline{r}_{C(f_2)}(Z, \alpha, \delta) + d.$$

Since d can be chosen arbitrary small we get

$$\overline{r}_{C(f_1)}(Z, \alpha, \delta) \leq \overline{r}_{C(f_2)}(Z, \alpha, \delta)$$

and

$$
\begin{aligned}
\overline{CP(f_1)}_Z(\delta) &= \inf\{\alpha : \overline{r}_{C(f_1)}(Z, \alpha, \delta) = 0\} \\
&\leq \inf\{\alpha : \overline{r}_{C(f_2)}(Z, \alpha, \delta) = 0\} = \overline{CP(f_2)}_Z(\delta).
\end{aligned}
$$

Passing with δ to zero we get

$$\overline{CP(f_1)}_Z = \lim_{\delta \to 0} \overline{CP(f_1)}_Z(\delta) \leq \lim_{\delta \to 0} \overline{CP(f_2)}_Z(\delta) = \overline{CP(f_2)}_Z.$$

In particular, if $f_1(x) \equiv f(x)$ and $e^{-x} \equiv g(x)$ then due to Corollary 1 for any $Z \subset X$ we get

$$\overline{CP(f)}_Z \leq \overline{CP(e^{-x})}_Z = h_{\text{top}}((H, H_1), Z).$$

In a similar way, if $f_1(x) \equiv e^{-x}$ and $f_2 \equiv g(x)$ then due to Corollary 1 for any $Z \subset X$ we get

$$h_{\text{top}}((H, H_1), Z) = \overline{CP(e^{-x})}_Z \leq \overline{CP(g)}_Z.$$

Finally, for any $Z \subset X$ we obtain

$$\overline{CP(f_1)}_Z \leq h_{\text{top}}((H, H_1), Z) \leq \overline{CP(f_2)}_Z. \qquad \square$$

Remark 1. The above construction provide an infinite class of lower and upper estimations of the topological entropy of finitely generated pseudogroups (and groups). Also, it suggests a construction of other entropy like invariants for groups and pseudogroups.

Remark 2. A pseudogroup is an extension of the group concept, therefore Theorem 1 and Theorem 2 are still true if we replace a finitely generated pseudogroup (H, H_1) of local homeomorphisms of a metric space X by a finitely generated group (H, H_1) of homeomorphisms of X.

5. Estimations of geometric entropy of foliations

A *p-dimensional C^r-foliation* F on a manifold M of dimension m is a decomposition of M into connected submanifolds $\{L_\alpha\}$, called *leaves*, of dimension p such that for any $x \in M$ there exists a C^r-differentiable chart $\phi = (\phi', \phi'') : U \to \mathbb{R}^p \times \mathbb{R}^{m-p}$ defined on a neighbourhood U of x and satisfying the following condition: for any leaf L_α the connected components of $L_\alpha \cap U$, which are called *plaques*, are described by the equation $\phi'' = \text{const}$. Any chart satisfying the above condition is said to be *distinguished by the foliation F*, an atlas built of distinguished charts is said to be *foliated*.

A foliated atlas A on a foliated manifold (M, F) is said to be *nice* if the covering $\{D_\phi : \phi \in A\}$ by domains of charts is locally finite, for any $\phi \in A$ the image $\phi(D_\phi)$ is an open cube in $\mathbb{R}^m$, and for charts $\phi_1, \phi_2 \in A$ with $D_{\phi_1} \cap D_{\phi_2} \neq \emptyset$ there exists a chart ϕ_3 distinguished by F satisfying two conditions:

(1) $\phi_3(D_{\phi_3})$ is an open cube in $\mathbb{R}^m$,
(2) $\overline{D_{\phi_1} \cup D_{\phi_2}} \subset D_{\phi_3}$ and $\phi_1 = \phi_3|_{D_{\Phi_1}}$.

It is known that for any foliation F on any manifold M there exists a nice covering of M. Nice coverings on compact manifolds are finite.

Ghys, Langevin and Walczak [5] introduced a notion of a geometric entropy $h_{\text{geom}}(F, g)$ of a compact foliated manifold (M, F) endowed with a Riemannian structure, which is closely related to geometry of (M, F). It is known that $h_{\text{geom}}(F, g)$ of codimension-one foliation is positive if and only if the foliation F has some resilient leaf. The topological entropies of holonomy pseudogroups $(H(U), H_1(U))$, determined by nice coverings U, of a compact foliated manifold (M, F) with Riemannian structure g on M and the geometric entropy $h_{\text{geom}}(F, g)$ of F are interrelated.

Proposition 2 (Theorem 3.4.1 in [25]). *For any nice covering U of a compact foliated manifold (M, F) endowed with a Riemannian structure g, denote by* $\text{diam}(U)$ *the maximum of the diameters of the plaques of U measured with respect to the Riemannian structures induced on the leaves. Then*

$$h_{\text{geom}}(F, g) = \sup \left\{ \frac{1}{\text{diam}(U)} h_{\text{top}}(H(U), H_1(U)) : \right.$$

$$\left. U \text{ is a nice covering of } M \right\}.$$

In Section 4 we have fixed a pseudogroup (H, H_1) acting on a compact metric space X and for any $f \in E$ we have constructed the upper capacity $\overline{CP(f)}_Z$ of a subset $Z \subset X$. Here, we write $\overline{CP(f)(H, H_1)}_Z$ instead of $\overline{CP(f)}_Z$ to emphasize the pseudogroup (H, H_1) involved in the construction of the upper capacity. A compact foliated manifold (M, F) with fixed nice covering U determines a finitely generated holonomy pseudogroup $(H(U), H_1(U))$ acting on the transversal T_U. Here, the finite generating set $H_1(U)$ consists of elementary holonomy maps corresponding to overlapping charts of U.

Theorem 2 yields upper and lower estimations of the geometric entropy of foliations from Corollary 2 formulated in Section 1.

Acknowledgment

The author is grateful to the referee for all the comments. The part of this article was written while the author was visiting the Erwin Schrödinger International Institute for Mathematical Physics. The author thanks the organizers of Teichmüller Theory Programme for their hospitality.

References

[1] L. Barreira, J. Schmeling, *Sets of "non-typical" points have full topological entropy and full Hausdorff dimension*, Israel J. Math., **116** (2000), 29–70.

[2] A. Biś, *An analogue of the Variational Principle for group and pseudogroup actions*, to appear in Annales de l'Institut Fourier.

[3] R. Bowen, *Topological entropy for noncompact sets*, Trans. of the AMS, **184** (1973), 125–136.

[4] K. Falconer, Techniques in Fractal Geometry, John Wiley and Sons, 1997.

[5] E. Ghys, R. Langevin and P. Walczak, *Entropie géométrique des feuilletages*, Acta Math., **160** (1988), 105–142.

[6] Ji-Hua Ma, Zhi-Ying Wen, *A Billingsley type theorem for Bowen entropy*, C. R. Acad. Sci. Paris, Serie I, **346** (2008), 503–507.

[7] D. Mauldin and M. Urbański, Graph Directed Markov Systems (Geometry and Dynamics of Limit Sets), Cambridge University Press, 2003.

[8] P. Mattila, Geometry of Sets and Measures in Euclidean Spaces, Cambridge University Press, 1995.

[9] M. Misiurewicz, *On Bowen's definition of topological entropy*, Disc. Cont. Dyn. Syst., **10**(3) (2004), 827–833.

[10] Ya. Pesin, Dimension Theory in Dynamical Systems, Chicago Lectures in Mathematics, The University of Chicago Press, Chicago, 1997.

[11] Ya. Pesin, *Dimension Type Characteristics for Invariant Sets of Dynamical Systems*, Russian Math. Surveys, **43** (1988), 111–151.

[12] Ya. Pesin, B.S. Pitskel, *Topological pressure and the variational principle for noncompact sets*, Funct. Anal. Appl., **18** (1984), 307–318.

[13] P. Walczak, Dynamics of Foliations, Groups and Pseudogroups, Monografie Matematyczne, **64**, Birkhäuser, Basel 2004.

Received January 18, 2013.

FOLIATIONS 2012
ed. by Paweł WALCZAK et al.
World Scientific, Singapore, 2013
pp. 41–66

The sutured Thurston norm

JOHN CANTWELL

Department of Mathematics, St. Louis University
St. Louis, MO 63103, USA
e-mail: cantwelljc@slu.edu

LAWRENCE CONLON

Department of Mathematics, Washington University
St. Louis, MO 63103, USA
e-mail: lc@math.wustl.edu

1. Introduction

For sutured three-manifolds M, there is a *sutured* Thurston norm x^s due to
M. Scharlemann [13]. Actually, this is one of a whole class of "generalized
Thurston norms" for sutured manifolds introduced in [13]. Here, we show
how depth one foliations of M can be useful tools for computing this norm.
This uses the relation of these foliations with fibrations of DM (the double
of M along the manifold $R \subset \partial M$ given by the sutured structure). We also
prove and use the fact that a natural doubling map $D_* : H_2(M, \partial M) \to
H_2(DM, \partial DM)$ is "norm doubling" with respect to the norms x^s and x on
$H_2(M, \partial M)$ and $H_2(DM, \partial DM)$, respectively. All of this implies significant
relations between the foliation cones of [6] and the sutured norm but, in
general, these relations are difficult to pin down.

Recent work of S. Friedl, A. Juhász and J. Rasmussen [8] produces
a sutured Floer norm and, using Example 2 of the present paper, they
show that it is not always equal to the sutured Thurston norm. This is
somewhat surprising since the Floer norm for closed 3-manifolds is equal to

the Thurston norm. This sutured Floer norm is obtained by symmetrizing a nonsymmetric norm defined by Juhász [12]. The nonsymmetric norm has a nonsymmetric polyhedral unit ball (the "sutured Floer polytope") and certain top dimensional faces of this ball subtend the foliation cones introduced by the authors in [6]. This is a result of I. Altman [1], who shows that these "foliated faces" correspond to those spin^c-structures that support infinite cyclic sutured Floer homology.

2. Cones of fibrations and foliations

If M is a compact 3-manifold, Thurston [14] defines a (pseudo)norm x on the real vector space $H_2(M, \partial M)$ (coefficients $\mathbb{R}$ will be understood throughout), with unit ball polyhedral, and proves:

Theorem 1. *The fibrations of M over the circle that are transverse to ∂M correspond up to isotopy to the rational rays in the open cones subtended by the interiors of certain top dimensional faces (called fibered faces) of the unit ball of the Thurston norm.*

Here, by rational rays in $H_2(M, \partial M)$, we mean the rays issuing from the origin that meet nontrivial elements of the integer lattice $H_2(M, \partial M; \mathbb{Z})$. The cones subtended by fibered faces of the Thurston ball will be called *fibration cones*. This is slightly misleading since the classes lying in the interior of fibration cones correspond to foliations without holonomy, "most" of which are dense-leaved. Such classes lie on rays not meeting the integer lattice except at $\mathbf{0}$.

Let (M, γ) be a compact, connected, oriented, sutured 3–manifold [10]. Write

$$\partial M = \partial_\tau M \cup \partial_\pitchfork M.$$

This notation, introduced in earlier papers of ours and in [2], anticipates a foliation tangent to $\partial_\tau M$ and transverse to $\partial_\pitchfork M$. Wherever these parts of ∂M meet, M has a convex corner. This notation relates to the standard sutured manifold notation as follows:

$$\partial_\pitchfork M = \gamma = A(\gamma) \cup T(\gamma),$$
$$\partial_\tau M = R(\gamma) = R_+ \cup R_-.$$

Here, $A(\gamma)$ is a union of annuli and $T(\gamma)$ is a union of tori, while $R_\pm$ are, respectively, the outwardly and inwardly oriented portions of $R(\gamma)$. The choice of orientations is part of the sutured structure and each component of R_- is separated from components of R_+ by components of $A(\gamma)$. Finally, each suture is a closed curve in the interior of a component of $A(\gamma)$, parallel to and oriented with the boundary curves of this annulus, these curves being oriented as boundaries of R_+ and R_- respectively. The union of the sutures is denoted by s or $s(\gamma)$.

We will be interested in taut foliations of M, hence will require that M be irreducible and, as a sutured manifold, *taut*. This latter requirement means that each component of $\partial_\tau M$ is norm-minimizing in $H_2(M, \partial_\pitchfork M)$. In particular, if $\sigma \subset \partial_\tau M$ is an imbedded loop bounding a disk in M, it also bounds a disk in $\partial_\tau M$.

In [6], we proved the following analog of Theorem 1 for depth one foliations.

Theorem 2. *Let (M, γ) be a compact, connected, oriented, irreducible, taut, sutured 3–manifold. There are finitely many closed, convex, polyhedral cones $\mathfrak{C}_1, \ldots, \mathfrak{C}_r$ in $H_2(M, \partial M)$, called* foliation cones, *having disjoint interiors and such that the taut, transversely oriented, depth one foliations of (M, γ) that are transverse to $\partial_\pitchfork M$ and have the components of $\partial_\tau M$ as sole compact leaves correspond up to isotopy to the rational rays in the interior of these foliation cones.*

Remark 1. Set $M_0 = M \smallsetminus \partial_\tau M$ and remark that a depth one foliation as above restricts to a fibration of M_0 over the circle. The classes in the interior of a foliation cone $\mathfrak{C}_i$ that are not on rational rays correspond up to isotopy to foliations "almost without holonomy" with each leaf in M_0 dense in M. This is proven in [7], where Theorem 2 is generalized to higher depths and has an improved proof. All classes in int $\mathfrak{C}_i$ are represented by "foliated forms", these being closed, nonsingular 1-forms ω on M_0 that "blow up nicely" at $\partial_\tau M$ (meaning that ω becomes unbounded near $\partial_\tau M$ in such a way that the 2-plane field ker ω extends smoothly to a 2-plane field on M tangent to $\partial_\tau M$).

Remark 2. In contrast to Thurston's result, the cones in Theorem 2 are generally not defined by a norm. Indeed, the set of cones is not generally symmetric with respect to multiplication by -1. By the work of Altman, cited in the introduction, they are defined by the nonsymmetric Floer norm of Juhász. For this result certain restrictions are placed on the sutured manifold. Most of these restrictions are mild, but one, the requirement that $H_2(M) = 0$, will hopefully be removed in the future.

3. Doubling

There are three basic topics to be treated here, namely: the doubling map in singular homology, the Thurston norm in sutured manifolds and their doubles, and the process of inducing fibrations in the double DM from certain depth one foliations on M.

3.1. *The doubling map*

If M is a smooth, connected, oriented, sutured manifold, we form the double DM along $\partial_\tau M$ (assumed to be nonempty). This is defined in complete analogy with the usual definition of the double of a manifold along its full boundary. Thus DM is an oriented manifold formed by taking a second copy of M, but with opposite orientation, and gluing the two together via the identity map on $\partial_\tau M$. We write

$$DM = M \cup (-M)/\sim .$$

There is a standard way to put a smooth, oriented structure on DM so that ∂DM is also smooth and so that the natural reflection map $\rho : DM \to DM$ is an orientation–reversing diffeomorphism. This map interchanges the corresponding points of M and $-M$, hence has $\partial_\tau M$ as its set of fixed points.

Let $S \subset M$ be a smooth, properly imbedded, oriented surface. Reversing orientations gives $-S \subset -M$. The double $DS = S \cup (-S) \subset DM$ can be viewed as an oriented, properly imbedded submanifold of DM. There is a technical problem that, if $S \cap \partial_\tau M \neq \emptyset$, smoothness of DS might fail along this set. To avoid this, one introduces a ρ–invariant Riemannian metric on DM. There is a ρ–invariant normal neighborhood U of $\partial_\tau M$ in DM and an isotopy of S makes $S \cap U$ saturated by the normal fibers of $U \cap M$. Now DS is a smooth, ρ–invariant subsurface of DM. Of course, if $S \cap \partial_\tau M = \emptyset$, DS is the disjoint union of S and $-S$. Note also that $\rho|DS$ is an orientation–reversing diffeomorphism of this surface.

A smooth triangulation of S determines a smooth triangulation of DS, producing a singular cycle mod the boundary in M and DM respectively. The corresponding classes $[S] \in H_2(M, \partial M)$ and $[DS] \in H_2(DM, \partial DM)$ are well defined, independently of the choice of triangulation. We will define a canonical "doubling" map

$$D_* : H_2(M, \partial M) \to H_2(DM, \partial DM)$$

such that $D_*[S] = [DS]$ and show that this map is "norm doubling".

At the level of singular chains, the map $\rho|M : M \to DM$ induces a linear map

$$\rho_\# : C_\#(M, \partial M) \to C_\#(DM, \partial_\tau M \cup \partial DM)$$

commuting with the singular boundary operator $\partial_\#$. Thus, we can define

$$D_\#(c) = c - \rho_\#(c), \quad \forall c \in C_\#(M, \partial M),$$

noting that this also commutes with $\partial_\#$. At this point, there is a small technical problem. The map D_* induced by $D_\#$ takes its image in the space

$$H_*(DM, \partial_\tau M \cup \partial DM),$$

whereas we want to interpret it as a map into the space $H_*(DM, \partial DM)$. The crucial property to notice is that, if the singular chain c is supported in $\partial_\tau M$, then $D_\#(c) = 0$.

Consider the open cover $\Phi = \{U, V\}$ of DM, where $U = \text{int } DM$ and V is a normal neighborhood of ∂DM with normal fibers along $\partial(\partial_\tau M)$ lying entirely within $\partial_\tau M$. Let $A = \partial_\tau M \cap V$ and note that ∂DM is a deformation retract of $A \cup \partial DM$. By abuse of notation, we also let Φ denote the induced open cover on any suspace of DM and we compute singular homology on DM and any of its subspaces using the Φ–small singular chain complex $C_\#^\Phi$. That is, each singular simplex in a chain $c \in C_\#^\Phi$ is supported either in U or in V. It is standard that the Φ–small homology H_*^Φ is canonically equal to the ordinary singular homology H_*, the equality being induced by $C_\#^\Phi \subset C_\#$.

If $c \in C_\#^\Phi(\partial M)$, then, since $D_\#$ annihilates all singular simplices in $\partial_\tau M$, $D_\#(c)$ is a chain on $A \cup \partial DM$. We obtain homomorphisms

$$D_\# : C_\#^\Phi(M) \to C_\#^\Phi(DM)$$
$$D_\# : C_\#^\Phi(\partial M) \to C_\#^\Phi(A \cup \partial DM),$$

of chain complexes, hence a chain homomorphism

$$D_\# : C_\#^\Phi(M, \partial M) \to C_\#^\Phi(DM, A \cup \partial DM).$$

This defines

$$D_* : H_*(M, \partial M) \to H_*(DM, A \cup \partial DM) = H_*(DM, \partial DM),$$

the desired doubling map.

Remark 3. The above supposes that $\partial_\tau M$ meets $\partial_\pitchfork M$. Otherwise, $\partial_\pitchfork M = T(\gamma)$ and the proof that $D_* : H_*(M, \partial M) \to H_*(DM, \partial DM)$ is even easier, not requiring the use of Φ-small homology.

Lemma 1. *If $S \subset M$ is a properly imbedded surface, then $D_*[S] = [DS]$.*

Proof. Indeed, if $c_S \in Z_2(M, \partial M)$ is a fundamental cycle for S obtained by a smooth triangulation, it is an elementary consequence of the orientation–reversing property of $\rho : DS \to DS$ that $c_S - \rho_\#(c_S) \in Z_2(DM, \partial DM)$ is a fundamental cycle for DS. □

Consider the inclusion map $i : M \hookrightarrow DM$ and the induced homomorphism

$$i^* : H^1(DM) \to H^1(M)$$

in real cohomology. Using Lefschetz duality, we view this as

$$i^* : H_2(DM, \partial DM) \to H_2(M, \partial M).$$

Lemma 2. *The composition $i^* \circ D_*$ is equal to the identity on $H_2(M, \partial M)$. In particular, the doubling map is injective on $H_2(M, \partial M)$.*

Proof. It will be enough to prove this for elements $[S] \in H_2(M, \partial M)$, where S is a properly imbedded, oriented surface in M. Indeed, these constitute the integer lattice in $H_2(M, \partial M)$. By Lemma 1, we must show that $i^*[DS] = [S]$. The Lefschetz dual of $[DS]$ is represented by a 1–form ω as follows. Fix a normal neighborhood V of DS in DM. This can be chosen so that $V \cap \partial_\tau M$ is saturated by normal fibers, as is $V \cap \partial DM$. The closed form ω is supported in V and has integral along each normal fiber equal to 1. Evidently, $V \cap M$ is a normal neighborhood of S and ω restricts in M to a representative of the Lefschetz dual of $[S]$. □

Remark 4. It is easy to give a geometric definition of

$$i^* : H_2(DM, \partial DM) \to H_2(M, \partial M)$$

on each element $[\Sigma]$ of the integer lattice. Represent this class by a properly imbedded surface $\Sigma \subset DM$ that is transverse to $\partial_\tau M$ and note that $\Sigma_+ = \Sigma \cap M$ is a properly imbedded surface in M. Then $i^*[\Sigma] = [\Sigma_+]$.

3.2. *The Thurston norm*

Roughly speaking, we define the Thurston norm in a sutured manifold by doubling along $\partial_{\text{th}} M$, computing the Thurston norm in the doubled manifold, and dividing by two. This is half the norm defined by Scharlemann in [13, Definition 7.4].

More precisely, let S be properly imbedded as usual and connected. By a small isotopy, ∂S can be assumed to be transverse to $\partial \partial_{\pitchfork} M$ and we compute $\chi_-^{\mathrm{s}}(S)$ by doubling along $\partial_{\pitchfork} M$, computing the usual χ_- of the doubled surface and dividing by two. (The superscript s stands for "sutured".) One can give an intrinsic formula for this number as follows.

The components of $S \cap \partial_{\pitchfork} M$ are circles and/or properly imbedded arcs in annular components of $\partial_{\pitchfork} M$. These circles need not be essential and some of the arcs might also fail to be essential in the sense that they start and end on the same boundary component of an annular component in $\partial_{\pitchfork} M$. We will see that these inessential arcs and circles can be eliminated, but for the moment they are allowed. Let $n(S)$ denote the number of arc components of $S \cap \partial_{\pitchfork} M$. Then the reader can verify that the formula for χ_-^{s} is

$$\chi_-^{\mathrm{s}}(S) = \begin{cases} -\chi(S) + \frac{1}{2}n(S), & \text{if this number is positive,} \\ 0, & \text{otherwise.} \end{cases}$$

As usual, if S is not connected, one defines $\chi_-^{\mathrm{s}}(S)$ as the sum of the values on each component. If z is an element of the integer lattice in $H_2(M, \partial M)$, $x^{\mathrm{s}}(z)$ is defined to be the minimum value of $\chi_-^{\mathrm{s}}(S)$ taken over all surfaces $S \in z$. Continuing to follow Thurston's lead, we extend x^{s} canonically to a pseudonorm on the vector space $H_2(M, \partial M)$ and call this the *sutured Thurston norm*.

Remark 5. Instead of computing the sutured norm by doubling in $\partial_{\pitchfork} M$, one can equally well double in $\partial_\tau M$. Again the components of $S \cap \partial_\tau M$ are properly imbedded arcs and/or circles and the number of arc components is the same number $n(S)$. One then notes that $2\chi_-^{\mathrm{s}}(S) = \chi_-(DS)$, where $\chi_-(DS)$ is defined as for the ordinary Thurston norm.

We further remark that, by a χ_-^{s}–reducing homology and/or isotopy, S can be assumed to meet each annular component of $\partial_{\pitchfork} M$ only in essential arcs, each crossing the suture once, or in essential circles, each parallel to the suture and disjoint from it. It can be assumed also that S meets each toral component only in essential circles, although this remark is not particularly consequential. At any rate, $n(S)$ is now just the number of times that ∂S crosses the sutures and it is elementary that this number is even. Thus, $\chi_-^{\mathrm{s}}(S)$ is an integer, as is $x^{\mathrm{s}}[S]$.

Example. A decomposing disk Δ in the sense of Gabai [10] has $\chi_-^{\mathrm{s}}(\Delta) = 0$ if it meets the sutures twice, $\chi_-^{\mathrm{s}}(\Delta) = 1$ if it meets them four times, etc.

Theorem 3. *The map*

$$D_* : H_2(M, \partial M) \to H_2(DM, \partial DM)$$

is norm–doubling, where the sutured Thurston norm is used on the first space and the usual Thurston norm is used on the second. Thus, if B is the Thurston ball of M and B^* that of DM, then $D_*(B/2) = B^* \cap D_*(H_2(M, \partial M))$.

Proof. It is enough to prove this on elements of the integer lattice. Let $[S]$ be represented by a χ^s_-–minimal surface S. We have already noted that $\chi_-(DS) = 2\chi^s_-(S)$, hence it will be enough to show that DS is a χ_-–minimal representative of $[DS] = D_*[S]$. If not, let $\Sigma \in [DS]$ have $\chi_-(\Sigma) < \chi_-(DS)$. Isotope Σ smoothly to be transverse to $\partial_\tau M$ and let $\Sigma_+ = \Sigma \cap M$ and $\Sigma_- = \Sigma \cap (-M)$. If no component of $\Sigma_\pm$ has positive Euler characteristic, one verifies the relation

$$\chi_-(\Sigma) = \chi^s_-(\Sigma_+) + \chi^s_-(\Sigma_-). \qquad (*)$$

The only possible components with positive Euler characteristic are spheres or disks. In the first case, irreducibility of M permits elimination of the offensive component. In the second, there will be no problem if the boundary of the disk Δ meets $\partial_\tau M$ in arcs. Otherwise, $\partial\Delta$ is a simple closed loop either in $\partial_\pitchfork M$ or $\partial_\tau M$. In the first case, Δ is also a component of Σ in DM and has zero Thurston norm. In M it has zero sutured norm, so this case also causes no problem. In the remaining case, $\partial\Delta \subset \partial_\tau M$ and tautness of the sutured manifold structure, together with irreducibility, yields an isotopy of Σ pulling the disk Δ through $\partial_\tau M$, hence eliminating it as a component of $\Sigma_\pm$. Thus $(*)$ can be assumed to hold. Interchanging the roles of M and $-M$, if necessary, we can then assume that $\chi^s_-(\Sigma_+) < \chi^s_-(S)$. But

$$[S] = i^*[DS] = i^*[\Sigma] = [\Sigma_+],$$

contradicting χ^s_-–minimality of S in $[S]$. $\qquad\square$

3.3. *Inducing fibrations on DM*

In this subsection, we assume that M, as a sutured manifold, is not a product $\partial_\tau M \times I$. This insures that $\partial_\tau M$ cannot be a fiber in a fibration of DM over the circle. We sketch some facts that are treated in greater detail in [4, 5, 6] and [7].

Let $\mathcal{F}$ be a smooth, depth one foliation of M, transverse to $\partial_{\pitchfork} M$ and having the components of $\partial_\tau M$ as sole compact leaves. A depth one leaf $L \subset M_0$ determines an element $\lambda(\mathcal{F}) \in H^1(M; \mathbb{Z})$ of the integer lattice in the real cohomology space $H^1(M)$ via the intersection product with loops in M_0. This class can also be represented by a foliated form ω on M_0 (cf. the first remark on page 43). The form ω defines $\mathcal{F}|M_0$, hence also determines $\mathcal{F}$, and its cohomology class can be viewed as a class on M via the homotopy equivalence $M_0 \hookrightarrow M$ (the natural inclusion map). For any positive constant a, the form $a\omega$ also defines $\mathcal{F}$, so we obtain a "foliated ray" $\langle \mathcal{F} \rangle \subset H^1(M)$ corresponding to $\mathcal{F}$. This ray, in turn, determines $\mathcal{F}$ up to an isotopy that is smooth in M_0 and continuous on M [4, Theorem 1.1]. We often think of a foliated ray as an isotopy class of foliations. These foliated rays are exactly the rational rays in the interiors of the foliation cones of Theorem 2.

Remark 6. Poincaré duality identifies $H^1(M) = H_2(M, \partial M)$.

The leaves of $\mathcal{F}|M_0$ spiral in a well–understood way on each component F of $\partial_\tau M$, giving rise to a nondivisible cohomology class

$$\nu : \pi_1(F) \to \mathbb{Z}$$

called the *juncture* of the spiral (cf. [4, §3]). The juncture on F depends only on the class $\lambda(\mathcal{F})$ [4, Lemma 3.1]. It can be represented by a compact, properly imbedded, oriented, nonseparating 1–manifold $N \subset F$ which need not be connected [4, pp. 159–160] and each component is assigned an integer weight.

If there is a depth one foliation $\mathcal{G}$ such that $\lambda(\mathcal{G}) = -\lambda(\mathcal{F})$, we will denote $\mathcal{G}$ by $-\mathcal{F}$ and call this the *opposite* foliation to $\mathcal{F}$. Remark that this is not the foliation defined by the form $-\omega$, even up to isotopy, since this foliation would require that the outwardly oriented components of $\partial_\tau M$ become inwardly oriented and vice versa. These orientations are part of the given sutured structure on M and may not be reversed. While, in many cases, $-\mathcal{F}$ exists, examples show that it may not. Indeed, the three vertices in Figure 2 of Section 6 are not foliated classes, but they are the negatives of foliated classes. Of course, at the cohomology level, $[-\omega] = \lambda(-\mathcal{F})$. By the ideas in the proof of [4, Lemma 3.1], the juncture for $-\mathcal{F}$ can be represented by $-N$, the manifold obtained by reversing the orientation of N. Intuitively, the foliations $\mathcal{F}$ and $-\mathcal{F}$ spin in "opposite directions" along F, appearing to be "mirror images" of one another in a small normal neighborhood of F in M.

Suppose that $\mathcal{F}$ admits an opposite foliation $-\mathcal{F}$. We can produce a taut foliation $\mathcal{F} \cup -\mathcal{F}$ on DM by using $\mathcal{F}$ in M and $-\mathcal{F}$ in $-M$, the components of $\partial_\tau M$ being the sole compact leaves. Since the foliation is taut, each of the compact leaves is a properly imbedded, incompressible surface in DM.

If F is one of these compact leaves, it inherits an orientation so that it is inwardly oriented with respect to M or $-M$ and outwardly oriented with respect to the other. Thus the junctures in F for the respective foliations can be taken to be physically the same submanifold of F, but with opposite orientations. It follows that the procedure in [5, pp. 379–381] applies, allowing us to erase these compact leaves by deleting their "spiral ramp" neighborhoods and fitting the resulting foliations together, matching convex corners of one to concave corners of the other and vice versa (cf. [5, Fig. 4]). Actually, our situation is a bit more complicated than that envisioned in [5] because our juncture need not be connected, but essentially the same construction goes through. In this way we erase all leaves that are components of $\partial_\tau M$. The resulting foliation of DM, denoted by $D\mathcal{F}$, has only compact leaves since the construction amputates the finitely many ends of all leaves and joins together their compact cores. Thus, $D\mathcal{F}$ is a fibration of DM over the circle, the fibers being transverse to ∂DM. The reader should be warned that $D\mathcal{F}$ is not uniquely determined by $\mathcal{F}$ and $-\mathcal{F}$. The topology of the fiber depends on the choices of spiral ramp neighborhoods of the components F of $\partial_\tau M$. With a little care, this construction can be carried out so that the following is true.

Lemma 3. *If the depth one foliation $\mathcal{F}$ admits an opposite foliation, then there are associated fibrations $D\mathcal{F}$ of DM over the circle with fibers transverse to ∂DM. Furthermore, there is a smooth, one–dimensional foliation $\mathcal{L}$ of DM, tangent to ∂DM and transverse both to $\mathcal{F} \cup -\mathcal{F}$ and $D\mathcal{F}$.*

While each component F of $\partial_\tau M$ fails to be a leaf of $D\mathcal{F}$, it remains an incompressible surface in DM with a special relationship to $D\mathcal{F}$.

Lemma 4. *The surface F is isotopic through properly imbedded surfaces in DM to a surface that has only positive saddle tangencies with $D\mathcal{F}$.*

Proof. The tangent bundles $\tau = \tau(\mathcal{F} \cup -\mathcal{F})$ and $\tau_0 = \tau(D\mathcal{F})$ are both transverse to $\mathcal{L}$ and transversely oriented so that both induce the same orientation along $\mathcal{L}$. It follows that τ and τ_0 are homotopic as oriented 2–plane bundles, hence have the same (relative) Euler class $e(\tau) = e(\tau_0) \in$

$H^2(M, \partial M)$. Thus

$$\int_F e(\tau_0) = \int_F e(\tau) = \chi(F).$$

We can assume, via a small isotopy near ∂DM, that each component of ∂F is either transverse to $D\mathcal{F}$ or lies in a fiber of $D\mathcal{F}$. The two possibilities correspond, respectively, to the cases in which the component of ∂F does or does not meet the juncture for $\mathcal{F}$. Thus, Thurston's general position result [14, Theorem 4] allows us to perform an isotopy of F, putting it in a position so that all tangencies with $D\mathcal{F}$ are saddles. (The possibility that F could be isotoped onto a fiber is eliminated by our assumption that M is not a product.) If some tangency is not positive (that is, the orientations of $\tau(F)$ and $\tau(D\mathcal{F})$ at the tangency are opposite), it would follow that $\int_F e(\tau_0) \neq \chi(F)$, a contradiction. □

Remark 7. Lemma 4 can also be proven more directly by a Morse theoretic argument.

Proposition 1. *If the depth one foliation $\mathcal{F}$ admits an opposite foliation and if $K \subset DM$ is a properly imbedded surface having only positive saddle tangencies with $D\mathcal{F}$, then $[K] \in H_2(DM, \partial DM)$ lies in the cone over a fibered face of the Thurston ball and K is a norm minimizing representative of $[K]$.*

Proof. Let $C \subset H_2(DM, \partial DM)$ be the cone over a top dimensional face of the Thurston ball, the interior of which contains contains the "fibered ray" $\langle D\mathcal{F} \rangle$ associated to $D\mathcal{F}$ as in Theorem 1. Let $[D\mathcal{F}] \in \langle D\mathcal{F} \rangle \smallsetminus \{0\}$. Then, by a standard argument of Thurston [14], the fact that the tangencies are positive saddles implies that the convex combination $t[D\mathcal{F}] + (1-t)[K] \in \operatorname{int} C$, $0 < t \leq 1$. Consequently, $[K] \in C$. The norm x is linear in C, coinciding there with the linear functional $-e(\tau(D\mathcal{F})) : H_2(DM, \partial DM) \to \mathbb{R}$, and so

$$x([K]) = -e(\tau(D\mathcal{F}))([K]) = -\chi(K).$$

This latter equality is due to the fact that the tangencies are positive saddles [14] (see also [3, Lemma 10.1.13]). □

Corollary 1. *If the depth one foliation $\mathcal{F}$ admits an opposite foliation and if $F \subset DM$ is as in Lemma 4, then $[F]$ lies in the cone over a lower dimensional face of a fibered face of the Thurston ball and F is norm minimizing in $[F]$.*

Proof. Indeed, by Proposition 1 and Lemma 4, F is norm minimizing in $[F]$ and that class lies in the cone over a fibered face. It cannot be in the interior of that cone since F is not the fiber of a fibration of DM.   □

Corollary 2. *If the depth one foliation* $\mathcal{F}$ *admits an opposite foliation and if* $S \subset M$ *is a properly imbedded surface such that* DS *is smooth and has only positive saddle tangencies with* $D\mathcal{F}$, *then* $x^{\mathrm{s}}[S] = -\frac{1}{2}\chi(DS)$ *and* $S \in [S]$ *realizes this minimal sutured norm.*

Proof. Indeed, by Proposition 1, DS is norm minimizing in $[DS]$. The assertion follows by Lemma 1 and Theorem 3.   □

4. Sutured handlebodies

Lemma 5. *There is a canonical decomposition*

$$H_2(DM, \partial DM) = H_2(M, \partial M) \oplus \ker i^*,$$

where $H_2(M, \partial M)$ *is imbedded as the image of* D_*.

Proof. Since $i^* \circ D_*$ is the identity on $H_2(M, \partial M)$, this is immediate.   □

Lemma 6. $\ker i^* \cong H_2(M, \partial_\pitchfork M)$.

Proof. By the long exact cohomology sequence of the pair (DM, M)

$$H^0(DM) \xrightarrow{i^*} H^0(M) \xrightarrow{\partial^*} H^1(DM, M) \to H^1(DM) \xrightarrow{i^*} H^1(M) \cdots$$

and the fact that $i^* : H^0(DM) \to H^0(M)$ is an isomorphism, it follows that $\partial^*(H^0(M)) = 0$. Thus, the kernel of $i^* : H^1(DM) \to H^1(M)$ is isomorphic to $H^1(DM, M)$. By excision and homotopy invariance, this space is isomorphic to $H^1(-M, \partial_\tau(-M))$. There is no harm in dropping the minus sign and employing Lefschetz duality to identify this space with $H_2(M, \partial_\pitchfork M)$. Here, the version of Lefschetz duality we are using is the seldom quoted one proven in [11, Theorem 3.43].   □

Let M be a sutured handlebody of genus n. We will let γ_i, $1 \le i \le m$, denote the sutures and also the homology class each suture represents in $H_1(M)$. Let $\{\gamma_i'\}_{i=1}^m$ denote the basis of $H_1(\partial_\pitchfork M)$ represented by these sutures. Let $X \subset M$ be a bouquet of circles $\alpha_j \subset M$, $1 \le j \le n$, that is a deformation retract of M. Viewing α_j as representing a homology class in $H_1(M)$ as well as a curve, one obtains a basis $\{\alpha_j\}_{j=1}^n$ of $H_1(M)$.

Consider the map

$$W : H_1(\partial_{\pitchfork} M) \to H_1(M)$$

induced by the inclusion $\partial_{\pitchfork} M \hookrightarrow M$.

Lemma 7. *The vector space $H_2(M, \partial_{\pitchfork} M)$ is canonically imbedded in the vector space $H_1(\partial_{\pitchfork} M)$ as $\ker W$.*

Proof. This follows from the long exact sequence

$$\cdots \to 0 = H_2(M) \to H_2(M, \partial_{\pitchfork} M) \xrightarrow{\partial} H_1(\partial_{\pitchfork} M) \xrightarrow{W} H_1(M) \to \cdots . \quad \square$$

Remark 8. In the above long exact sequence, the map W can be represented by the $n \times m$ matrix

$$\mathbf{W} = \begin{bmatrix} w_{11} & \cdots & w_{1m} \\ \vdots & & \vdots \\ w_{n1} & \cdots & w_{nm} \end{bmatrix}.$$

Here, we coordinatize $H_1(\partial_{\pitchfork} M)$ by the basis $\{\gamma_i'\}_{i=1}^m$ and $H_1(M)$ by $\{\alpha_j\}_{j=1}^n$. The columns of $\mathbf{W}$ are the vectors γ_i, $1 \le i \le m$. The column rank r of this matrix is the rank of the linear map W and the dimension of the kernel of W is $d = m - r$.

Theorem 4. $H_2(DM, \partial DM) \cong H_2(M, \partial M) \oplus \mathbb{R}^d$.

Proof. Indeed,

$$\begin{aligned} H_2(DM, \partial DM) &\cong H_2(M, \partial M) \oplus \ker i^* && \text{(Lemma 5)} \\ &\cong H_2(M, \partial M) \oplus H_2(M, \partial_{\pitchfork} M) && \text{(Lemma 6)} \\ &\cong H_2(M, \partial M) \oplus \ker W && \text{(Lemma 7)} \quad \square \end{aligned}$$

Let c be the number of components of $\partial_\tau M = R_+ \cup R_-$.

Theorem 5. *One has $d \ge c - 1$, with equality if and only if the linear map W has rank $m - c + 1$ if and only if the identification in Lemma 5 is*

$$H_2(DM, \partial DM) = H_2(M, \partial M) \oplus \mathbb{R}^{c-1}.$$

If $d = c - 1$, the factor $\mathbb{R}^{c-1}$ is generated by the classes represented by any $c - 1$ of the components of $R_+ \cup R_-$.

Proof. The first equivalence follows since the rank of W equals $m - d$ while the second equivalence is immediate by Theorem 4. By Lemma 5, the factor $\mathbb{R}^{c-1}$ is identified in $H_2(DM, \partial DM)$ as $\ker i^*$ and it is clear that each component N_i of $R_+ \cup R_-$ determines a homology class $\nu_i = [N_i] \in \ker i^*$. Thus, it will be sufficient to show that any $c - 1$ of these classes are linearly independent. This will also show that $d \geq c - 1$.

First note that the classes determined by the components of R_+ are linearly independent, as are those determined by the components of R_-. Indeed, there is a loop in DM having intersection number 1 with any given component of R_+ and intersection number 0 with all others. The same argument works for the components of R_-, proving that there is no nontrivial linear relation between the classes corresponding to the components of one of $R_{\pm}$.

Next, choosing the indexing appropriately, let $\{\nu_i = [N_i]\}_{i=1}^{c-1}$ be a choice of $c - 1$ of the classes and let $\nu_c = [N_c]$ be the omitted one. For definiteness, suppose that N_c is a component of R_+. We consider a linear relation

$$0 = \sum_{i=1}^{c-1} a_i \nu_i$$

and show that each a_i is forced to be zero. For each component N_i of R_-, there is an arc in M from N_c to N_i and this doubles to a loop in DM that has intersection number a_i with the right hand side of the above relation. Thus, $a_i = 0$ whenever N_i is a component of R_-. The above relation, therefore, involves only terms corresponding to components of R_+. As already observed, there is no such nontrivial relation. An entirely similar argument works when N_c is a component of R_-. □

Corollary 3. *The linear map W has rank $m - 1$ if and only if the identification in Lemma 5 is*

$$H_2(DM, \partial DM) = H_2(M, \partial M) \oplus \mathbb{R}.$$

In this case, the factor $\mathbb{R}$ is generated by $[R_+] = [R_-]$ and both R_+ and R_- are connected.

Let g be the genus of $R_+ \cup R_-$.

Theorem 6. $m - c + 1 + g = n$.

Proof. The disjoint union of R_+ and R_- has genus g. The proof consists of sequentially pasting together adjoining components of the disjoint union of R_+ and R_- along a common suture. This operation either reduces the number of components by one or adds a handle. The totality of such pastings produces a surface homeomorphic to ∂M, a connected surface of genus n. Since there are c components, $c - 1$ of the pastings along sutures reduce the number of components and the remaining $m - (c - 1)$ pastings add handles to give a total of $m - c + 1 + g$ handles. The assertion follows. $\square$

5. Computing the sutured Thurston norm

Our goal in this section is to state and prove a proposition that can often be used to find top dimensional faces of the Thurston ball. It applies to all the examples at the end of [6] and Example 2 of Section 6. We let $[\mathbf{a}_1, \ldots, \mathbf{a}_n]$ denote the closed, convex hull of a set of points $\{\mathbf{a}_1, \ldots, \mathbf{a}_n\}$ in $H_2(M, \partial M)$ or $H_2(DM, \partial DM)$ and we let $\langle \mathbf{a}_1, \ldots, \mathbf{a}_n \rangle$ be the cone with base $[\mathbf{a}_1, \ldots, \mathbf{a}_n]$ and cone point $\mathbf{0}$.

Definition 1. A *simple disk decomposition* of M is a complete disk decomposition of M in which all the disks are disjoint proper disks in M.

That is we can assume all the disk are there at the beginning when we do the disk decomposition rather than having to do the disk decomposition sequentially.

The following lemmas are consequences of Gabai's procedure of disk decomposition [9]. Let $\{D_1, \ldots, D_n\}$ be the oriented disks of a simple disk decomposition. The disk oppositely oriented to D_i will be denoted by $-D_i$. Sometimes we will use $+D_i$ for D_i to emphasize that there is a preferred orientation. We will denote the class $[+D_i] \in H_2(M, \partial M) = H^1(M)$ by $\mathbf{e}_i$. If $\mathcal{F}$ is a depth one foliation defined by the foliated form $\omega_\mathcal{F}$, we let $[\mathcal{F}] = [\omega_\mathcal{F}] \in H^1(M)$ and we denote by $\mathfrak{C}_\mathcal{F}$ the foliation cone containing $[\mathcal{F}]$.

Lemma 8. *If $\{+D_1, \ldots, +D_n\}$ is a simple disk decomposition of M giving the depth one foliation $\mathcal{F}$, then each D_i, $1 \le i \le n$, meets $\mathcal{F}$ in positive saddles. Furthermore, $\langle \mathbf{e}_1, \ldots, \mathbf{e}_n \rangle$ is a subcone of the foliation cone $\mathfrak{C}_\mathcal{F}$ and $\langle \mathcal{F} \rangle \smallsetminus \{\mathbf{0}\} \subset \mathrm{int}\, \langle \mathbf{e}_1, \ldots, \mathbf{e}_n \rangle$.*

Proof. By Gabai's construction [9], each disk $+D_i$ is transverse to $\mathcal{F}$ except at a single positive saddle tangency. By the fact that the decomposition is simple, one sees that there are positive closed transversals σ_i to $\mathcal{F}$ meeting D_i precisely at the saddle tangency with $\int_{\sigma_i} \alpha_j = \delta_{ij}$, $1 \le i, j \le n$. Here, α_j is the usual Poincaré dual form to D_j, supported in a small normal neighborhood of D_j, $1 \leq j \leq n$. Remark that $[\alpha_j] = [\mathbf{e}_j]$ and so

$$[\mathcal{F}] = \sum_{i=1}^{n} c_i[\mathbf{e}_i],$$

with all $c_i > 0$, and $[\mathcal{F}] \in \text{int} \langle [\mathbf{e}_1], \ldots, [\mathbf{e}_n] \rangle$. Then, since the saddle tangencies are positive, one constructs a smooth, nonvanishing vector field v on M such that

$$(t\omega_{\mathcal{F}} + (1-t)\alpha_i)(v) > 0$$

on M, $0 < t \leq 1$, $1 \leq i \leq n$. This can be done so that the 1-dimensional foliation $\mathcal{L}$ tangent to v is transverse both to $\mathcal{F}$ and D_i and, even though the disks meet the annular components of $\partial_\pitchfork M$, one can arrange that $\mathcal{L}$ fibers these components over S^1. If $\mathcal{Z}$ is the core lamination of $\mathcal{L}$, it follows that $t[\omega_{\mathcal{F}}] + (1-t)[\alpha_i]$ takes strictly positive values on all nontrivial structure cycles of $\mathcal{Z}$, $0 < t \leq 1$, $1 \leq i \leq n$. Thus, these cycles do not bound and, by [7, Theorem 5.4], the interior of the cone $\mathfrak{C}_{\mathcal{Z}} \subset H^1(M)$, dual to the cone of structure cycles of $\mathcal{Z}$, consists of foliated classes. Furthermore, since $\mathfrak{C}_{\mathcal{Z}}$ is closed, one takes the limit as $t \to 0$, concluding that $\mathbf{e}_i = [\alpha_i] \in \mathfrak{C}_{\mathcal{Z}}$, $1 \leq i \leq n$. Thus, $\langle \mathbf{e}_1, \ldots, \mathbf{e}_n \rangle \subset \mathfrak{C}_{\mathcal{Z}}$. It is evident that $\langle \mathcal{F} \rangle \smallsetminus \{\mathbf{0}\} \subset \text{int} \langle \mathbf{e}_1, \ldots, \mathbf{e}_n \rangle$ and our assertions follow from the maximality of $\mathfrak{C}_{\mathcal{F}}$ [7, Subsec. 6.4]. □

Lemma 9. *If $\{+D_1, \ldots, +D_n\}$ is a simple disk decomposition of M giving the foliation $\mathcal{F}$, then $\{+D_1, \ldots, +D_n\}$ is a simple disk decomposition of $-M$ giving the foliation $\mathcal{F}$. Each $D_i \subset -M$, $1 \leq i \leq n$, meets $\mathcal{F}$ in positive saddles. Furthermore, $\langle \mathbf{e}_1, \ldots, \mathbf{e}_n \rangle$ is a subcone of a foliation cone of $-M$.*

Proof. Each D_i, $1 \leq i \leq n$, and $\mathcal{F}$ have the opposite transverse orientation in $-M$ as in M, as does $R(\gamma)$. □

Lemma 10. *If $\{-D_1, \ldots, -D_n\}$ is a simple disk decomposition of M giving the foliation $\mathcal{F}$, then each $-D_i$ meets $\mathcal{F}$ in positive saddles and so the cone*

$$\langle -\mathbf{e}_1, \ldots, -\mathbf{e}_n \rangle = - \langle \mathbf{e}_1, \ldots, \mathbf{e}_n \rangle$$

is a subcone of a foliation cone of both M and $-M$.

Proof. Apply Lemmas 8 and 9. □

In the following, a boundary component of a properly imbedded surface S is said to cross the sutures essentially if its intersections with annular components of $\partial_\pitchfork M$ are essential arcs. Indeed, a small isotopy of S removes any inessential intersections of ∂S with sutures. When $S = D$ is a disk of a disk decomposition, the term "essentially" is redundant by Gabai's definition of disk decomposition, but we will use it anyway for emphasis.

Proposition 2. *If $\{D_1, \ldots, D_n\}$ and $\{-D_1, \ldots, -D_n\}$ are simple disk decompositions of M, then there is a fibration DF of DM over the circle such that the surfaces $D_i \cup -D_i$, $1 \leq i \leq n$, and R_+ have only positive saddle tangencies with the fibration. Further the Thurston norm of $D_* e_i = [D_i \cup -D_i] \in H_2(DM, \partial DM)$ is the number of times minus 2 that ∂D_i essentially crosses the sutures and the sutured Thurston norm of e_i is half this number.*

Proof. The disk decomposition $\{D_1, \ldots, D_n\}$ (respectively $\{-D_1, \ldots, -D_n\}$) gives the subcone $\langle e_1, \ldots, e_n \rangle$ of a foliation cone of M (respectively, it gives the subcone $\langle -e_1, \ldots, -e_n \rangle$ of a foliation cone of $-M$). If $\langle F \rangle \subset \text{int} \langle e_1, \ldots, e_n \rangle$, then $\langle -F \rangle \subset \text{int} \langle -e_1, \ldots, -e_n \rangle$. Then by Lemma 3, F and $-F$ can be matched up across $\partial_\tau M$ to give a fibration DF. Further Lemmas 8 and 10 imply that $D_i \cup -D_i$, $1 \leq i \leq n$, has only positive saddle tangencies with DF while Lemma 4 implies that (after a small isotopy of DM moving DF and all $D_i \cup -D_i$) R_+ has only positive saddle tangencies with DF.

Let b_i be the number of times ∂D_i essentially crosses the sutures. Then the surface $D_i \cup -D_i$ is a punctured sphere with b_i boundary components and thus $-\chi(D_i \cup -D_i) = b_i - 2$. Since this surface has only positive saddle tangencies with the fibration, Proposition 1 implies that $x(D_* e_i) = b_i - 2$ and Corollary 2 implies that $x^s(e_i) = x(D_* e_i)/2$. $\square$

In the examples we are interested in, the matrix $\mathbf{W}$ of Section 4 has rank $m - 1$ so, by Corollary 3, $\partial_\tau M$ has one positive component R_+ and one negative component R_- and $H_2(DM, \partial DM) = H_2(M, \partial M) \oplus \mathbb{R}$ where the $\mathbb{R}$ factor is generated by $\mathbf{R} = [R_+] = [R_-]$, and, without loss, we can assume

$$D_*(H_2(M, \partial M)) = H_2(M, \partial M) \oplus \{0\} \subset H_2(DM, \partial DM).$$

In the following corollary, the integer m and the matrix $\mathbf{W}$ are as in Section 4.

Corollary 4. *All four of the cones*

$$\langle D_* e_1, \ldots, D_* e_n, \pm \mathbf{R} \rangle \text{ and } \langle D_*(-e_1), \ldots, D_*(-e_n), \pm \mathbf{R} \rangle$$

are subcones of fibration cones (full-dimensional if rank $\mathbf{W} = m - 1$*) and thus each lies in a cone over a fibered face of the Thurston ball of DM. Also, the sutured Thurston norm is linear on both the cones* $\langle \mathbf{e}_1, \ldots, \mathbf{e}_n \rangle$, $\langle -\mathbf{e}_1, \ldots, -\mathbf{e}_n \rangle \subset H_2(M, \partial M)$ *and both of these cones are full-dimensional subcones of foliation cones and are contained in cones over top dimensional faces of the Thurston ball of M.*

Proof. Since $\{D_1, \ldots, D_n\}$ and $\{-D_1, \ldots, -D_n\}$ are simple disk decompositions of M and $-M$ respectively, Proposition 2 gives a fibration $D\mathcal{F}$ meeting the surfaces $D_i \cup -D_i$, $1 \leq i \leq n$, and R_+ in positive saddles. Thus, $\langle D_*\mathbf{e}_1, \ldots, D_*\mathbf{e}_n, \mathbf{R} \rangle$ is a subcone of a fibration cone (Proposition 1). If rank $\mathbf{W} = m - 1$, this cone is full-dimensional by Corollary 3 and the fact that $\mathbf{R}$ is not in the image of D_*. Similarly, since $\{-D_1, \ldots, -D_n\}$ and $\{D_1, \ldots, D_n\}$ are simple disk decompositions of M and $-M$ respectively, one sees that the cone $\langle D_*(-\mathbf{e}_1), \ldots, D_*(-\mathbf{e}_n), \mathbf{R} \rangle$ is a subcone of a fibration cone (full-dimensional if rank $\mathbf{W} = m - 1$). One obtains the other two fibration cones because the Thurston ball and its fibered faces are symmetric under multiplication by -1.

We prove the second part of the corollary for $\langle \mathbf{e}_1, \ldots, \mathbf{e}_n \rangle$. The proof for the cone $- \langle \mathbf{e}_1, \ldots, \mathbf{e}_n \rangle$ is identical. We must show that if $\mathbf{p} = u \cdot \mathbf{p}_1 + v \cdot \mathbf{p}_2$, with $\mathbf{p}, \mathbf{p}_1, \mathbf{p}_2 \in \langle \mathbf{e}_1, \ldots, \mathbf{e}_n \rangle$ and $u, v \in \mathbb{R}$ then $x^s(\mathbf{p}) = u \cdot x^s(\mathbf{p}_1) + v \cdot x^s(\mathbf{p}_2))$. Suppose on the contrary that $x^s(\mathbf{p}) \neq u \cdot x^s(\mathbf{p}_1) + v \cdot x^s(\mathbf{p}_2)$. Then, by Theorem 3, $x(D_*\mathbf{p}) \neq u \cdot x(D_*\mathbf{p}_1) + v \cdot x(D_*\mathbf{p}_2)$. This contradicts the linearity of the Thurston norm over faces of the Thurston ball of DM.

Since the sutured Thurston norm is linear on $\langle \mathbf{e}_1, \ldots, \mathbf{e}_n \rangle$, this is an (obviously full-dimensional) subcone of the cone over a fibered face of the Thurston ball. It is also a subcone of a foliation cone by Lemma 8.   □

6. Examples

In many case we can figure out the Thurston ball of knot or link complements cut apart along the Seifert surface using the methods of Section 5. The methods of Example 1 can be used to make rigorous the computations of the Thurston norm in [6, §7].

Example 1. Let M be the complement of the pretzel link $(2, 2, 2)$ cut apart along its Seifert surface as in [6, §7, Example 1] (see Figure 1).

One can do disk decompositions using disks $\{D_i, -D_j\}$ as long as $i \neq j \in \{0, 1, 2\}$. These disk decompositions are extremely easy to do using Gabai's graphical algorithm in [9, Theorem 6.1]. Since each of the ∂D_i's

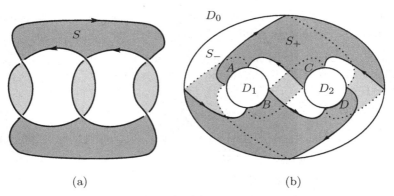

(a) (b)

Figure 1. (a) A Seifert surface for $(2, 2, 2)$ (b) The sutured manifold M obtained from $(2, 2, 2)$

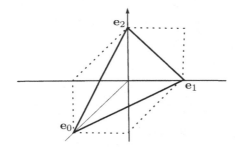

Figure 2. Thurston ball and foliation cones for $(2, 2, 2)$

essentially crosses the sutures 4 times, it follows from Proposition 2 that $x(D_*\mathbf{e}_i) = 2$ and $x^s(\mathbf{e}_i) = 1$. By Corollary 4, it follows that the Thurston ball is the dotted hexagon B of Figure 2.

The Markov process argument of [6, §7, Example 1 or Example 2] shows that $\langle \mathbf{e}_1, \mathbf{e}_2 \rangle$, $\langle \mathbf{e}_2, \mathbf{e}_0 \rangle$ and $\langle \mathbf{e}_0, \mathbf{e}_1 \rangle$ are the foliation cones.

Suitably labelling the sutures, we have that $\gamma_1 = -\alpha_1 + \alpha_2$, $\gamma_2 = \alpha_1 + \alpha_2$, and $\gamma_3 = \alpha_1 - \alpha_2$ in $H_1(M)$ (notation as in §4). The matrix

$$\mathbf{W} = \begin{bmatrix} -1 & 1 & 1 \\ 1 & 1 & -1 \end{bmatrix}$$

has rank 2. Further, since $x(\pm \mathbf{R}) = 1$, Corollary 4 allows us to conclude:

Proposition 3. *The Thurston ball of DM is the double cone (suspension) over $D_*(B/2)$ with cone points $\pm\mathbf{R}$.*

Remark 9. Similarly, if M is any of the sutured manifolds in [6, §7], the Thurston ball of DM is the double cone over $D_*(B/2)$ with cone points $\pm\mathbf{R}/x(\mathbf{R})$.

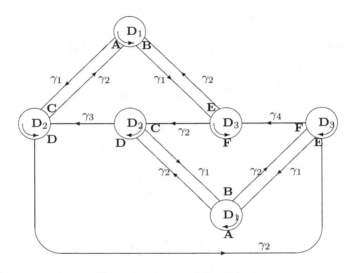

Figure 3.   A sutured handlebody

Example 2. Regard Figure 3 as drawn on S^2, the boundary of a solid ball $\mathcal{B}$. Paste D_1 to D_1 so that A (respectively B) on one copy of D_1 is matched to A (respectively B) on the other copy of D_1 and the sutures match up, paste D_2 to D_2 so that C (respectively D) on one copy of D_2 is matched to C (respectively D) on the other copy of D_2 and the sutures match up, and paste D_3 to D_3 so that E (respectively F) on one copy of D_3 is matched to E (respectively F) on the other copy of D_3 and the sutures match up. Then Figure 3 represents a sutured handlebody M of genus 3 with sutures $\gamma_1, \gamma_2, \gamma_3, \gamma_4$. Clearly, $H_2(M, \partial M) = \mathbb{R}^3$.

The arrows on the disks in Figure 3 define the positive orientation of the disks. Let α be a simple closed curve in Figure 3 going once around D_1, D_2, and D_3 in the negative sense and essentially crossing the sutures γ_2 twice and γ_3 and γ_4 once each. Then α bounds an oriented disk in the solid ball $\mathcal{B}$ which we will denote D_0. In $H_2(M, \partial M)$, $\mathbf{e}_0 + \mathbf{e}_1 + \mathbf{e}_2 + \mathbf{e}_3 = 0$.

6.1. *The Thurston ball*

Consider the compact, convex polyhedron depicted in Figure 4. One easily checks that the vertices of the quadrilateral faces really are coplanar. Two of these faces will present special problems in the following analysis.

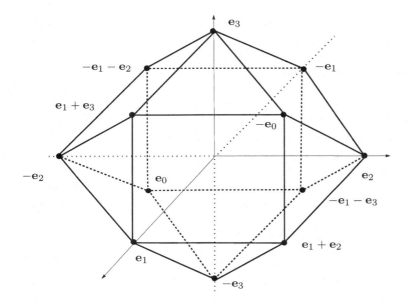

Figure 4.   The Thurston ball B

Definition 2. The two quadrilateral faces $Q^{\pm} = \pm[\mathbf{e}_2, \mathbf{e}_3, -\mathbf{e}_0, -\mathbf{e}_1]$ will be called the *exceptional faces*.

Lemma 11. *Each of the vertices in Figure 4 is represented by an oriented properly imbedded disk in M, the boundary of which essentially crosses the sutures four times.*

Proof. This is clear for $\mathbf{e}_1, \mathbf{e}_2, \mathbf{e}_3$ and has already been observed for $\mathbf{e}_0$. For $\mathbf{e}_1 + \mathbf{e}_2$, draw a closed, positively oriented curve on ∂B meeting the suture γ_1 once, γ_2 twice, and γ_3 once. This bounds the desired disk in M. One argues similarly for $\mathbf{e}_1 + \mathbf{e}_3$, obtaining a disk with boundary meeting γ_1 once, γ_2 twice, and γ_4 once. The negatives of these classes are represented by the respective oppositely oriented disks.   □

Lemma 12. *The vertices in Figure 4 all have sutured Thurston norm* 1 *and the sutured norm is identically equal to* 1 *on each of the nonexceptional faces.*

Proof. If $\{p_1, p_2, p_3\}$ are any three vertices of a nonexceptional face with corresponding representative disks $\{\Delta_1, \Delta_2, \Delta_3\}$, then these disks and their negatives give simple disk decompositions and each of the disks has boundary that essentially crosses the sutures 4 times. Verifying these disk decompositions by Gabai's algorithm is routine but tedious. The lemma then follows by Proposition 2 and Corollary 4.   □

Let $\pm D_i$ denote the disk representing $\pm e_i$, $0 \leq i \leq 3$. Then there are simple disk decompositions $\{-D_1, D_2, D_3\}$ and $\{-D_0, D_2, D_3\}$ and simple disk decompositions $\{D_0, D_1, -D_2\}$ and $\{D_0, D_1, -D_3\}$. There can be no pairs of simple disk decompositions $\{\Delta_1, \Delta_2, \Delta_3\}$ and $\{-\Delta_1, -\Delta_2, -\Delta_3\}$ that can be used in Corollary 4 to show that $Q^{\pm}$ are faces. Instead we will show that $x^s(e_2 + e_3) = 2$, which proves, by convexity of the sutured Thurston ball, that Q^+ is a face. Of course, the norm of $-e_2 - e_3$ is also 2 and Q^- is a face.

In the following, $\partial(D_2 \cup D_3)$ and the sutures γ_i are viewed as 1-cycles on ∂M.

Lemma 13. *The intersection numbers of $\partial(D_2 \cup D_3)$ with the sutures is given by:* $\gamma_1 \cdot \partial(D_2 \cup D_3) = -2$, $\gamma_2 \cdot \partial(D_2 \cup D_3) = 4$, $\gamma_3 \cdot \partial(D_2 \cup D_3) = -1$, $\gamma_4 \cdot \partial(D_2 \cup D_3) = -1$.

Proof. Let n be an exterior normal to ∂M and use a right hand rule to define the intersection number $\gamma_i \cdot D_j$, i.e. $\gamma_i \cdot D_j = \pm 1$ depending on whether (γ_i, D_j, n) is a right or left handed system $1 \leq i, j \leq 3$. One can compute the intersection numbers:

$$\gamma_1 \cdot \partial D_2 = -1 \quad \gamma_2 \cdot \partial D_2 = +2 \quad \gamma_3 \cdot \partial D_2 = -1 \quad \gamma_4 \cdot \partial D_2 = 0$$
$$\gamma_1 \cdot \partial D_3 = -1 \quad \gamma_2 \cdot \partial D_3 = +2 \quad \gamma_3 \cdot \partial D_3 = 0 \quad \gamma_4 \cdot \partial D_3 = -1.$$

The lemma follows.   □

Lemma 14. *If D is a properly embedded disk in M and ∂D crosses the sutures essentially at most twice, then D is boundary compressible. If S is a properly embedded, connected surface in M which is not a boundary compressible disk and whose boundary crosses the sutures essentially (and does so cross some sutures), then $\chi_-(DS) \geq 2$.*

Proof. Suppose D is a properly embedded disk with ∂D meeting the sutures at most twice. Put D into general position with respect to D_1, D_2, and D_3. The points of intersection of D with D_1, D_2, and D_3 will consist of circles and arcs. Assume the ends of the arcs do not lie on sutures.

By an innermost circle on D argument, we can get rid of all circles of intersection.

Similarly, by an innermost arc argument on D we can get rid of all arcs of intersection without increasing the number of intersections of ∂D with the sutures. In fact, choose an arc of intersection α in D having endpoints x and y such that there exists an arc $\beta \subset \partial D$ having endpoints x and y with $\alpha \cup \beta$ bounding a disk $D' \subset D$ such that int D' meets none of the arcs in the innermost arc argument. Since there are at least two such α and β and since ∂D meets the sutures at most twice, we can assume α and β chosen so that β meets the sutures at most once. The arc α will be a properly embedded arc in D_{i_0}, some $1 \leq i_0 \leq 3$. Thus, there is an arc $\delta \subset \partial D_{i_0}$ with endpoints x and y, such that $\alpha \cup \delta$ bounds a disk $D'' \subset D_{i_0}$. Since ∂D_{i_0} meets the sutures four times and there are two possible choices of δ, we can assume δ meets the sutures at most twice. Thus $\delta \cup \beta$ is a simple closed curve in ∂M meeting the sutures at most three times, therefore never or twice. Therefore $\delta \cup \beta$ bounds a disk $D''' \subset \partial M$ (D''' lies on the sphere represented in Figure 3 and D''' contains none of $\pm D_j$, $1 \leq j \leq 3$) and a suture meets δ if and only if it meets β. Since M is irreducible, the sphere $D' \cup D'' \cup D'''$ bounds a ball that can be used to give an isotopy of D removing the arc of intersection α. Indeed, D' can be moved onto D'', keeping α fixed, and then an arbitrarily small isotopy pulls this image of D' free of D_{i_0}. Since a suture meets δ if and only if it meets β, the isotopy does not change the number of intersections of ∂D with the sutures. After finitely many isotopies, we may assume that D does not meet D_i, $1 \leq i \leq 3$ and that ∂D meets the sutures at most twice. Cut M apart along D_1, D_2, and D_3 to give the solid ball $\mathcal{B}$ with boundary S^2 (see Figure 3). Clearly, D is boundary compressible in the solid ball $\mathcal{B}$ and so in M.

Thus if S has boundary meeting the sutures and S is not a boundary compressible disk with ∂S meeting the sutures twice, then either S is a disk with ∂S meeting the sutures 4 or more times, or S has genus $g \geq 1$, or S has at least 2 boundary components and S has genus $g = 0$. In the first case $\chi_-(DS) \geq 4-2 = 2$ and, in the second case, $\chi_-(DS) \geq 2+4g-2 = 4g > 2$. The third case falls into two subcases. If only one boundary component meets $\partial_\tau M$, then DS has genus 0 and at least four boundary components, in which case $\chi_-(DS) \geq 4+0-2 = 2$. If at least two boundary components of S meet $\partial_\tau M$, then DS has genus at least 1 and at least two boundary components, hence $\chi_-(DS) \geq 2 + 2 - 2 = 2$. □

Lemma 15. $x^s(\mathbf{e}_2 + \mathbf{e}_3) = 2$ and so $x^s \equiv 1$ on each of the exceptional faces $Q^\pm$.

Proof. The double of $S = D_2 \cup D_3$ consists of two four times punctured spheres with Euler characteristic $2 \cdot (2 - 4) = -4$. Dividing by two we see that

$$x^s(\mathbf{e}_2 + \mathbf{e}_3) \leq \chi_-^s(S) \leq |-2| = 2.$$

Let S be a surface representing $[D_2 \cup D_3]$ in $H_2(M, \partial M)$. Thus $\chi_-^s(S) = \frac{1}{2}\chi_-(DS)$. By Lemma 13,

$$\gamma_1 \cdot \partial S = -2, \gamma_2 \cdot \partial S = 4, \gamma_3 \cdot \partial S = -1, \gamma_4 \cdot \partial S = -1.$$

Therefore, ∂S must meet the sutures at least eight times. If S has only one component S_1 whose boundary meets the sutures, then

$$\chi_-^s(S) \geq \chi_-^s(S_1) \geq \frac{1}{2}\chi_-(DS_1) \geq \frac{1}{2}(8 + 4g - 2) \geq 3,$$

where g is the genus of S_1. Otherwise S has at least two components, S_1 and S_2, whose boundaries meet the sutures. Thus, by Lemma 14,

$$\chi_-^s(S) \geq \frac{1}{2}\chi_-(DS_1) + \frac{1}{2}\chi_-(DS_2) \geq 2.$$

In any event, $x^s(\mathbf{e}_2 + \mathbf{e}_3) \geq 2$ and equality holds.

For the last assertion, the fact that $x^s = 1$ on $\pm(\mathbf{e}_2 + \mathbf{e}_3)/2$ and on each vertex of $Q^\pm$, together with convexity of the unit ball, implies that $x^s|Q^\pm \equiv 1$. □

Theorem 7. *The polyhedron B in Figure 4 is the unit ball of x^s.*

Indeed, by Lemma 12 and Lemma 15, $x^s \equiv 1$ on each of the faces.

6.2. *The foliation cones*

Bases of the foliation cones are given in Figure 5 and can be found by doing the four simple disk decompositions using the disks $\{D_1, D_2, D_3\}$, $\{D_0, D_2, D_3\}$, $\{D_1, D_0, D_3\}$, and $\{D_1, D_2, D_0\}$. Thus every lattice point in the four open cones of Figure 5 correspond to depth one foliations. The foliation cones obtained this way are seen to be maximal by the Markov processes argument of [6, §7].

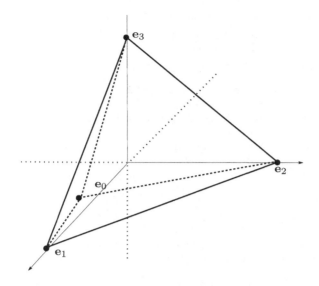

Figure 5.   Foliation cones

Remark 10. The face Q^+ (respectively Q^-) meets the interior of both $\langle \mathbf{e}_1, \mathbf{e}_2, \mathbf{e}_3 \rangle$ and $\langle \mathbf{e}_0, \mathbf{e}_2, \mathbf{e}_3 \rangle$ (respectively $\langle \mathbf{e}_0, \mathbf{e}_1, \mathbf{e}_3 \rangle$ and $\langle \mathbf{e}_0, \mathbf{e}_1, \mathbf{e}_2 \rangle$). Thus none of the foliation cones can be the union of cones over faces of the Thurston ball.

Remark 11. In this example it is not true that the Thurston ball of DM is the double cone (suspension) of $B/2$. The dimension of $H_2(DM, \partial DM)$ is 4 and $x(\pm \mathbf{R}/2) = 1$ but the two exceptional faces, coned with $\pm \mathbf{R}/2$ do not give faces of the unit ball. The cones over the other faces are faces of the unit ball.

References

[1] I. Altman, *The sutured Floer polytope and taut depth one foliations*, arXiv:1205.0442v1, 1–28.

[2] A Candel and L Conlon, *Foliations I*, Graduate Studies in Mathematics, **23**, Amer. Math. Soc., Providence, 2001.

[3] A. Candel and L. Conlon, *Foliations II*, Graduate Studies in Mathematics, **60**, Amer. Math. Soc., Providence, 2004.

[4] J. Cantwell and L. Conlon, *Isotopy of depth one foliations*, Proceedings of the

International Symposium and Workshop on Geometric Study of Foliations, Tokyo, World Scientific, November 1993, 153–173.

[5] J. Cantwell and L. Conlon, *Surgery and foliations of knot complements*, J. Knot Th. Ramific., **2** (1993), 369–397.

[6] J. Cantwell and L. Conlon, *Foliation cones*, Geometry and Topology Monographs, Proceedings of the Kirbyfest, **2**, 1999, 35–86.

[7] J. Cantwell and L. Conlon, *Open saturated sets without holonomy*, preprint.

[8] S. Friedl, A. Juhász and J. Rasmussen, *Decategorification of sutured Floer homology*, J. Topol., **4** (2011), 431–478.

[9] D. Gabai, *Foliations and Genera of Links*, Topology, **23** (1984), 381–394.

[10] D. Gabai, *Foliations and the topology of 3–manifolds*, J. Diff. Geom., **18** (1983), 445–503.

[11] A. Hatcher, Algebraic Topology, Cambridge University Press, Cambridge, 2002.

[12] A. Juhász, *The sutured Floer homology polytope*, Geometry and Topology, **14** (2010), 1303–1354.

[13] M. Scharlemann, *Sutured manifolds and generalized Thurston norms*, J. Diff. Geom., **29** (1989), 557–614.

[14] W. Thurston, *A norm on the homology of three–manifolds*, Mem. Amer. Math. Soc., **59** (1986), 99–130.

Received June 22, 2012.

FOLIATIONS 2012
ed. by Paweł WALCZAK et al.
World Scientific, Singapore, 2013
pp. 67–102

Foliations of $\mathbb{S}^3$ by Dupin cyclides

RÉMI LANGEVIN*

Institut de Mathématiques de Bourgogne, U.M.R. C.N.R.S 5584
Université de Bourgogne, France
e-mail: Remi.Langevin@u-bourgogne.fr

JEAN-CLAUDE SIFRE

Institut de Mathématiques de Bourgogne, U.M.R. C.N.R.S 5584
Université de Bourgogne, France
e-mail: jean-claude.sifre@orange.fr

1. Introduction

The restriction imposed to the geometry of the leaves of a foliation by the geometry of the ambient manifold is widely studied. For example non-existence theorem of totally geodesic foliations of hyperbolic spaces are given in [9, 13, 14]. Here we study codimension one foliations of $\mathbb{S}^3$ satisfying globally some weaker local constraint, and we accept some singularities.

More precisely, we impose that the principal curvature k_i of the leaves of the foliation satisfy

$$X_i(k_i) \equiv 0,$$

where X_i is a unit vector field tangent to the line of curvature associated to the curvature k_i. This is a conformal property of the leaves which implies that they are Dupin cyclides, that is surfaces which are in two different ways envelope of a one-parameter family of spheres (see [3] or [9]). This condition

---
*The author was supported by the Polish NSC grant N 6065/B/H03/2011/40

is still too strong to obtain examples of foliations of 3-dimensional space-forms but for the foliations of flat 3-dimensional tori by flat planes, cylinders or tori (see [9]).

One can relax even more the condition on the local geometry of the leaves as in [10]. Here we keep the Dupin condition, but accept a smooth curve of singular points.

Our construction starts from an observation of computational graphics (see [2] and [6]): Dupin cyclides maybe tangent along curves which are not characteristic circles (see Figure 1). We prove here that the one-parameter families of Dupin cyclides constructed in [2] and [6] form a foliation of the complement of the tangency curve in $\mathbb{S}^3$. Moreover we show that the paradigm of such foliations is the Hopf foliation defined in Subsection 3.1, where the Dupin cyclide leaves are tangent along a Villarceau circle.

Finally, in Subsection 6.3, we classify our foliations up to action of the group of global conformal transformations of $\mathbb{S}^3$.

Figure 1.   Two Dupin cyclides tangent along a curve.

2. Classical foliation of $\mathbb{S}^3$ by tori and Hopf fibration

2.1. *A foliation of $\mathbb{S}^3$ by tori*

Let us define a foliation of $\mathbb{S}^3$ by a one-parameter family of tori. To use complex coordinates, we consider $\mathbb{S}^3$ as the unit sphere of $\mathbb{C}^2$; this endows it with the usual constant curvature $+1$ metric.

The torus $T_{a,b} \subset \mathbb{S}^3$, $a^2 + b^2 = 1$, is given by the equations: $|z_1|^2 = a^2$, $|z_2|^2 = b^2$. It is the boundary of a tubular neighborhood in $\mathbb{S}^3$ of the geodesic $C_1 = (\mathbb{C} \times \{0\}) \cap \mathbb{S}^3$ and also the boundary of a tubular neighborhood in $\mathbb{S}^3$ of the geodesic $C_2 = (\{0\} \times \mathbb{C}) \cap \mathbb{S}^3$.

The tori $T_{a,b}$, $a^2 + b^2 = 1$, form a foliation of $\mathbb{S}^3 \setminus (C_1 \cup C_2)$ where $C_1 = \{z_2 = 0\} \cap \mathbb{S}^3$ and $C_2 = \{z_1 = 0\} \cap \mathbb{S}^3$.

Figure 2 shows the images of some of these tori by a stereographic projection from $\mathbb{S}^3 \subset \mathbb{R}^4$ onto $\mathbb{R}^3$. Our stereographic projection *Stereo* is obtained choosing the "North pole" $North = (0,0,0,1)$ on the geodesic C_2, and the "south pole" $South = -North$ (also on C_2). The projection is defined on $\mathbb{S}^3 \setminus \{South\}$ and sends a point of $\mathbb{S}^3 \setminus \{South\}$ to the intersection $Stereo(z)$ of the line joining $North$ to $z \in \mathbb{S}^3 \setminus \{South\}$ with the plane $\mathbb{R}^3 \times \{0\}$ (parallel to the 3-plane tangent to $\mathbb{S}^3$ at $North$).

The image of $T_{a,b}$ by this stereographic projection is a torus of revolution; the axis of revolution is the line image of C_2 by the stereographic projection (see Figure 2).

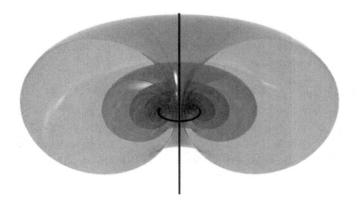

Figure 2. Foliation of $\mathbb{S}^3$ by nested tori.

2.2. Tori in $\mathbb{S}^3$ and Hopf fibration

Recall that $\mathbb{S}^3 \subset \mathbb{C}^2$ is equipped with a fibration by geodesic circles, called the *Hopf fibration*, the fibers of which are the intersection of $\mathbb{S}^3$ with the complex lines of $\mathbb{C}^2$. In other words, considering the complex projective space $\mathbb{CP}^1$, set of complex lines of $\mathbb{C}^2$, the Hopf map *Hopf* sends $(z_1, z_2) \in \mathbb{S}^3 \subset \mathbb{C}^2$ to the point $[z_1, z_2] \in \mathbb{CP}^1$.

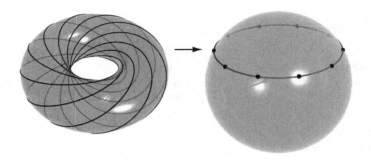

Figure 3.   Some inverse images by Hopf of the points of Hopf($T_{a,b}$).

The *Hopf fibers* are the inverse images of the points of $\mathbb{CP}^1$ by the Hopf map; the two complex lines $z_2 = 0$ and $z_1 = 0$ provide two antipodal points A and B of the 2-sphere $\mathbb{CP}^1$.

The orbits of the action of the unit circle of the complex plane, $\mathcal{U} \subset \mathbb{C}$, on $\mathbb{S}^3 \subset \mathbb{C}^2$, $(u, (z_1, z_2)) \to (uz_1, uz_2)$, are the fibers of the Hopf map. They are geodesics (i.e. great circles) of $\mathbb{S}^3$.

Lemma 1. *The image of a torus $T_{a,b} \subset \mathbb{S}^3 \subset \mathbb{C}^2$ by the Hopf map $\mathbb{S}^3 \to \mathbb{CP}^1$ is a circle centered at A (and B).*

Proof. The torus $T_{a,b}$ is given by $|z_1|^2 = a, |z_2|^2 = b; a^2 + b^2 = 1, |z_1|^2 + |z_2|^2 = 1$, equivalently, it is defined by $|\frac{z_1}{z_2}| = \lambda = \frac{a}{b}, |z_1|^2 + |z_2|^2 = 1$. In the chart $t = \frac{z_1}{z_2}$ of $\mathbb{CP}^1$, the image of $T_{a,b}$ by the Hopf map is the circle given by the equality $|t| = \lambda$. This chart maps circles of $\mathbb{C}$ centered at the origin to circles of $\mathbb{CP}^1 \simeq \mathbb{S}^2$ centered at A (and B) (for the geodesic distance in $\mathbb{CP}^1 \simeq \mathbb{S}^2$).   □

The reader guessed that the inverse map of this chart can be seen as a stereographic projection from $\mathbb{CP}^1 \simeq \mathbb{S}^2$ to $\mathbb{C}$.

Then the foliation of $\mathbb{S}^3$ by the tori $T_{a,b}$ is the inverse image by the Hopf map of the pencil of circles in the 2-sphere $\mathbb{CP}^1$ with antipodal limit points A and B.

For each of these circles, the inverse images (i.e. the Hopf fibers) of the points of this circle form a one-parameter family of Villarceau circles on the torus $T_{a,b}$. Figure 3 shows the stereographic projection of $T_{a,b}$ (instead of $T_{a,b}$ itself), the image of the torus by $Hopf$ and some of the fibers.

The other family of Villarceau circles is obtained exchanging i and $-i$ in one of the factors $\mathbb{C}$ of $\mathbb{C}^2$.

Corollary 1. *The image of a tubular neighborhood of a Hopf fiber is a disc of* $\mathbb{CP}^1$

Proof. With no loss of generality, we can suppose the Hopf fiber is the orbit Γ_0 of the point $(0,1) \in \mathbb{S}^3 \subset \mathbb{C}^2$ by the action of the unit circle $\mathcal{U} \subset \mathbb{C}$, that is $\Gamma_0 = \{0, z_2\}, |z_2|^2 = 1$. A tubular neighborhood of Γ_0 is a solid torus $ST_{a,b}$ given by $|z_1|^2 < a, |z_2|^2 > b; a^2 + b^2 = 1, |z_1|^2 + |z_2|^2 = 1$, or, equivalently by $\left|\frac{z_1}{z_2}\right| < \lambda = \frac{A}{b}, |z_1|^2 + |z_2|^2 = 1$. In the chart $t = \frac{z_1}{z_2}$ of $\mathbb{CP}^1$, the image of $ST_{a,b}$ by the Hopf map is the disc given by the inequality $|t| < \lambda$. $\qquad\square$

Definition 1. We call the tori, boundary of tubular neighborhoods of Hopf fibers *Hopf tori*.

Hopf tori are the inverse images of the circles of $\mathbb{CP}^1$ by the Hopf map.

Remark 1. As there are geodesics on $\mathbb{S}^3 \subset \mathbb{C}^2$ which are not Hopf fibers, there are tubular neighborhoods of geodesics which are not Hopf tori. We shall consider them later in this text.

Corollary 2. *The image of a sphere* $\Sigma \subset \mathbb{S}^3$ *by the Hopf map is a disc* $\Delta \subset \mathbb{CP}^1$ *(or* $\mathbb{CP}^1$ *if the sphere is totally geodesic).*

Proof. The orbit of Σ by the action of the unit circle $\mathcal{U} \subset \mathbb{C}$ is a tubular neighborhood of the orbit of its center by the action of the unit circle $\mathcal{U}$. If the sphere is totally geodesic, its image by the action of $\mathcal{U}$ is $\mathbb{S}^3$. $\qquad\square$

Remark 2. The boundary δ of the disc Δ is the image by the Hopf map of the points of Σ where it is tangent to the Hopf fibration.

3. Villarceau foliations of $\mathbb{S}^3$

3.1. *Hopf foliation of* $\mathbb{S}^3$

Now, let us replace in $\mathbb{CP}^1$ the pencil of circles with limit points A and B seen above by a tangential pencil (see Figure 4).

Proposition 1. *Consider a pencil of tangent circles at a point* $A \in \mathbb{CP}^1$. *Then the tori inverse images of the circles of the pencil by the Hopf map form a foliation* $\mathcal{F}_0$ *of* $\mathbb{S}^3$ *by Hopf tori, with singular locus the fiber of* A. *We call it* Hopf foliation.

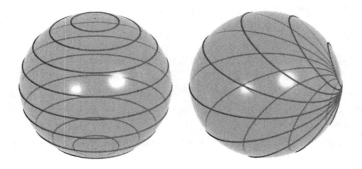

Figure 4.   Pencils with two limit points and tangential pencil in $\mathbb{CP}^1$.

As all the circles of the pencil are tangent at A, all the leaves are tangent along the Hopf fiber of A, which is a Villarceau circle on all the leaves (compare Figure 2 and Figure 5).

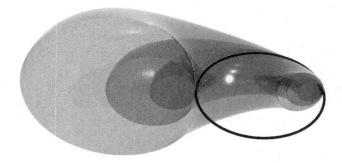

Figure 5.   Image by a stereographic projection of the Hopf foliation by Hopf tori with their common Villarceau circle.

The foliation $\mathcal{F}_0$ described in Proposition 1 will in some sense model all the other foliations by Dupin cyclides we will construct in this article.

Another way to construct this foliation by Hopf tori in $\mathbb{S}^3$ is to consider them as envelopes.

The images of a sphere $\Sigma \subset \mathbb{S}^3$ under the action of the unit circle $\mathcal{U} \subset \mathbb{C}$ form a one parameter family of spheres. When Σ is not totally geodesic, the envelope of the spheres orbit of Σ is a Hopf torus (when Σ is totally geodesic, the envelope degenerates into a geodesic circle). All the spheres of the family are tangent to all the Hopf fibers drawn on this torus, which are Villarceau circles on this torus. We call this torus T_Σ the *Hopf torus tangent to the sphere* Σ.

Consider now a fixed Hopf fiber and a pencil of spheres tangent to this Hopf fiber at some fixed point. For each sphere Σ of the pencil, the Hopf torus T_Σ contains the given Hopf fiber. Moreover, all the Hopf tori T_Σ are tangent along the initial Hopf fiber, which is a common Villarceau circle of the Hopf tori (see Figure 6).

Figure 6. Two tori, envelopes of the orbits of two tangent spheres, are tangent along a common Villarceau circle.

3.2. *Dupin cyclides and Villarceau foliations*

In $\mathbb{R}^3$, the three possible shapes for a torus of revolution are shown in Figure 7.

If we compose the inclusion $\mathbb{R}^3 \subset \mathbb{S}^3$ (by the inverse of a stereographic projection) with a conformal transformation of $\mathbb{S}^3$, the image of a torus of revolution in $\mathbb{R}^3$ is a *Dupin cyclide* in $\mathbb{S}^3$. Thus the three kinds of tori of revolution give three kinds of Dupin cyclides, *regular cyclides*, *singular cyclides* with two conical singularities and the *cuspidal* cyclides. The regular cyclides are also the images of the tori $T_{a,b} \subset \mathbb{S}^3$ by conformal transformations. The images of cones in $\mathbb{R}^3$ by (embedding $\mathbb{R}^3$ into $\mathbb{S}^3$ followed by) a conformal transformation in $\mathbb{S}^3$ have two conical singularities. Images of cylinders in $\mathbb{R}^3$ have a cuspidal singular point.

A detailed description of the different Euclidean views of cyclides of

Figure 7.   The three possible shapes of tori of revolution in $\mathbb{R}^3$.

different types may be found in [1] pp. 152–159.

Another important property of the Dupin cyclides is that they are the envelopes of the spheres tangent to three given spheres in $\mathbb{S}^3$. It was their original definition by Dupin.

In this paper, we shall construct foliations of $\mathbb{S}^3$ by Dupin cyclides. Novikov theorem implies that there is no regular foliation of $\mathbb{S}^3$ by regular Dupin cyclides. Here, by a *foliation of* $\mathbb{S}^3$ *by cyclides* we mean a one parameter families of Dupin cyclides which covers $\mathbb{S}^3$ and such that, outside a regular (compact) curve, the cyclides are the leaves of a foliation of the complement of the curve.

A first example is the Hopf foliation of Proposition 1 (see Figure 5). We will call it, and all its images by conformal maps of $\mathbb{S}^3$, *Villarceau foliations*.

4. The space of spheres and its completion

It will be convenient for us to realize both our ambient space $\mathbb{S}^3$ and the set of oriented spheres as subsets of the Lorentz space $\mathbb{R}_1^5$, that is $\mathbb{R}^5$ endowed with the Lorentz quadratic form $\mathbb{L}(x) = \mathbb{L}(x_0, x_1, x_2, x_3, x_4) = -x_0^2 + \sum_{i=1}^4 x_i^2$.

The *light-cone* $\mathcal{L}i$ is the set $\mathbb{L}(x) = 0$. Its generatrices are called *light-rays*. We also call affine lines parallel to a generatrix of the light-cone light-rays.

The light-cone separates vectors of $\mathbb{R}^5 \setminus \mathcal{L}i$ in two types: space-like vectors, such that $\mathbb{L}(v) > 0$ and time-like vectors, such that $\mathbb{L}(v) < 0$.

Definition 2. A parameterized curve $\gamma = \{\gamma(t)\}$ is space-like if its tangent vector $\dot{\gamma}(t)$ is everywhere space-like.

4.1. *The space of spheres*

The space of oriented 2-dimensional spheres in $\mathbb{S}^3$ may be parameter-ized by the *de Sitter quadric* $\Lambda^4 \subset \mathbb{R}^5$ defined as the set of points $\sigma = (x_0, x_1, x_2, x_3, x_4)$ such that $\mathcal{L}(\sigma) = 1$, in the following way. The hyper-plane $\sigma^\perp$ orthogonal to σ (for the Lorentz quadratic form $\mathcal{L}$) cuts the affine hyperplane $H_0 = \{\, x_0 = 1 \,\}$ along a 3-dimensional oriented affine hyper-plane, which cuts the unit sphere $\mathbb{S}^3 \subset H_0$ along a 2-dimensional sphere Σ. Let us orient the sphere Σ as boundary of the ball $B_\sigma = \mathbb{S}^3 \cap \{\mathbb{L}(x, \sigma) \geq 0\}$.

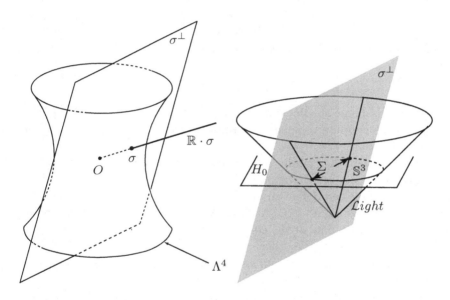

Figure 8.   The correspondence between points of Λ^4 and spheres in $\mathbb{S}^3$.

This correspondence between points σ of Λ^4 and oriented spheres $\Sigma \subset \mathbb{S}^3 \subset H_0$ is bijective.

Geometric properties of spheres have a counterpart in Λ^4. For exam-ple, two oriented spheres Σ and Σ' in $\mathbb{S}^3$ are positively (i.e. respecting the orientation) tangent if and only if the corresponding points σ and σ' in Λ^4 verify: $\mathcal{L}(\sigma, \sigma') = 1$. In that case, the points σ and σ' are joined by a segment of light-ray contained in Λ^4. In fact the oriented spheres tangent to Σ correspond to the points of the 3-dimensional cone $T_\sigma \Lambda^4 \cap \Lambda^4$ which is a union of (affine) light-rays.

The tangent space $T_\sigma \Lambda^4$ is parallel to the hyperplane $\mathbb{R} \cdot \sigma)^\perp$. It is therefore of index $(3, 1)$. This means it contains space-like, time like and light-like vectors.

Then, the spheres corresponding to the points of a space-like path (see Definition 2) $\gamma \subset \Lambda^4$ admit an envelope called a *canal surface* (see for example [8] for a proof of this fact).

A Dupin cyclide $C \subset \mathbb{S}^3$ is in two ways the envelope of a one parameter family of spheres Σ (resp Σ^*), which are represented, when the cyclide is oriented, by two space-like curves of Λ^4. From the original definition of Dupin, the spheres Σ of the first family are exactly all the spheres tangent to *three* spheres Σ_1^*, Σ_2^* and Σ_3^*, arbitrarily chosen, of the second family. In Λ^4, the points σ, corresponding to the oriented spheres Σ are tangent to Σ_1^*, Σ_2^* and Σ_3^* (with the same orientation of the common tangent plane) if and only if

$$\mathcal{L}(\sigma, \sigma_1^*) = \mathcal{L}(\sigma, \sigma_2^*) = \mathcal{L}(\sigma, \sigma_2^*) = 1.$$

The equations define an affine plane $P = T_{\sigma_1^*} \Lambda^4 \cap T_{\sigma_2^*} \Lambda^4 \cap T_{\sigma_3^*} \Lambda^4$. The points σ of Λ^4 corresponding to spheres of the first family form the intersection $\mathcal{C} = \Lambda^4 \cap P \subset \mathbb{R}^5$.

Similarly, the other family of spheres Σ^* tangent to the given cyclide corresponds to the set $\mathcal{C}^*$ of points $\sigma^* \in \Lambda^4$ such that, for all $\sigma \in \mathcal{C}$, $\mathcal{L}(\sigma^*, \sigma) = 1$. One can define an affine plane P^* using three spheres of the first family: $P^* = T_{\sigma_1} \Lambda^4 \cap T_{\sigma_2} \Lambda^4 \cap T_{\sigma_3} \Lambda^4$. The second family of spheres correspond to the points of the conic $\mathcal{C}^* = P^* \cap \Lambda^4$. Therefore each of the curves $\mathcal{C}, \mathcal{C}^*$ determines the other.

The conics $\mathcal{C}$ and $\mathcal{C}^*$ are called *sister conics*. They play a symmetric role in defining the Dupin cyclide as an envelope.

Notice that an affine plane P may miss Λ^4. If it intersects it along two time-like curves, the intersection $P \cap \Lambda^4$ corresponds to a family of nested spheres which does not define an envelope. A space-like plane intersects Λ^4 along a space-like curve, but the fact that the curve $\mathcal{C} = P \cap \overline{\Lambda^4}$ is space-like does not imply that P is space-like.

Let us translate in Λ^4 the definition of C as the envelope of a one-parameter family of spheres. Two nearby spheres intersect in a circle which tends to a *characteristic circle* when the second spheres approaches the first. The Dupin cyclide C is the union of these characteristic circles. Let us parameterize C by (Lorentz) arclength: $s \mapsto \sigma_s$. The derivative $\sigma_s' = \dfrac{d\sigma_s}{ds}$

is thus in Λ^4. Then the intersection in $\mathbb{S}^3$ of Σ_s and Σ'_s is a characteristic circle of the cyclide C, and C is the union of these circles.

Let us show that the two sister conics give a natural parameterization of C.

At each point of C, the cyclide is tangent to one (and only one) sphere Σ of the first family, and to one sphere Σ^* of the second family. Using $\mathcal{C}$ and $\mathcal{C}^*$, we see that C is the set of contact points of the spheres Σ corresponding to points σ in $\mathcal{C}$ with the spheres Σ^* corresponding to points in $\mathcal{C}^*$. The parameterization thus obtained of C associates to any couple $(\sigma, \sigma^*) \in \mathcal{C} \times \mathcal{C}^*$ the contact point $\Sigma \cap \Sigma^*$.

A *contact element* (or simply a *contact*) in $\mathbb{S}^3$ is a pair (m, h), where $m \in \mathbb{S}^3$ and h is a vector plane $h \subset T_m \mathbb{S}^3$. The set of contact conditions is of dimension 5. To each contact element (m, h) corresponds a pencil of spheres tangent to h at m. Orienting $h \subset T_m \mathbb{S}^3$ allows to orient the spheres of the pencil, and distinguishes one of the light-rays of Λ^4 corresponding to spheres of the pencil.

Reciprocally, each light-ray contained in Λ^4 defines a *contact element* in $\mathbb{S}^3$. Precisely, the intersection of the direction of the light-ray ℓ with H_0 is a point m_ℓ of $\mathbb{S}^3$ and the spheres Σ associated to the points $\sigma \in \ell$ are the spheres having a common oriented contact $h \subset m_\ell$ at the point m_{ell}. We can now observe that the quadric Λ^4 is ruled by a 5-dimensional family of (affine) light-rays.

We see now that C is the set of intersections with $\mathbb{S}^3 \subset H_0$ of the directions of the light rays $(\sigma\, \sigma^*)$, where σ belongs to $\mathcal{C}$ and σ^* belongs to $\mathcal{C}^*$.

Remark 3. More generally F. Klein (see [5]) considers the set $V(M) \subset \Lambda^4$ of all oriented spheres tangent to an oriented surface $M \subset \mathbb{S}^3$. It is a singular 3-dimensional submanifold immersed in Λ^4. When the surface is a Dupin cyclide, the singularities of $V(M)$ are the two conics $\mathcal{C}$ and $\mathcal{C}^*$.

4.2. *The projective completion of* Λ^4

To avoid going to and coming back from infinity, it is convenient to work in a projective setting.

Let us embed as usual $\mathbb{R}^5$ in $\mathbb{RP}^5$ by: $(x_0, x_1, \ldots, x_4) \mapsto [1, x_0, x_1, \ldots, x_4]$. The *Lie quadric* $\overline{\Lambda^4}$, projective completion of Λ^4, is the *set of lines* of the isotropic cone $C_\mathbb{L} \subset \mathbb{R}^6$ of the *Lie quadratic form* $\mathbb{L}$:

$$\mathbb{L}(x_{-1}, x_0, x_1, x_2, x_3, x_4) = -x_{-1}^2 - x_0^2 + \sum_{i=1}^4 x_i^2.$$

The intersection of $\overline{\Lambda^4}$ with the projective hyperplane $\mathbb{H}_\infty$ defined by $x_{-1} = 0$ is a 3-dimensional sphere $\mathbb{S}^3_\infty$ defined by the projective equations:

$$x_{-1} = 0, \quad x_0^2 = \sum_{i=1}^{4} x_i^2.$$

The points named σ are now in $\overline{\Lambda^4}$. When two subspaces of $\mathbb{R}^6$ are orthogonal for $\mathbb{L}$, we say that the corresponding projective subspaces of $\mathbb{RP}^5$ are conjugate. We denote $P^\perp$ the conjugate subspace of maximal dimension of P.

In particular, for each $\sigma \in \overline{\Lambda^4}$ not in the hyperplane $\mathbb{H}_\infty$, the conjugate projective hyperplane $\sigma^\perp \in \mathbb{RP}^5$ cuts $\mathbb{H}_\infty$ in a 2-sphere $\Sigma \subset \mathbb{S}^3_\infty$ *associated to σ.*

If σ is in $\overline{\Lambda^4} \cap \mathbb{H}_\infty$, the projective hyperplane $\sigma^\perp$ is tangent to $\overline{\Lambda^4}$ along a light line and the intersection $\sigma^\perp \cap \mathbb{S}^3_\infty$ is σ itself. Therefore, the points σ of $\overline{\Lambda^4} \cap \mathbb{H}_\infty$ correspond to the spheres reduced to a point. It means that $\overline{\Lambda^4}$ is a model for the compact set of *sphere-or-points* of $\mathbb{S}^3$.

This view point comes from the dissertation of Sophus Lie. It is developed in [1] (where $\overline{\Lambda^4}$ is named Q^4, [1] page 14) under the name of *Lie sphere geometry.*

Notice also that

Proposition 2. *The 2-spheres Σ and Σ' (in the new $\mathbb{S}^3$) associated respectively to the points σ and σ' of $\overline{\Lambda^4}$ are positively tangent if and only if σ and σ' are conjugate for the Lie quadratic form $\mathbb{L}$.*

Proof. The relation $\mathcal{L}\big((x_0, \ldots x_4), (x_0', \ldots x_4')\big) = 1$ is equivalent to

$$\mathbb{L}\big((1, x_0, \ldots x_4), (1, x_0', \ldots x_4')\big) = 0. \qquad \square$$

The point $\Sigma \cap \Sigma^*$ of C becomes just the intersection of $\mathbb{S}^3_\infty$ with the *projective line* joining σ to σ^*, which is the projective completion of the light-ray $(\sigma \, \sigma^*) \subset \Lambda^4$.

The projective completion of a light-ray $\ell \subset \mathbb{R}^5$ is the set of lines of a *totally isotropic plane* for $\mathbb{L}$ in $\mathbb{R}^6$. We call the projective lines of $\mathbb{RP}^5$ whose underlying plane in $\mathbb{R}^6$ is totally isotropic for $\mathbb{L}$ *light lines*. As $\mathbb{R}^6$ does not contain any 3-dimensional subspace totally isotropic for $\mathbb{L}$, there does not exist light projective subspaces of higher dimension. As the restriction of $\mathbb{L}$ to the underlying hyperplane $\mathcal{H}$ of $\mathbb{H}_\infty$ has the signature $(4, 1)$, $\mathbb{H}_\infty$ does not contain any (projective) light line, so that all the light-rays in $\mathbb{RP}^5$ are the projective completions of the affine light lines in $\mathbb{R}^5$. All these projective light lines intersect transversely $\mathbb{H}_\infty$.

4.3. Sister conics in $\overline{\Lambda^4}$

We can now give a characterization of projective planes $P \subset \mathbb{RP}^5$ such that the intersection $\mathcal{C} = P \cap \overline{\Lambda^4}$ defines a Dupin cyclide in $\mathbb{S}^3$.

Observe first that the affine condition $\mathcal{L}(\sigma, \sigma^*) = 1$ which characterized the sister conics $\mathcal{C}$ and $\mathcal{C}^*$ in the affine context, are now translated by the conjugation of σ and σ^*, which means that $(\sigma\sigma^*)$ is a projective light line.

Proposition 3. *Let P be a 2-dimensional projective plane P in $\mathbb{RP}^5$ not contained in $\mathbb{H}_\infty$, and not tangent to $\overline{\Lambda^4}$. Then the intersection $\mathcal{C} = P \cap \overline{\Lambda^4}$ is a conic associated to a cyclide C in $\mathbb{S}^3_\infty$ if and only if the restriction of $\mathbb{L}$ to the 3-dimensional underlying subspace $P_{vect} \subset \mathbb{R}^6$ has signature $(2, 1)$.*

Proof. If $\mathcal{C}$ is the set of points in $\overline{\Lambda^4}$ associated to a family of spheres enveloped by a cyclide $C \subset \mathbb{S}^3_\infty$, then the underlying set of vectors in $\mathbb{R}^6$ is a cone. A priori, the signature of $\mathbb{L}$ restricted to P_{vect} is $(2, 1)$ or $(1, 2)$. It cannot be $(1, 2)$, since the signature of the restriction of $\mathbb{L}$ to its conjugate subspace $(P^\perp)_{vect}$ would be $(3, 0)$ and the sister conic $\mathcal{C}^* = P^\perp \cap \overline{\Lambda^4}$ would be empty. Then it is $(2, 1)$.

Reciprocally, if the signature of $\mathbb{L}$ restricted to P_{vect} is $(2, 1)$, then the signature of $\mathbb{L}$ restricted to $(P^\perp)_{vect}$ is also $(2, 1)$. Thus $\mathcal{C}$ and $\mathcal{C}^* = P^\perp \cap \overline{\Lambda^4}$ are conjugate conics, that is *sister conics*, for which the set of the intersection points $m = (\sigma\,\sigma^*) \cap \mathbb{S}^3_\infty$, $\sigma \in \mathcal{C}$ is a cyclide in $\mathbb{S}^3_\infty$. □

4.4. Action of $\mathcal{U}$ on $\overline{\Lambda^4}$ by Lie sphere transformations

A *Lorentz isometry* is a linear mapping $f\colon \mathbb{R}^5 \to \mathbb{R}^5$ which preserves the Lorentz quadratic form $\mathcal{L}$. It induces a unique homography $\overline{f} : \mathbb{RP}^5 \to \mathbb{RP}^5$: defined by $[x_{-1}, v] \mapsto [x_{-1}, f(v)]$. The linear mapping $(x_{-1}, v) \mapsto (x_{-1}, f(v))$ from $\mathbb{R}^6$ to $\mathbb{R}^6$ preserves the quadratic form $\mathbb{L}(x_{-1}, v) = \mathcal{L}(v) - x_{-1}^2$. Then $\overline{f}$ is a *Lie sphere transformation* (see [1] chapter 3), which means a projective mapping $\mathbb{RP}^5 \to \mathbb{RP}^5$ leaving $\overline{\Lambda^4}$ globally invariant.

The Lie sphere transformation $\overline{f}$ sends the hyperplane at infinity $\mathbb{H}_\infty \approx \mathbb{RP}^4$ onto itself, and sends $\mathbb{S}^3_\infty \subset \mathbb{H}_\infty$ onto itself, therefore $\overline{f}$ induces a conformal mapping $\underline{f}$. Indeed, for each $\sigma \in \overline{\Lambda^4}$, the conjugate subspace $\sigma^\perp$ cuts $\mathbb{S}^3_\infty \subset \mathbb{H}_\infty$ along a sphere or a point, and $\overline{f}$ preserves conjugacy for $\mathbb{L}$. Then $\underline{f}$ sends spheres onto spheres and is conformal.

If $\overline{\sigma}$ is a point of Λ^4, the *corresponding sphere* $\Sigma \subset \mathbb{S}^3_\infty$ is the set of points at infinity (i.e. in $\mathbb{H}_\infty$) of the conjugate subspace $\sigma^\perp$ of σ.

Proposition 4. *The sphere $\underline{f}(\Sigma)$ is also the sphere corresponding to $f(\sigma)$.*

Proof. As $\overline{f}$ preserves the quadratic form $\mathbb{L}$, for any $\sigma \in \overline{\Lambda^4}$, $\overline{f}(\sigma^\perp) = \overline{f}(\sigma)^\perp$.

If Σ is the sphere in $\mathbb{S}^3_\infty = \mathbb{H}_\infty \cap \overline{\Lambda^4}$ associated to $\sigma \in \overline{\Lambda^4}$, then

$$\underline{f}(\Sigma) = \overline{f}(\sigma^\perp \cap \overline{\Lambda^4} \cap \mathbb{H}_\infty) = \overline{f}(\sigma)^\perp \cap \overline{\Lambda^4} \cap \mathbb{H}_\infty.$$

It means that the image by $\underline{f}$ of the sphere associated to σ is the sphere associated to $\overline{f}(\sigma)$.   □

An important particular case where the construction of f from the conformal mapping $\underline{f}$ is direct is the following. Given an isometry g of $\mathbb{S}^3 \subset \mathbb{R}^4$, the linear mapping $f: (x_0, v) \mapsto (x_0, g(v))$ is a Lorentz isometry, and the induced mapping $\underline{f}: \mathbb{S}^3_\infty \to \mathbb{S}^3_\infty$ is: $[0, 1, v] \mapsto [0, 1, g(v)]$.

Some isometries are of importance for us. For each $u \in \mathcal{U}$, the group $\mathcal{U}$ of complex numbers of module 1 acts on $\mathbb{S}^3 \subset \mathbb{C}^2$ (see Subsection 3.1). The map $(z_1, z_2) \mapsto (uz_1, uz_2)$ is an isometry, and the Lorentz isometry associated above is: $(x_0, z_1, z_2) \mapsto (x_0, uz_1, uz_2)$.

The above construction of $\overline{f}$ from f, applied to each Lorentz isometry $(x_0, z_1, z_2) \mapsto (x_0, uz_1, uz_2)$ $(u \in \mathcal{U})$ defines an action of $\mathcal{U}$ on $\mathbb{RP}^5$ by Lie sphere transformations:

$$(u, [x_{-1}, x_0, z_1, z_2]) \mapsto u.[x_{-1}, x_0, z_1, z_2] = [x_{-1}, x_0, uz_1, uz_2].$$

This defines an action of $\mathcal{U}$ on $\overline{\Lambda^4}$ which will be denoted by: $(u, \sigma) \mapsto u.\sigma$. In this projective context, and considering now the points σ in $\overline{\Lambda^4}$, the proposition (4) says that if $\sigma \in \overline{\Lambda^4}$ corresponds to the sphere $\Sigma = \sigma^\perp \cap \mathbb{S}^3_\infty$, then the image $u.\sigma$ corresponds to $u.\Sigma$. The action of $\mathcal{U}$ on $\overline{\Lambda^4}$ and on $\mathbb{S}^3_\infty$ commutes with the correspondence $\sigma \mapsto \Sigma$.

5. Foliations by cyclides tangent along a curve

We will obtain our foliations transforming data in $\overline{\Lambda^4}$ constructed from a Villarceau foliation by a *Lie sphere transformation* which is a map from $\overline{\Lambda^4}$ to itself (which is not in general associated to a conformal map of $\mathbb{S}^3$).

5.1. *Contact elements and cyclides*

Recall that a contact element (or simply a *contact*) in $\mathbb{S}^3$ is a pair (m, h), where $m \in \mathbb{S}^3$ and h is a vector plane $h \subset T_m\mathbb{S}^3$. To each contact element (m, h) corresponds a pencil of spheres tangent to h at m. Orienting $h \subset T_m\mathbb{S}^3$ allows to orient the spheres of the pencil, and distinguishes one of the light-rays of Λ^4 corresponding to spheres of the pencil.

Observe that, to a singular point of a Dupin cyclide corresponds a circle in the space of contacts, as the plane h is now allowed to rotate, staying tangent to the cone tangent to the cyclide at m, or should contain the line tangent at m to the cyclide when the cyclide has a unique singular point which is then a cuspidal point.

A triplet (p_1, p_2, p_3) of contacts is called *generic* when there is no (2-dimensional) sphere positively tangent to at least two of the three contacts. It is equivalent to say that the corresponding light lines in $\overline{\Lambda^4}$ are pairwise disjoint.

Remark 4. When the three contacts are tangent to a cyclide C, this genericity may be related to the characteristic circles of C. If C is regular, it means that no two of the three base points m_1, m_2, m_3 of the contacts are on the same characteristic circle.

Indeed, if the light lines $\ell_1 = (\sigma_1 \sigma_1^*)$ and $\ell_2 = (\sigma_2 \sigma_2^*)$ intersect at a point $\sigma \in \overline{\Lambda^4}$, then σ must be on C or C^*, otherwise σ_1, σ_1^*, σ_2 and σ_2^* would be contained in a 2-dimensional projective plane, which contradicts the fact that the union of the two sister conics are not contained in any strict projective subspace of $\mathbb{RP}^5$.

Then the two contacts are on a common characteristic circle.

If C has a singularity at m_1 and m_2 is not singular, the genericity imposes that the light-rays associated to $m_2, T_{m_2} C$ has no sphere in common with any of the light-rays associated to contacts m_1, h accepted at m_1.

5.2. *Contacts and cyclides*

Theorem 1 (Three contacts theorem). *For a generic choice of three contacts p_1, p_2, p_3 on a cyclide C, there is a one parameter family of cyclides tangent to C at these points. The cyclides are then tangent along a common regular curve Γ. This family of cyclides is obtained from a Villarceau foliation by a Lie sphere transformation of $\overline{\Lambda^4}$, the projective completion of Λ^4.*

Notice that the existence of the one-parameter family of cyclides tangent along a curve containing the three points of contact was already proved in [6].

The three contact condition, when it can be satisfied, implies contact along a curve.

Theorem 2 (Foliation by tangent cyclides). *The cyclides of each family constructed in Theorem 1 form a foliation of $\mathbb{S}^3$ singular only along* *the common curve tangency* Γ.

In Subsection 5.3, we prove the three contact theorem (Theorem 1), and in Subsection 5.4, the foliation by tangent cyclides theorem (Theorem 2).

The proofs will imply that the eventual singularities of the cyclides of the family should belong to the curve Γ.

In the Subsections 6.1 and 6.2, we describe more precisely which cyclides of the family are singular.

Corollary 3. *The foliations obtained are of two kinds.*

- *Villarceau foliations.*
- *Foliations by cyclides both regular and singular.*

The family of cyclides C_σ is parameterized by a circle Q divided in two arcs. On the first arc, all the cyclides are regular. On the second, all the cyclides have a conic singularity. For the two points of Q separating the two arcs, the two corresponding cyclides are cuspidal.

5.3. *Proof of the three contact theorem*

The Hopf foliation $\mathcal{F}_0$ of Proposition 1 will be our reference foliation.

Let us choose a Hopf torus T_0 and three contact elements p_1, p_2, p_3 on the same Hopf fiber $\Gamma_0 \subset T_0$. Recall that this fiber Γ_0 is a Villarceau circle of T_0.

We denote by $\mathcal{T}_0$ and $\mathcal{T}_0^*$ the two sister conics associated to T_0 and by σ'_1, σ'_2, σ'_3 (resp. σ'^*_1, σ'^*_2, σ'^*_3) the intersections with $\mathcal{T}_0$ (resp. $\mathcal{T}_0^*$) of the light lines corresponding to the contact elements p_i.

Let $\mathcal{W}_0$ be the union of the light lines ℓ associated to the contacts of the Hopf torus T_0 at the points of Γ_0. Thus $\mathcal{W}_0$ contains the three light lines $(\sigma'_i \sigma'^*_i)$.

Lemma 2. *The set $\mathcal{W}_0$ is a regular surface ruled by light lines.*

Proof. The action of $\mathcal{U}$ on $\mathbb{S}^3 \subset \mathbb{C}^2 \subset \mathbb{H}_\infty$ extends to an action of $\mathcal{U}$ on $\mathbb{RP}^5$. This action of $\mathcal{U}$ on $\mathbb{RP}^5 = \mathbb{P}(\mathbb{R}^2 \times \mathbb{C}^2)$, given by the formula

$$(u, [x_{-1}, x_0, z_1, z_2]) \mapsto [x_{-1}, x_0, uz_1, uz_2],$$

defines a family of Lie sphere transformations which leave $\overline{\Lambda^4}$ invariant (see Section 4.4). It sends light lines onto light lines. We denote by $u.\ell$ the image of ℓ under the action of $u \in \mathcal{U}$. In particular, the light lines corresponding to the spheres tangent (with the same orientation as the cyclide) to contacts of T_0 along the Hopf fiber Γ_0 are the light lines $u.\ell_0, u \in \mathcal{U}$, where ℓ_0 is the light line corresponding to a contact of T_0 at some point m_0 of Γ_0.

These light lines $u.\ell_0$ are disjoint, as an oriented sphere containing a Villarceau circle of a cyclide is tangent to the cyclide at a unique point if we impose that the orientations of the sphere and the cyclide coincide; therefore, their union $\mathcal{W}_0$ is regular. □

We can now show three contacts satisfying the admissibility condition of the three contact theorem (Theorem 1) gives rise to a foliation which is in some sense a pull-back of the foliation $\mathcal{F}_0$ of Proposition 1.

Let us again start from a generic triplet (p_1, p_2, p_3) of oriented contacts on a Dupin cyclide C_0.

We denote by $\mathcal{C}_0$ and $\mathcal{C}_0^*$ the sister conics of $\overline{\Lambda^4}$ associated to the spheres of envelope C_0. The three oriented contacts are associated to three light lines ℓ_1, ℓ_2, ℓ_3 which cut $\mathcal{C}_0$ in $\sigma_1, \sigma_2, \sigma_3$, and $\mathcal{C}_0^*$ in $\sigma_1^*, \sigma_2^*, \sigma_3^*$. As the projective planes containing $\mathcal{C}_0$ and $\mathcal{C}_0^*$ are conjugate, the six projective points σ_i, σ_j^* are projectively independent in $\mathbb{RP}^5$.

Corollary 4 of the appendix implies that there exists a Lie sphere transformation, which we denote by Φ, which sends σ_i onto σ_i' and σ_i^* onto $\sigma_i'^*$. It sends $\mathcal{C}_0$ onto $\mathcal{T}_0$ and $\mathcal{C}_0^*$ onto $\mathcal{T}_0^*$. It sends also each light line ℓ_i onto the light line $(\sigma_i' \sigma_i'^*)$.

The inverse image $\mathcal{W} = \Phi^{-1}\mathcal{W}_0$ of $\mathcal{W}_0$ is a regular surface and will give both sister conics in $\overline{\Lambda^4}$ corresponding to the one-parameter family of Dupin cyclides we are looking for. The conics $\mathcal{T} \subset \overline{\Lambda^4}$ are associated to the Hopf tori T tangent to T_0 along their common Villarceau circle Γ_0 (see Subsection 5.3). Let us prove that the common tangency curve $\Gamma = \mathcal{W} \cap \mathbb{H}_\infty$ of the Dupin cyclides associated to the conics $\Phi^{-1}(\mathcal{T})$ is regular.

The ruled surface $\mathcal{W}$ is the union of the light lines $\Phi^{-1}(u.\ell_0), u \in \mathcal{U}$. It cuts transversally $\mathbb{H}_\infty$, since each point of $\Phi^{-1}(\mathcal{W}_0) \cap \mathbb{H}_\infty$ contains a light line transverse to $\mathbb{H}_\infty$. Therefore, the intersection $\Gamma = \mathcal{W} \cap \mathbb{H}_\infty$ is a regular curve Γ.

The surface $\mathcal{W}_0$ is the union of the conics $\mathcal{T}$. Indeed, each Hopf torus T is tangent to all the contacts $\{u.\ell_0\}$ along Γ_0 which form a ruling by light lines of $\mathcal{W}_0$. In other words, each contact $u.\ell_0$ along Γ_0 is common to all the Hopf tori of the Villarceau foliation $\mathcal{F}_0$ of Proposition 1.

Then, $\mathcal{W}$ is also the union of the conics $\mathcal{C} = \Phi^{-1}(\mathcal{T}), \mathcal{T} \subset \mathcal{W}_0$ associated to the Hopf tori $T \in \mathcal{F}_0$. We notice that, as Φ preserves conjugation, the sister conic of $\Phi^{-1}(\mathcal{T})$ is $\Phi^{-1}(\mathcal{T}^*)$.

All these circles $\mathcal{C}$ are space-like, since the signature of $\mathbb{L}$ restricted to the 3-space $P_{vect} \subset \mathbb{R}_2^4$ which defines the projective plane containing C is, as the signature of a subspace is preserved by any Lie sphere transformation (see proposition 3), the same as the signature of $\mathbb{L}$ restricted to the subspace π_{vect} which defines the projective plane containing the conic $\mathcal{T} = \Phi(\mathcal{C})$. The light lines $\Phi^{-1}(u.\ell_0)$, $u \in \mathcal{U}$, cut any pair of sister conics $\mathcal{C}, \mathcal{C}^*$ contained in $\mathcal{W}$. Then all the cyclides C associated to the conics $\mathcal{C} = \Phi^{-1}(\mathcal{T})$ are tangent along Γ. This proves the three contacts theorem (Theorem 1). $\quad\square$

5.4. *Proof of the foliation by tangent cyclides theorem* (*Theorem 2*)

We keep the notations of Subsection 5.3. We have to prove that the cyclides associated to the conics $\mathcal{C} = \Phi^{-1}(\mathcal{T})$ constructed therein form a regular foliation of $\mathbb{S}^3 \setminus \Gamma$. We shall use for that a singular foliation of $\overline{\Lambda^4}$ by singular varieties $V(\mathcal{C})$ constructed upon the $\mathcal{C}$'s.

For any space-like conic $\mathcal{C} \in \overline{\Lambda^4}$, we denote by $V(\mathcal{C})$ the union of the light lines $(\sigma\sigma^*)$, $\sigma \in \mathcal{C}$, $\sigma^* \in \mathcal{C}^*$. It represents, when the cyclide is regular, the set of all the spheres tangent to the cyclide C.

When the cyclide is singular, $V(\mathcal{C})$ is the closure of the set of all the spheres tangent to contacts tangent to the cyclide C at regular points; this means that $V(\mathcal{C})$ contains also the spheres tangent to contacts formed by a singular point of the cyclide C and a plane tangent to the cone tangent to C at the singular point. When C admits a unique singular point, this tangent cone degenerates into a line, and the plane defining an acceptable contact should contain this line.

The set $V(\mathcal{C})$ is a 3-dimensional singular manifold, containing $\mathcal{C} \cup \mathcal{C}^*$, singular at the points of $\mathcal{C} \cup \mathcal{C}^*$, and regular at its other points. By construction, the cyclide associated to $\mathcal{C}$ is the intersection $C = V(\mathcal{C}) \cap \mathbb{H}_\infty$.

Now, we consider the union of the $V(\mathcal{C})$'s, where $\mathcal{C}$ goes over all the conics $\mathcal{C} = \Phi^{-1}(\mathcal{T})$ constructed in the subsection 1. For each $\mathcal{C}$, the set of singular points of $V(\mathcal{C})$ is $\mathcal{C} \cup \mathcal{C}^*$, which is contained in $\mathcal{W} \subset V(\mathcal{C})$.

Lemma 3. *The open set* $\overline{\Lambda^4} \setminus \mathcal{W}$ *is regularly foliated by the regular* 3-*manifolds* $V(\mathcal{C}) \setminus \mathcal{W}$.

Proof. The Lie sphere transformation Φ restricted to $\overline{\Lambda^4}$ is a diffeomorphism, therefore it is sufficient to prove that the open set $\overline{\Lambda^4} \setminus \mathcal{W}_0$ is regularly foliated by the smooth 3-manifolds $V(\mathcal{T}) \setminus \mathcal{W}_0$, where $\mathcal{T}$ goes over all the conics of $\overline{\Lambda^4}$ associated to the Hopf tori T.

First, let us show hat the regular 3-manifolds $V(\mathcal{T}) \setminus \mathcal{W}$ form a partition of $\overline{\Lambda^4} \setminus \mathcal{W}$ (recalling that $V(\mathcal{T}) \setminus \mathcal{W} = V(\mathcal{T}^*) \setminus \mathcal{W}$).

We recall that $\Gamma_0 \subset (\mathbb{H}_\infty \cap \overline{\Lambda^4})$ is one of the conics $\mathcal{T}$, which allows us to consider the singular 3-manifold $V(\Gamma_0)$, union of the light line joining points of Γ_0 to points of Γ_0^*; it corresponds the set of spheres tangent to the circle $\Gamma_0 \subset \mathbb{S}_\infty^3 \subset \mathbb{H}_\infty$. $V(\Gamma_0)$ contains $\mathcal{W}_0$ and the regular 3-manifold $V(\Gamma_0) \setminus \mathcal{W}_0$.

Let us now prove that for each point $\sigma \in \overline{\Lambda^4}$ not in $V(\Gamma_0)$ there exists a unique Hopf torus T of $\mathcal{F}_0$ such that $\sigma \in V(\mathcal{T}) \setminus \mathcal{W}$.

This assertion will be proved by translating it into a property of spheres in $\mathbb{S}^3$, and, using the Hopf map $\mathbb{S}^3 \to \mathbb{CP}^1$, into a property of circles in $\mathbb{CP}^1 \simeq \mathbb{S}^2$.

We just observed that σ is not in $V(\Gamma_0)$ if and only if the corresponding sphere Σ is not tangent to the circle $\Gamma_0 \subset \mathbb{S}^3$. Now we have to prove that such a sphere Σ is positively tangent to exactly one torus T of $\mathcal{F}_0$. The sphere $\Sigma \subset \mathbb{S}^3$ is not tangent to Γ_0 if and only if its image by the Hopf projection is a disc Δ bounded by a circle δ of $\mathbb{S}^2$ (see Corollary 2) which does not contains the projection q_0 of the Hopf fiber Γ_0. Then the boundary δ is positively tangent to a unique circle in $\mathbb{S}^3$ projection of some Hopf torus T of our family (see Figure 9).

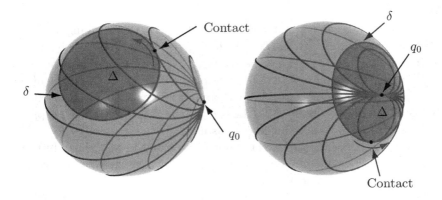

Figure 9.   Contact of Δ with a projection of a Hopf torus of $\mathcal{F}_0$.

Therefore the 3-manifolds $V(\mathcal{T}) \setminus \mathcal{W}_0$ form a partition of $\overline{\Lambda^4} \setminus \mathcal{W}_0$. It remains to prove that this partition is in fact a foliation.

For any element $\sigma \in \ell_0$, the conic $\mathcal{T}_\sigma$ is the orbit $\mathcal{U}.\sigma$ of σ under the action

$$(u, [x_{-1}, x_0, z_1, z_2]) \mapsto [x_{-1}, x_0, uz_1, uz_2]$$

of $\mathcal{U}$ on $\overline{\Lambda^4}$ constructed in the Section 4.4. Moreover, the mapping η_σ: $u \mapsto u.\sigma$ is a homography from $\mathcal{U}$ to $\mathcal{T}_\sigma$.

The proof that the 3-manifolds $V(\mathcal{T}_\sigma) \setminus \mathcal{W}_0$ foliate $\overline{\Lambda^4}$ is local, and may be done around a point $\sigma' \in \overline{\Lambda^4} \setminus \mathcal{W}_0$.

If σ' is not in $V(\Gamma_0)$, the corresponding sphere Σ' is not tangent to Γ_0. One sees on Figure 9 that the mapping associating to σ'' (near σ') the unique Hopf torus T_σ to which Σ'' (associated to σ'') is positively tangent is a submersion with fibers (locally) the 3-manifolds $V(\mathcal{T}_\sigma) \setminus \mathcal{W}_0$. Then near σ', the 3-manifolds $V(\mathcal{T}_\sigma) \setminus \mathcal{W}_0$ foliate $\overline{\Lambda^4}$.

Unfortunately, around a point σ' of $V(\Gamma_0) \setminus \mathcal{W}_0$, the geometric construction of Figure 9 does not give immediately such a submersion.

We shall construct a Lie sphere transformation ρ which sends $V(\Gamma_0)$ onto another 3-manifold $V(\mathcal{T}_{\sigma_0})$, such that the inverse image of the local foliation of $\overline{\Lambda^4}$ around $\rho(\sigma') \in V(\mathcal{T}_{\sigma_0}) \setminus \mathcal{W}_0$ by the 3-manifolds $V(\mathcal{T}_\sigma) \setminus \mathcal{W}_0$ gives a local foliation of $\overline{\Lambda^4}$ around σ' by the 3-manifolds $V(\mathcal{T}_{\rho^{-1}(\sigma)}) \setminus \mathcal{W}_0$.

Let us chose from now on (in this proof) a point $m_0 \in \Gamma_0$ and let ℓ_0 be the light line corresponding to the contact element $(m_0, T_{m_0}T_0)$. We have seen that the other light lines generating $\mathcal{W}_0$ are of the form $u.\ell_0$, $u \in \mathcal{U}$ (see Lemma 2).

For any two points σ_1 and σ_2 on the light line ℓ_0, the mapping $\eta_{\sigma_2} \circ \eta_{\sigma_1}^{-1}$ is a homography from $\mathcal{T}_{\sigma_1}$ to $\mathcal{T}_{\sigma_2}$ which may be writes $u.\sigma_1 \mapsto u.\sigma_2$.

If $\sigma \in \ell_0$, let σ^* be the unique element of $\ell_0 \cap \mathcal{T}_\sigma^*$. Its geometric meaning in $\mathbb{S}_\infty^3$ is the following: if $\Sigma \subset \mathbb{S}_\infty^3$ corresponds to σ, σ^* is the point of $\overline{\Lambda^4}$ corresponding to the sphere Σ^* of the second family of spheres tangent to the torus T_σ at the point of T_σ defined by the contact ℓ_0. The mapping $\sigma \mapsto \sigma^*$ from ℓ_0 to itself is involutive, as it reflects sister-hood.

For any σ_1 and σ_2 on ℓ_0, the lemma 9 (proved in the appendix) implies that there exists a unique orientation preserving Lie sphere transformation whose restriction to $\mathcal{T}_{\sigma_1}$ is $u.\sigma_1 \mapsto u.\sigma_2$ and whose restriction to $\mathcal{T}_{\sigma_1^*}$ is $u.\sigma_1^* \mapsto u.\sigma_2^*$. This Lie sphere transformation will be denoted by $\rho_{\sigma_1, \sigma_2}$. It leaves ℓ_0 invariant, since it sends σ_1 onto σ_2 and σ_1^* onto σ_2^*. Using the lemma 9, we can see that $\rho_{\sigma_1, \sigma_2}$ commutes with the action of $\mathcal{U}$ on $\overline{\Lambda^4}$. For any $\sigma \in \ell_0$, it sends $\mathcal{T}_\sigma$ onto $\mathcal{T}_{\rho_{\sigma_1, \sigma_2}(\sigma)}$ and $V(\mathcal{T}_\sigma) \setminus \mathcal{W}_0$ onto $V(\mathcal{T}_{\rho_{\sigma_1, \sigma_2}(\sigma)}) \setminus \mathcal{W}_0$.

We notice that the mappings $\rho_{\sigma_1, \sigma_2}$ satisfy

- $\rho_{\sigma_1, \sigma_1}$ is the identity,

$$- \rho_{\sigma_2,\sigma_3} \circ \rho_{\sigma_1,\sigma_2} = \rho_{\sigma_1,\sigma_3}.$$

This implies that they form a one parameter group G of Lie sphere transformations which permute the conics associated to the Dupin cyclides of our family.

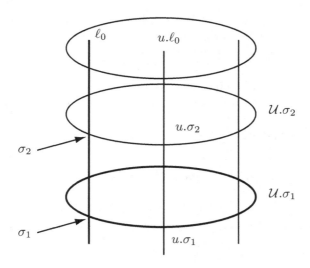

Figure 10. $\mathcal{W}_0$.

Let us choose $\sigma_0 \in \ell_0$ is different from $m_0 = \ell_0 \cap \Gamma_0$ and $m_0^* = \ell_0 \cap \Gamma_0^*$. The Lie transformation $\rho = \rho_{\sigma_0,m_0}$ is a diffeomorphism which preserves globally the 3-manifolds $V(\mathcal{T})$ and maps a neighborhood of σ' to a foliated neighborhood of $\rho(\sigma')$. $\qquad \square$

Let us get back to the proof of the foliation by tangent cyclide theorem (Theorem 2). We denote by $\mathcal{C}_\sigma$ the conics $\mathcal{C}_\sigma = \Phi^{-1}(\mathcal{T}_\sigma)$, $\sigma \in \ell_0$.

Recall that $\mathbb{S}^3 = \mathbb{H}_\infty \cap \overline{\Lambda^4}$ and $\Gamma = \mathbb{H}_\infty \cap \mathcal{W}$. Then at each point $m \in \mathbb{S}^3 \setminus \Gamma = \mathbb{S}^3 \setminus \mathcal{W}$, the manifold $V'(\mathcal{C}_\sigma) = V(\mathcal{C}_\sigma) \setminus \mathcal{W}$ which contains m cuts transversally $\mathbb{S}^3$, since $V(\mathcal{C}_\sigma)$ contains a light line passing through m and transverse to $\mathbb{H}_\infty$. Then the intersections $V'(\mathcal{C}_\sigma) \cap (\mathbb{S}^3 \setminus \Gamma)$, $\sigma \in \ell_0$, foliate $\mathbb{S}^3 \setminus \Gamma$. $\qquad \square$

6. Detailed description of the foliations

As in the previous subsections, we denote by $\mathcal{T}_\sigma$ conics associated to Hopf tori forming a Hopf foliation $\mathcal{F}_0$ (see Proposition 1 and Subsection 5.3). The conic $\mathcal{T}_\sigma$ is the orbit $\mathcal{U}.\sigma$, of a point $\sigma \in \ell_0$. The conic $\mathcal{C}$ is the image $\mathcal{C}_\sigma = \Phi^{-1}(\mathcal{T}_\sigma)$, its associated cyclide is $C_\sigma \subset \mathbb{S}^3$. The foliation by the C_σ's is denoted by $\mathcal{F}$.

In this section, we prove Corollary 3. Let us prove that, when $\mathbb{H}_\infty$ contains some conic $\mathcal{C}_\sigma$, the foliation is a Villarceau foliation, and that, otherwise the leaves are both singular and regular cyclides.

Recall that a singular point of a Dupin cyclide C correspond to an intersection point of $\mathbb{H}_\infty$ and $\mathcal{C} \cup \mathcal{C}^*$. The reader will find in [9] an affine version of this result: when one of the conics $\mathcal{C}$ or $\mathcal{C}^*$ (in Λ^4) is an hyperbola, the two asymptotes correspond to two singular points, and when one of the conics is a parabola, the asymptotic direction corresponds to a unique cuspidal singular point.

6.1. *Singularities of the leaves: the Villarceau case*

Proposition 5. *When $\mathbb{H}_\infty$ contains some conic $\mathcal{C}_{\sigma_1}$ of the family $\mathcal{C}_\sigma$, the foliation $\mathcal{F}$ by the Dupin cyclides envelopes of spheres corresponding to points of the conics $\mathcal{C}_\sigma$ is a Villarceau foliation. The singular curve Γ of $\mathcal{F}$ is a Villarceau circle of all the cyclides of the foliation and $\Gamma = \mathcal{C}_{\sigma_1} \subset \mathbb{S}^3 \subset \mathbb{H}_\infty$*

Proof. Let us first prove the

Lemma 4. *The singular curve $\Gamma \subset \mathbb{S}^3_\infty \subset \overline{\Lambda^4}$ of the foliation $\mathcal{F}$ is the conic $\mathcal{C}_{\sigma_1}$. Moreover, all the other Dupin cyclides C_σ are regular, and Γ is the common Villarceau circle of these Dupin cyclides.*

Proof. When $\mathbb{H}_\infty$ contains some $\mathcal{C}_{\sigma_1}$, the cyclide C_{σ_1} degenerates into the circle $C_{\sigma_1} \subset \mathbb{S}^3_\infty \subset \mathbb{H}_\infty$. The sister conic of $\mathcal{C}_{\sigma_1}$ is associated to the pencil of spheres containing $C_{\sigma_1} \subset \mathbb{S}^3_\infty$.

Let us show that, in this case, $\mathcal{C}_{\sigma_1}$ is the intersection $\mathcal{W} \cap \mathbb{H}_\infty = \Gamma$. First, $\mathcal{C}_{\sigma_1} \subset \mathcal{W}$ is contained in $\Gamma = \mathcal{W} \cap \mathbb{H}_\infty$. We must show that Γ contains no point out of $\mathcal{C}_{\sigma_1}$. By construction, each Hopf torus $\mathcal{T}_\sigma \subset \mathcal{W}_0$ cuts each generating light line $\ell \subset \mathcal{W}_0$. The surface $\mathcal{W}$, which is, keeping the notation of the proof of Lemma 2, the image of $\mathcal{W}_0$ by a Lie sphere transformation Φ^{-1}, inherits this property. Then if m is a point of Γ not in $\mathcal{C}_{\sigma_1}$, the light line ℓ_m of $\mathcal{W}$ which contains m meets $\mathcal{C}_{\sigma_1} \subset \mathbb{H}_\infty$ at a point $\xi \neq m$. It then should be contained in $\mathbb{H}_\infty$, which is impossible. Therefore $\mathcal{C}_{\sigma_1} = \Gamma$, and Γ is a circle, intersection of a plane and $\mathbb{S}_\infty^3$.

Moreover, as the conics $\mathcal{C}_\sigma$ are disjoint and as $\mathcal{W} \cap \mathbb{H}_\infty = \Gamma$, the only conic $\mathcal{C}_\sigma$ which meets $\mathbb{H}_\infty$ is $\mathcal{C}_{\sigma_1} = \Gamma$. It means that except $\Gamma = \mathcal{C}_{\sigma_1}$, all the cyclides of the foliation are regular. As Γ cannot be a characteristic circle of a cyclide $\mathcal{C}_\sigma$ disjoint from $\mathbb{H}_\infty$, Γ is a Villarceau circle common to all the cyclides of the foliation. $\square$

We chose now one of the regular cyclides of the family, say $\mathcal{C}_{\sigma_2}$; Γ is a Villarceau circle on $\mathcal{C}_{\sigma_2}$. We may construct from the pair $(\mathcal{C}_{\sigma_2}, \Gamma)$ a Villarceau foliation which we denote by $\mathcal{F}_2$. Our goal will be to prove that our foliation $\mathcal{F}$ is the pull back $\mathcal{F}_2$ of the Hopf foliation $\mathcal{F}_0$ that we constructed in Proposition 1.

Lemma 5. *For any couple (C, Γ), C a regular Dupin cyclide $C \subset \mathbb{S}_\infty^3$ and $\Gamma \subset C$ a Villarceau circle, there exists a unique Villarceau foliation $\mathcal{F}_2$ made of leaves which are Dupin cyclides, one of them C, tangent along the curve Γ.*

Proof. The existence is easy to deal with. Let us chose an arbitrary Hopf torus T_0 with the same conformal invariant as C, and a Hopf fiber Γ_0 on it. There exists a conformal transformation ψ of $\mathbb{S}_\infty^3$ which sends C onto T_0 and Γ onto Γ_0. The inverse image $\mathcal{F}_2$ of the Hopf foliation $\mathcal{F}_0$ of Proposition 1 is a Villarceau foliation which satisfies our requirements.

A priori the Villarceau foliation $\mathcal{F}_2$ depends on the conformal mapping ψ chosen.

Let us consider another conformal map $\psi' : \mathbb{S}^3 \to \mathbb{S}^3$ which sends C onto T_0 and Γ onto Γ_0. The composition $\psi' \circ \psi^{-1}$ is a conformal map of $\mathbb{S}^3 \subset \mathbb{C}^2$ which preserves globally T_0 and Γ_0. We let the reader prove that it is a map of the form

$$(z_1, z_2) \mapsto (e^{i\theta} z_1, e^{i\theta} z_2).$$

The pull-backs of the foliation $\mathcal{F}_0$ by the maps ψ and ψ' are therefore the same. $\square$

We suppose now that $\mathcal{F}$ is a singular foliation of $\mathbb{S}^3$ with leaves Dupin cyclides tangent along a curve Γ. We want to prove that it is the Villarceau foliation $\mathcal{F}_2$ we constructed above. This is a consequence of the uniqueness of the Dupin cyclides satisfying three admissible contact conditions in the version proved in ([6]) of Theorem 1. We will give another proof of this result here, which does not use the "dynamical" argument of [6].

It is sufficient to prove that each cyclide C', leaf of $\mathcal{F}$ is a leaf of $\mathcal{F}_2$.

Let us choose a point $m_0 \in \Gamma \subset C'$. In each of the two families of spheres enveloped by C', there is one sphere positively tangent to C' at m_0. Let us choose one of the two possible spheres, say Σ; the *first* family of spheres enveloped by C' is the family containing Σ.

There exists exactly one cyclide C'' of $\mathcal{F}_2$ such that Σ is enveloped by C'', since, at m_0, Σ is in the pencil of spheres tangent to the imposed contact at this point. Let us prove that $C' = C''$.

The first family of spheres enveloped by C'' is for us again the family which contains the sphere Σ .

We recall that C' and C'' are tangent along the common Villarceau circle Γ.

Let us prove that the second families of spheres enveloped respectively by C' and by C'' are the same. This result relies on the following lemma (see Figure 11).

Lemma 6. *Let C' be a regular cyclide, Σ a sphere of one of the two families enveloped by C' (called the* first *family), and m a point of C' not on Σ.*

Then the (unique) sphere Σ' positively tangent to C' at m and positively tangent to Σ belongs to the second family of spheres enveloped by C'.

Proof. The (oriented) spheres positively tangent to C' at m form a pencil of spheres, and exactly one of them is positively tangent to Σ. Let us denote it by Σ'. Now, let Σ'' be the unique sphere of the second family of spheres enveloped by C' which is positively tangent to C' at m. All the spheres tangent to C at m contained in Σ'' are disjoint from C', and therefore from all the spheres of the first family of spheres enveloping C'. Then Σ'', which is positively tangent to all the spheres of the first family, has the property characterizing Σ': to be positively tangent to C' at m and to be positively tangent to Σ. As both definitions define a unique sphere, $\Sigma' = \Sigma''$, that is Σ' belongs to the second family enveloping C'.   □

We finish now the proof of Proposition 5. The Lemma 6 associates to each $m \in \Gamma \backslash \{m_0\}$ a sphere Σ', constructed using only the contact condition at m and Σ. These spheres Σ' form the second family of spheres enveloping C' (but for one of them, tangent at m_0 to C').

The same construction, from the contacts at the points $m \in \Gamma \setminus \{m_0\}$ and Σ, provides the spheres of the second family of spheres enveloped by

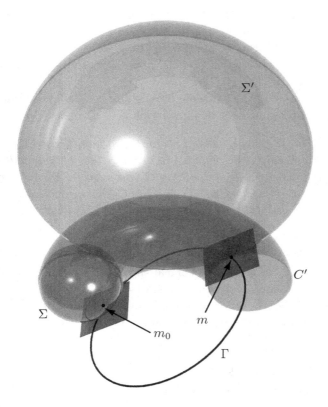

Figure 11.   The sphere Σ' tangent to C' at m and to Σ.

C''. Then the second families of spheres enveloping respectively C' and C'' are the same, and therefore $C' = C''$.

6.2. *Singularities of the leaves: the non-Villarceau case*

We have now to study the foliation obtained when $\mathbb{H}_\infty$ does not contain any conic $\mathcal{C}_\sigma$ of the family. We recall that for any $\sigma \in \ell_0$, we have $\mathcal{C}_\sigma = \Phi^{-1}(\mathcal{T}_\sigma)$, where $\mathcal{T}_\sigma$ is the conic associated to the Hopf torus T_σ of the Villarceau foliation $\mathcal{F}_0$.

This set of cyclides is already parameterized by $\ell_0 \colon \sigma \mapsto C_\sigma$, but this parameterization is not injective. Indeed, for $\sigma \in \ell_0$, let σ^* be the unique point of the sister conic $(\mathcal{T}_\sigma)^*$ on ℓ_0. Then $\mathcal{T}_\sigma^* = \mathcal{T}_{\sigma^*}$, and $\mathcal{C}_\sigma^* = \mathcal{C}_{\sigma^*}$, which implies that $C_\sigma = C_{\sigma^*}$.

Nevertheless, we see on the Hopf fibration $\mathcal{F}_0$ that the quotient Q of ℓ_0 by the identification $\sigma \sim \sigma^*$ may be taken to parametrize injectively the leaves of $\mathcal{F}_0$, and also the set of cyclides $\{C_\sigma\}$.

This set of parameters Q is a circle, as $\sigma \mapsto \sigma^*$ is an involutive homeomorphism of ℓ_0.

During the proof of the following proposition, we shall need (and prove) a more precise description of this involution $\sigma \mapsto \sigma^*$ of ℓ_0.

Proposition 6. *We suppose that no conic C_σ is contained in $\mathbb{H}_\infty$. Then the circle of parameters Q is decomposed in two arcs. For all parameters in the interior of the first arc, all the cyclides C_σ are regular. For all parameters in the interior of the second, all the cyclides C_σ have a conical singularity. The two limit points of the arcs correspond to two distinct cuspidal cyclides.*

We denote by $[\sigma]$ the class $\{\sigma, \sigma^*\}$ of σ in Q.

If the cyclide $C_\sigma \subset \mathbb{S}^3_\infty$ is singular, a singularity is a point of $\mathbb{H}_\infty \cap C_\sigma$ or $\mathbb{H}_\infty \cap C_\sigma^*$. Moreover, C_σ has two conical singularities when C_σ or C_{σ^*} cuts transversally $\mathbb{H}_\infty$, and a cuspidal singularity when C_σ or C_{σ^*} is tangent to $\mathbb{H}_\infty$. As the cyclides C_σ vary continuously with σ, the set of parameters for which C_σ cuts transversally $\mathbb{H}_\infty$ is open in Q. Similarly, the set of parameters for which neither C_σ nor C_{σ^*} cuts $\mathbb{H}_\infty$ is open. When σ varies, we pass from cyclides with conical singularities to non-singular ones. On the way we must encounter a cuspidal cyclide.

To prove the proposition, we will prove that there are exactly two parameters $[\sigma]$ for which C_σ is a cuspidal cyclide, and that in one of the arcs delimited by them, there is one cyclide with conical singularity, and on the other arc, there is a non-singular cyclide.

Proof of Proposition 6. We denote by $\mathbb{H}$ the image $\Phi(\mathbb{H}_\infty)$. For each $\sigma \in \ell_0$, the number of points of $\mathbb{H}_\infty \cap C_\sigma$ is the same as the number of points of intersection of $\mathbb{H}$ with $\mathcal{T}_\sigma = \Phi(C_\sigma)$, since Φ is a diffeomorphism.

We shall study this number when σ varies. The projective plane P_σ containing $\mathcal{T}_\sigma$ cuts $\mathbb{H}$ along a projective line Δ_σ, otherwise P_σ would be contained in $\mathbb{H}$ and C_σ would be contained in $\mathbb{H}_\infty$, which has been excluded from the hypothesis. Then $\mathbb{H} \cap \mathcal{T}_\sigma = \Delta_\sigma \cap \mathcal{T}_\sigma$.

A simple way to follow the evolution of the points $\Delta_\sigma \cap \mathcal{T}_\sigma$ while σ varies is to pull them back in the same plane, using the Lie sphere transformations $\rho_{\sigma_0, \sigma}$, which form a one-parameter group of Lie sphere transformations; in the proof of Lemma 3 we denoted it by G.

Let us recall the definition of these transformations $\rho_{\sigma_0, \sigma}$ $(\sigma \in \ell_0)$. We have fixed $\sigma_0 \in \ell_0$. Given $\sigma \in \ell_0$, the map $\rho_{\sigma_0,\sigma}$ is the unique orientation preserving Lie sphere transformation sending each $u.\sigma_0$ to $u.\sigma$, and sending $u.\sigma_0^*$ to $u.\sigma^*$, for all $u \in \mathcal{U}$, where σ^* is the point $\ell_0 \cap (\mathcal{T}_\sigma)^*$. The inverse images of the planes P_σ's (when σ varies on ℓ_0) by the respective maps $\rho_{\sigma_0,\sigma}$ is always the plane $P = P_{\sigma_0}$. The inverse images of all the $\mathcal{T}_\sigma$'s by $\rho_{\sigma_0,\sigma}$ is then the conic $\mathcal{T}_{\sigma_0}$.

Lemma 7. *The inverse images $\delta_\sigma = \rho_{\sigma_0,\sigma}^{-1}(\Delta_\sigma)$ of the lines Δ_σ form, in the plane $P = P_{\sigma_0}$, a pencil of lines (see Figure 12).*

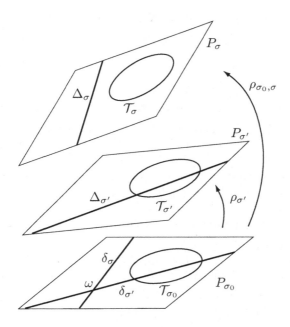

Figure 12. Different types of singularities seen on P_{σ_0}.

To prove Lemma 7, we need a precise projective description of the Lie sphere transformations $\rho_{\sigma_0,\sigma}$. We denote by $[X]$ the projective class of a vector $X \in \mathbb{R}^6$, and by $[F]$ the homography of $\mathbb{RP}^5$ associated to a linear automorphism F of $\mathbb{R}^6$.

Lemma 8 (Projective description of G). *Let $\mathcal{J}$ be the unique orientation preserving Lie sphere transformation which extends to $\mathbb{RP}^5$ the couple (h, h^*) of homographies $h: u.\sigma_0 \mapsto u.\sigma_0^*$ from $\mathcal{T}_0$ to $\mathcal{T}_0^*$ and $h^*: u.\sigma_0^* \mapsto u.\sigma$.*

Let J be a linear mapping $\mathbb{R}^6 \to \mathbb{R}^6$ such that $\mathcal{J} = [J]$. Then G is the set of homographies $[\alpha \operatorname{Id} + \beta J]$, where $[\alpha, \beta]$ goes over $\mathbb{RP}^1$.

The proof of this purely technical lemma is given in the appendix, Subsection 7.2.

Proof of Lemma 7. We denote by $\rho_{\alpha,\beta}$ the projective map $[R_{\alpha,\beta}]$.

Let ϕ be a linear form, equation of $\mathbb{H}$. Then, for $X \in P_{vect}$ (the underlying vector subspace of P), its class $[X]$ is sent by $\rho_{\alpha,\beta}$ in $P_\sigma \cap \mathbb{H}$ if and only if

$$\phi(R_{\alpha,\beta}(X)) = \alpha\phi(X) + \beta\phi(J(X)) = 0.$$

When $[\alpha, \beta]$ goes over $\mathbb{RP}^1$, this defines a pencil of (projective) lines in P. □

Let us now get back to the proof of Proposition 6.

When σ varies in ℓ_0 (which is topologically a circle covering twice Q), the lines δ_σ of Lemma 7 form a pencil in the projective plane P. Let ω be the base-point of this pencil. If ω were inside the disc bounded by $\mathcal{T}_{\sigma_0}$, all the lines δ_σ would cut $\mathcal{T}_{\sigma_0}$, and, for some σ, $\mathbb{H}$ would cut both $\mathcal{T}_\sigma$ and $\mathcal{T}_{\sigma^*}$ and would therefore contain a light line ℓ. This is impossible, since $\mathbb{H} = \Phi(\mathbb{H}_\infty)$, as $\mathbb{H}_\infty$ contains no light line and should contain $\Phi(\ell)$.

Then, for each $\sigma \in \ell_0$, at least one of the two lines δ_σ or δ_{σ^*} does not cut $\mathcal{T}_{\sigma_0}$. It implies that the common point ω of all the δ_σ's is outside the disc bounded by $\mathcal{T}_{\sigma_0}$.

The complement of the two lines δ_{σ_1} and δ_{σ_2} tangent to $\mathcal{T}_{\sigma_0}$ and the two lines $\delta_{\sigma_1^*}$ and $\delta_{\sigma_2^*}$ form four intervals of the pencil of lines of base point ω. Their images in Q form two open intervals I_1 and I_2. When δ_σ turns around the point ω, letting $[\sigma]$ sweep say I_1, suppose that one of the lines δ_σ or δ_{σ^*} cuts $\mathcal{T}_{\sigma_0}$. The cyclide $C_\sigma \subset \mathbb{S}^3_\infty$ is therefore singular. When $[\sigma]$ belongs to the complementary arc of Q, neither δ_σ nor δ_{σ^*} cut $\mathcal{T}_{\sigma_0}$, therefore the cyclide C_σ is regular (see Figure 13). There are exactly two limiting cases, corresponding to a line δ_σ tangent to $\mathcal{T}_0$; this occurs when $\Delta_\sigma \subset P_\sigma$ is tangent to $\mathcal{T}_\sigma$. This is precisely the case when the cyclide $C_\sigma \subset \mathbb{S}^3_\infty$ has a cuspidal singularity.

6.3. *Singularities of the leaves: portrait of the family*

One can follow in Figure 14 the evolution of the cyclides, when we take the two cuspidal cyclides as reference. We recall that the parameter set of

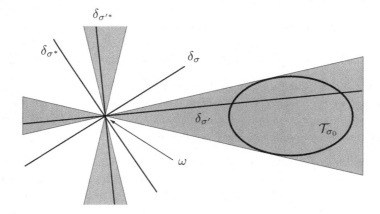

Figure 13.   Sectors for which C'_σ is singular.

Figure 14.   A sequence of cyclides C_σ.

the cyclides C_σ of the (singular) foliation $\mathcal{F}$ is the quotient Q of ℓ_0 by the identification $\sigma \sim \sigma^*$. The class of σ has been denoted by $[\sigma]$.

When $[\sigma]$ is in the arc Q for which C_σ is singular, C_σ is contained in the union of the two full crescents (made transparent to turn Figure 14 more readable) bounded by the two cuspidal cyclides. This can be seen on the first two pictures in Figure 14. When σ is in the other arc, in which case the cyclide C_σ is regular (made transparent for readability in the last two pictures of Figure 14), the open balls bounded by the two cuspidal cyclides are in different components of the complement of any regular cyclide.

A natural task is now the classification up to conformal equivalence of all the foliations constructed above. The following theorem provides it.

Theorem 3. *All the Villarceau foliations are conformally equivalent.*

The set of equivalence classes of non-Villarceau foliations constructed above modulo conformal equivalence form an open interval. Using a cross-ration it can be parameterized by the interval $]0, 1[$.

We shall use the two cuspidal cyclides of a non-Villarceau foliation, and see that, up to conformal equivalence, the possible pairs of tangent cuspidal cyclides (like in Figure 14) depend on only one real parameter, the cross-ratio of four points on a circle $C_\mathcal{F}$ naturally associated to the foliation.

Proof of Theorem 3. As usual, we denote by Σ the sphere corresponding to the point σ of $\overline{\Lambda^4}$.

Let us call a cuspidal cyclide in $\mathbb{S}^3$ a *crescent*.

We call the component of its complement which is homeomorphic to an open ball *interior* of the crescent.

The fact that all the Villarceau foliations are conformally equivalent is a consequence of Lemma 5.

Each non-Villarceau foliation $\mathcal{F}$ is determined by the two crescents C_1 and C_2 of the family, whose contact determines Γ, the contact elements along Γ (which will be called *the imposed contacts*), and thus $\mathcal{W}$ and the family (C_σ).

The crescent C_1 (associated to the conic $\mathcal{C}_1 \subset \overline{\Lambda^4}$) has a singular point m_1 with tangent direction L_1 at this point. This crescent is the boundary of a topological open ball union of the "interior" balls of the non-singular oriented spheres corresponding to the points of the conic $\mathcal{C}_1$.

In the same way, the crescent C_2 has a singular point m_2 with tangent direction L_2 at this point. This crescent bounds an open ball union of the "interior" balls of the non-singular oriented spheres corresponding to the points of the conic $\mathcal{C}_2$.

Two cases may occur.

1) There exists a circle containing the two points m_1 and m_2 and to which the two lines L_1 and L_2 are tangent.

2) There is a (unique) sphere Σ which contains both the circle c_1 tangent to L_1 at m_1 passing by m_2 and the circle c_2 ,tangent to L_2 at m_2 passing by m_1.

Let us begin by this second case.

The circle c orthogonal to Σ and passing by m_1 and m_2 is also orthogonal to the spheres Σ_2 (associated to a point of C_2) and Σ_1 (associated to a point of C_1) respectively tangent to L_1 at m_1 and to L_2 at m_2.

The circle c cuts Σ_2 at m_1 and at another point denoted by n_1, and cuts Σ_1 at m_2 and at another point denoted by n_2. The cross-ratio $[m_1, m_2, n_1, n_2]$ is a conformal invariant of the pair of crescents, and thus of the foliation (see Figure 15).

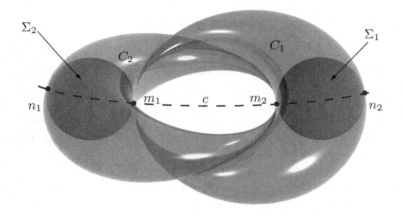

Figure 15. The points m_1, n_1 on C_2 and m_2 and n_2 on C_1.

Figure 16. Impossible configurations.

The crescents are tangent to the sphere Σ, C_1 along c_1 and C_2 along c_2. The spheres Σ_1 and Σ_2 should be on the same side of Σ otherwise the crescents would not be tangent along a curve, as the two possible tangency points would be m_1 and m_2, the two intersection points of c_1 and c_2. The crescents must be disjoint otherwise the topological balls bounded by the crescent would have an open set in common, and could not be tangent a long a curve. Figure 16 shows all the impossible configurations of the points m_1, m_2, n_1 and n_2 on c. The two conditions that the crescents should satisfy imply that the cross ratio $[m_1, m_2, n_1, n_2]$ is in $]0, 1[$.

Reciprocally, given four points on a circle respecting the above specified configuration (see Figure 15), we can construct the spheres Σ, Σ_1 and Σ_2 (the last two on the same side of Σ and disjoint).

Without loss of generality, we can suppose that m_1 and m_2 are antipodal. Then the sphere Σ is geodesic. Each choice of L_1 and L_2 determines (with the spheres Σ_1 and Σ_2) two crescents C_{L_1,Σ_1} and C_{L_2,Σ_2}: for $i = 1, 2$, the crescent C_{L_i,Σ_i} is the union of the circles tangent to L_i at m_i and tangent to Σ_i (see Figure 15). The two crescents may be disjoint, secant or tangent. Moreover, turning one of the lines L_i forces the crescent to turn by the same isometry of axis the circle c.

The two crescents are always disjoint when the circles c_1 and c_2 are orthogonal. In fact, there exists only one value of the angle between c_1 and c_2 which makes the two crescents C_1 and C_2 tangent. It suffices for that to turn one of the lines, say L_1, until the crescent C_1 becomes tangent to a sphere Σ_λ^2 (different from Σ_2), to create the third contact between the two Dupin cyclides which make them tangent along a curve, as is proved in the three contact Theorem 1. Notice that the two contacts at m_1 and m_2 are tangent to the same sphere but one has an orientation opposite to the orientation of the sphere. The corresponding light lines are therefore disjoint.

On the right of Figure 17, the rotating crescents are cut to show how the contact is obtained.

In case 1), the two crescents degenerate into a circle. In fact, when m_1 tends to n_1 and, simultaneously, m_2 tends to n_2 (see Figure 15) and the foliation "tends to" a Villarceau foliation.   □

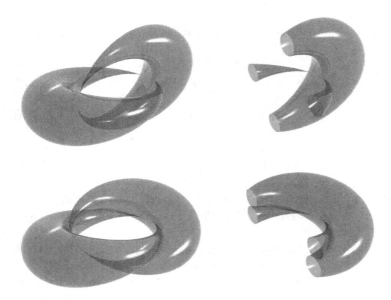

Figure 17.   Adjusting crescents.

7. Appendix

7.1. *Determination of a Lie sphere transformation by its restriction to conics*

We prove here a lemma which implies that a Lie sphere transformation operates transitively on the set of projective planes such that the signature of $\mathbb{L}$ restricted to the underlying 3-dimensional vector space is $(2, 1)$.

Recall that a Lie sphere transformation is an homography of $\mathbb{RP}^5$ which preserves $\overline{\Lambda^4}$.

Lemma 9. *Let C and C^* be two sister conics, carried respectively by the projective planes P and P^*. In the same way, let C' and C'^* be two sister conics carried respectively by the projective planes P' and P'^*.*

Let h be a homography $C = P \cap \overline{\Lambda^4} \to C' = P' \cap \overline{\Lambda^4}$, and h^* a homography (defined independently) $C^* = P^* \cap \overline{\Lambda^4} \to C'^* = P'^* \cap \overline{\Lambda^4}$.

Then there exist two Lie sphere transformations whose restriction to C is h and whose restriction to C^* is h^*. Only one of them preserves orientation.

Proof. The two 3-dimensional subspaces P_{vect} and P^*_{vect} of $\mathbb{R}^6$ carrying C and C^* are orthogonal and form a direct sum decomposition of $\mathbb{R}^6$. The same properties are satisfied by P'_{vect} and $P'^{*}{}_{vect}$.

The homographies h and h^{*} extend to maps that we still call h and h^{*}, $h : P \to P'$ and $h^{*} : P^{*} \to P'^{*}$. These two maps are respectively represented by linear maps $F_0 : P_{vect} \to P'_{vect}$ and $F_0^{*} : P_{vect}^{*} \to P'^{*}_{vect}$. As F_0 and F_0^{*} define projective maps which send respectively C to C' and C^{*} to C'^{*}, they satisfy $\mathbb{L}(F_0(X)) = \mu_0 \mathbb{L}(X)$ and $\mathbb{L}(F_0^{*}(X^{*})) = \mu_0^{*} \mathbb{L}(X^{*})$.

Let us show that the constants μ_0 and μ_0^{*} are positive. Recall that the *cone of positivity* (resp. of *negativity*) of $\mathbb{L}$ is the set of vectors $v \in \mathbb{R}^6$ such that $\mathbb{L}(v) > 0$ (resp $\mathbb{L}(v) < 0$). We define similarily cones of positivity or of negativity in the 3-dimensional spaces $P_{vect}, P'_{vect}, P_{vect}^{*}$ and P'_{vect}, restricting $\mathbb{L}$ to these subspaces. A priori, F_0 sends the whole cone of positivity of $\mathbb{L}$ restricted to P_{vect} into the cone of positivity or into the cone of negativity of $\mathbb{L}$ restricted to P'_{vect}. We assert that it is in fact the cone of positivity. Indeed, when the signature of $\mathbb{L}$ restricted to P_{vect} is $(2, 1)$, the cone of positivity of $\mathbb{L}$ restricted to P_{vect} is not topologically equivalent to the cone of negativity (the second one is not simply connected).

We look for multiples of F_0 and F_0^{*}, $F = aF_0$ and $F^{*} = bF_0^{*}$ such that the mapping $F \oplus F^{*}$ defined by $(F \oplus F^{*})(X + X^{*}) = aF_0(X) + bF_0^{*}(X^{*})$, $X \in P_{vect}, X^{*} \in P_{vect}^{*}$ preserves the light cone $C_{\mathbb{L}}$ (of image $\overline{\Lambda^4}$ in $\mathbb{RP}^5$). It means that there exists a constant μ' such that:

$$\mathbb{L}(F(X) + F^{*}(X^{*})) = \mu_0 a^2 \mathbb{L}(X) + \mu_0^{*} b^2 \mathbb{L}(X^{*}) = \mu' \mathbb{L}(X + X^{*})$$
$$= \mu'\big(\mathbb{L}(X) + \mathbb{L}(X^{*})\big).$$

This relation must be true for all $X \in P_{vect}$, and $X^{*} \in P_{vect}^{*}$. Then

$$\frac{\mu_0 a^2}{\mu_0^{*} b^2} = \frac{\mu'}{\mu'} = 1,$$

which gives : $\dfrac{a^2}{b^2} = \dfrac{\mu_0^{*}}{\mu_0}$.

This determines the quotient b^2/a^2, but not the sign of b/a. Then there are two Lie sphere transformations whose restriction to P is h and whose restriction to P^{*} is h^{*}.

Only one of them is orientation preserving. To see this fact, observe first that, if one of them is obtained from the action of $F \oplus F^{*}$ on $\mathbb{R}^6$, the other one is obtained from the action of $-F \oplus F^{*}$ on $\mathbb{R}^6$. Let us consider a basis of $\mathbb{R}^6$ union of three vectors of P_{vect} and three vectors of P_{vect}^{*}. The images of this basis by $F \oplus F^{*}$ and by $-F \oplus F^{*}$ have opposite orientations, as we replaced three vectors of a basis of $\mathbb{R}^6$ by opposite vectors.   □

Corollary 4. *With the same data P and P', let us choose three points σ_i,*

$i = 1, 2, 3$ on $\mathcal{C} = P \cap \overline{\Lambda^4}$, and three points σ_j^*, $j = 1, 2, 3$ on the sister conic $\mathcal{C}^*$. In the same way, let us choose three points σ_i', $i = 1, 2, 3$ on $\mathcal{C}'$, and three points $\sigma_j'^*$, $j = 1, 2, 3$ on $\mathcal{C}'^*$.

Then there exists two Lie sphere transformations sending each σ_i onto σ_i', and each σ_j^* onto $\sigma_j'^*$. These Lie sphere transformations send $\mathcal{C}$ onto $\mathcal{C}'$ and $\mathcal{C}^*$ onto $\mathcal{C}'^*$. Only one of them preserves orientation.

Proof. The group of homographies of a conic of $\mathbb{RP}^2$ acts transitively on the triplets of distinct points. Then there exists a unique homography h : $\mathcal{C} \to \mathcal{C}'$ sending each σ_i onto σ_i', and, independently, a unique homography $h^* : \mathcal{C}^* \to \mathcal{C}'^*$ sending each σ_j^* onto $\sigma_j'^*$. The previous lemma applied to h and h^* gives the desired Lie sphere transformations. $\square$

7.2. A one parameter group of Lie sphere transformations

In this subsection, we describe the group G of Lie sphere transformations which permute the conics associated to the Dupin cyclides of a family of Dupin cyclides tangent along a curve (see Subsection 6.2).

Proof of Lemma 8. We have seen in Section 4.4 that $\overline{\Lambda^4}$ is the set of classes modulo colinearity of points (x_{-1}, x_0, z_1, z_2) of $\mathbb{R}^2 \times \mathbb{C}^2$ such that $\mathbb{L}(x_{-1}, x_0, z_1, z_2) = -x_{-1}^2 - x_0^2 + |z_1|^2 + |z_2|^2 = 0$. Moreover, the action of $\mathcal{U}$ on $\mathbb{RP}^5$ is: $u.[x_{-1}, x_0, z_1, z_2] = [x_{-1}, x_0, uz_1, uz_2]$.

We may, without loss of generality, choose coordinates such that $\sigma_0 = [X_0]$, with $X_0 = (0, 1, 1, 0)$, and $\sigma_0^* = [X_0^*]$ with $X_0^* = [1, 0, 0, 1]$.

Let us show that the linear mapping J $(x_{-1}, x_0, z_1, z_2) \mapsto (x_0, -x_{-1}, -z_2, z_1)$ induces the Lie sphere transformation $\mathcal{J}$ of the statement of the lemma. Indeed, J preserves $\mathbb{L}$ and preserves orientation. For each $u \in \mathcal{U}$, the projective mapping $[J]$ exchanges $u.\sigma_0$ and $u.\sigma_0^*$. Lemma 9 implies that these properties characterize $\mathcal{J}$.

Each point $\sigma \in \ell_0$ may be written $\sigma = [\alpha X_0 + \beta X_0^*] = [\beta, \alpha, \alpha, \beta]$, with $\alpha^2 + \beta^2 = 1$.

Let us define $R_{\alpha, \beta} = \alpha \operatorname{Id} + \beta J$. One can see that $R_{\alpha, \beta}$ preserves $\mathbb{L}$ and preserves orientation. For each $u \in \mathcal{U}$, the projective mapping $[R_{\alpha, \beta}]$ sends $u.\sigma_0$ onto $u.\sigma$, and $u.\sigma_0^*$ onto $u.\sigma^*$. Again, Lemma 9 implies that these properties characterize $\rho_{\sigma_0, \sigma}$. $\square$

References

[1] T.E. Cecil, Lie sphere geometry with applications to submanifolds, 2nd edition, Springer 2008.

[2] L. Druoton, L. Garnier, R. Langevin, H. Marcellier and R. Besnard, *Les cyclides de Dupin et l'espace des sphères*, Refig (Revue Francophone d'Informatique Graphique), **5**(1) (2011), 41–59.

[3] C. Dupin, *Applications de Géométrie*, (1822), paragraph III p. 200

[4] R. Garcia, R. Langevin and P. Walczak, *Darboux Curves on Surfaces*, Preprint IMB, Dijon.

[5] F. Klein, Lectures on mathematics, Mac Millan and company (1894), new edition AMS, Chelsea, 2000.

[6] R. Langevin, J-C. Sifre, L. Druoton, L. Garnier and M. Paluszny, *Finding a cyclide given three contact conditions*, Preprint IMB, Dijon.

[7] R. Langevin and J. O'Hara, *Conformal arc-length via osculating circles*, Commentarii Mathematici Helvetici, **85** Issue 2, 273–312.

[8] R. Langevin and G. Solanes, *The geometry of canal surfaces and the length of curves in de Sitter space*, Advances in Geometry, **11** fasc. 4 (2011), 585–602.

[9] R. Langevin and P. Walczak, *Conformal geometry of foliations*, Geom. Dedicata, **132** (2008), 135–178.

[10] R. Langevin and P. G. Walczak, *Canal foliations of* $\mathbb{S}^3$, accepted for publication in the Journal of the Mathematical Society of Japan.

[11] B. O'Neill, *Semi-Riemannian geometry. With applications to relativity*, Pure and Applied Mathematics, 103. Academic Press, Inc., New York, xiii+468 pages (1983).

[12] Y. Villarceau. *Théorème sur le tore*, Nouvelles Annales de Mathématiques, Paris, Gauthier-Villars, 1ère série **7** (1848), 345–347.

[13] A. Zeghib, *Laminations et hypersurfaces géodésiques des variétés hyperboliques*, Ann. Sci. École Norm. Sup., **24** (1991), 171–188.

[14] A. Zeghib, *Sur les feuilletages géodésiques continus des variétés hyperboliques*, Invent. Math., **114** (1993), 193–206.

Received February 18, 2013.

FOLIATIONS 2012
ed. by Paweł WALCZAK *et al.*
World Scientific, Singapore, 2013
pp. 103–113

Transverse invariant measures extend to the ambient space

CARLOS MENIÑO COTÓN

Departamento de Xeometría e Topoloxía, Facultade de Matemáticas
Universidade de Santiago de Compostela, Spain
e-mail: carlos.menino@usc.es

1. Introduction

Transverse invariant measures of foliated spaces play an important role in the study of their transverse dynamics. They are measures on transversals invariant by holonomy transformations. There are many interpretations of transverse invariant measures; in particular, they can be extended to generalized transversals, which are defined as Borel sets that meet each leaf in a countable set [4]. Here, we show that indeed invariant measures can be extended to the σ-algebra of all Borel sets becoming an "ambient" measure (a measure on the ambient space). Precisely, the following result is proved.

Theorem 1. *Let X be a foliated space with a transverse invariant measure Λ. There exists a Borel measure $\widetilde{\Lambda}$ on X such that $\widetilde{\Lambda}(T) = \Lambda(T)$ for all generalized transversal T.*

This $\widetilde{\Lambda}$ is constructed as a "pairing" of the transverse invariant measure Λ with the counting measure on the leaves. Connes has proved that this pairing is coherent for generalized transversals but it can not be directly extended to any Borel set since the local projection of a Borel set is not necessarily a Borel set [6, 7]. The solution of this problem is the main difficulty of the proof. In fact, this result is proved in the more general setting of foliated measurable spaces [1, 2]. The uniqueness of this extension is also discussed. Remark that we work in the general setting of measure theory, without assuming any condition on the measure apart from its transverse invariance.

2. Measurable laminations

A *Polish space* is a completely metrizable and separable topological space. A *standard Borel space* is a measurable space isomorphic to a Borel subset of a Polish space. A *measurable topological space* or *MT-space* X is a set equipped with a σ-algebra and a topology. Usually, measure theoretic concepts will refer to the σ-algebra of X, and topological concepts will refer to its topology. Notice that the σ-algebra does not necessarily agree with the Borel σ-algebra associated with the topology. An *MT-map* between MT-spaces is a measurable continuous map. An *MT-isomorphism* is a map which is a measurable isomorphism and a homeomorphism, simultaneously.

Let T be a standard Borel space. On $T \times \mathbb{R}^n$, we consider the σ-algebra generated by products of Borel subsets of T and $\mathbb{R}^n$, and the product of the discrete topology on T and the Euclidean topology on $\mathbb{R}^n$. $T \times \mathbb{R}^n$ will be endowed with the structure of MT-space defined by this σ-algebra and this topology.

A *foliated measurable chart* on X is an MT-isomorphism $\varphi : U \to T \times \mathbb{R}^n$, where U is open and measurable in X. A *foliated measurable atlas* on X is a countable family of foliated measurable charts whose domains cover X. The sets $\varphi^{-1}(\{*\} \times \mathbb{R}^n)$ are the *plaques* of the foliated chart φ and the sets $\varphi^{-1}(T \times \{*\})$ are called *transversals* induced by φ. A *foliated measurable space* is an MT-space that admits a foliated measurable atlas. Observe that we always consider countable atlases. The connected components of X are called its *leaves*. An example of foliated measurable space is a foliated space with its Borel σ-algebra and the leaf topology. According to this definition, the leaves are second countable connected manifolds but they may not be Hausdorff.

A measurable subset $T \subset X$ is called a *generalized transversal* if its intersection with each leaf is countable; these are slightly more general than the transversals of [1], which are required to have discrete and closed intersection with each leaf. Let $\mathcal{T}(X)$ be the set of generalized transversals of X. This set is closed under countable unions and intersections, but it is not a σ-algebra.

A *measurable holonomy transformation* is a measurable isomorphism $\gamma : T \to T'$, for $T, T' \in \mathcal{T}(X)$, which maps each point to a point in the same leaf. A *transverse invariant measure* on X is a σ-additive map $\Lambda : \mathcal{T}(X) \to [0, \infty]$ which is invariant by measurable holonomy transformations. The classical definition of transverse invariant measure in the context of foliated spaces is a measure on usual transversals invariant by usual holonomy transformations [3]. Both definitions are equivalent in the case of foliated measurable spaces induced by foliated spaces [4].

3. Case of a product foliated measurable space

In this section, we take foliated measurable spaces of the form $T \times P$, where T is a standard measurable space and P a connected, separable and Hausdorff manifold. Indeed, the results of this section hold when P is any Polish space. We assume that a new topology is given in this space as follows. All standard Borel spaces are Borel isomorphic to a finite set, $\mathbb{Z}$ or the interval $[0, 1]$ (see [7, 8]). Identify $T \times P$ with $\mathbb{Z} \times P$, $[0, 1] \times P$ or $A \times P$ (A finite), via a Borel isomorphism. We work with two topologies in $T \times P$. On the one hand, the topology of the MT-structure is the product of discrete topology on T and the topology of P. On the other hand, the second topology is the product of the topology of $[0, 1]$, $\mathbb{N}$ or P with the topology of P; the term "open set" is used with this topology. The σ-algebra of the MT-structure on $T \times P$ is generated by these "open sets". Let $\pi : T \times P \to T$ be the first factor projection.

Proposition 1 (R. Kallman [5]). *If $B \subset T \times P$ is a Borel set such that $B \cap (\{t\} \times P)$ is σ-compact for all $t \in T$, then $\pi(B)$ is a Borel set. Moreover there exists a Borel subset $B' \subset B$ such that $\#(B' \cap (\{t\} \times P)) = 1$ if $B \cap (\{t\} \times P) \neq \emptyset$, and $\#(B' \cap (\{t\} \times P)) = 0$ otherwise.*

For any measurable space $(X, \mathcal{M}, \Lambda)$, the *completion* of $\mathcal{M}$ with respect to Λ is the σ-algebra

$$\mathcal{M}_\Lambda = \{ Z \subset X \mid \exists A, B \in \mathcal{M}, \ A \subset Z \subset B, \ \Lambda(B \setminus A) = 0 \}.$$

The measure Λ extends in a natural way to $\mathcal{M}_\Lambda$ by defining $\Lambda(Z) = \Lambda(A) = \Lambda(B)$ for Z, A, B as above.

Now, let Λ be a Borel measure on T. Define

$$\pi(\mathcal{B}^*, \mathcal{B}_\Lambda) = \{ B \in \mathcal{B}^* \mid \pi(B \cap U) \in \mathcal{B}_\Lambda \ \forall \text{ open } U \subset T \times P \},$$

where $\mathcal{B}$ and $\mathcal{B}^*$ are the Borel σ-algebras of T and $T \times P$, respectively.

Proposition 2. $\pi(\mathcal{B}^*, \mathcal{B}_\Lambda)$ *is closed under countable unions.*

Proof. This is obvious since, for any countable family $\{B_n\} \subset \mathcal{B}^*$, we obtain $\bigcup_n B_n \in \mathcal{B}^*$ and

$$\pi\left(\left(\bigcup_n B_n\right) \cap U\right) = \bigcup_n \pi(B_n \cap U) \in \mathcal{B}_\Lambda$$

for any open subset $U \subset T \times P$. $\hfill\square$

Remark 1. If Λ is σ-finite (*i.e.*, T is a countable union of Borel sets with finite Λ-measure), then $\pi(\mathcal{B}^*, \mathcal{B}_\Lambda) = \mathcal{B}^*$: by Exercise 14.6 in [6], any set in $\mathcal{B}^*$ projects onto an analytic set, which is Λ-measurable since Λ is σ-finite [7, Theorem 4.3.1].

Remark 2. If $B \in \pi(\mathcal{B}^*, \mathcal{B}_\Lambda)$ and U is an open set, then $B \cap U \in \pi(\mathcal{B}^*, \mathcal{B}_\Lambda)$. By Proposition 1, $\pi(\mathcal{B}^*, \mathcal{B}_\Lambda)$ contains the Borel sets with σ-compact intersection with the plaques $\{t\} \times P$.

Now, we want to extend Λ to all Borel sets satisfying the conditions of a measure. Let $\mathcal{B}^{**}$ denote the Borel σ-algebra of $T \times P \times T \times P$, $\tilde{\pi}$ the natural projection $T \times P \times T \times P \to T \times T$, and $\langle \mathcal{B}_\Lambda \times \mathcal{B}_\Lambda \rangle$ the σ-algebra generated by sets of the form $A \times B$ for $A, B \in \mathcal{B}_\Lambda$.

Lemma 1. *If* $B, B' \in \pi(\mathcal{B}^*, \mathcal{B}_\Lambda)$, *then* $B \times B' \in \tilde{\pi}(\mathcal{B}^{**}, \langle \mathcal{B}_\Lambda \times \mathcal{B}_\Lambda \rangle)$.

Proof. Since $B, B' \in \mathcal{B}^*$, we have $B \times B' \in \mathcal{B}^{**}$. Observe that every open set $U \subset T \times P$ is a countable union of products of open sets. Write $U = \bigcup_{n=1}^\infty (U_n \times V_n)$ with U_n and V_n open subsets of T and P, respectively. Then

$$\tilde{\pi}((B \times B') \cap U) = \tilde{\pi}\left((B \times B') \cap \bigcup_{n=1}^\infty (U_n \times V_n)\right)$$

$$= \tilde{\pi}\left(\bigcup_{n=1}^\infty ((B \cap U_n) \times (B' \cap V_n))\right) = \bigcup_{n=1}^\infty \tilde{\pi}((B \cap U_n) \times (B' \cap V_n))$$

$$= \bigcup_{n=1}^\infty (\pi(B \cap U_n) \times \pi(B' \cap V_n)),$$

which is in $\langle \mathcal{B}_\Lambda \times \mathcal{B}_\Lambda \rangle$. $\hfill\square$

Definition 1. For $B \in \pi(\mathcal{B}^*, \mathcal{B}_\Lambda)$, let

$$\tilde{\Lambda}(B) = \int_T \#(B \cap (\{t\} \times P))\, d\Lambda(t) = \int_T \left(\int_{\{t\} \times P} \chi_{B \cap (\{t\} \times P)}\, d\nu\right) d\Lambda(t),$$

where ν denotes the counting measure and χ_X the characteristic function of a subset $X \subset \{t\} \times P$.

Remark 3. A measure on T induces a transverse invariant measure on $T \times P$. When B is a generalized transversal, $\widetilde{\Lambda}(B)$ is the value of this transverse invariant measure on B. Therefore Definition 1 defines an extension of this transverse invariant measure to a map $\widetilde{\Lambda} : \pi(\mathcal{B}^*, \mathcal{B}_\Lambda) \to [0, \infty]$.

Proposition 3. *On* $B \in \pi(\mathcal{B}^*, \mathcal{B}_\Lambda)$, $\widetilde{\Lambda}$ *is well defined and satisfies the following properties:*

(a) $\widetilde{\Lambda}(\emptyset) = 0$.
(b) $\widetilde{\Lambda}(\bigcup_{n \in \mathbb{N}} B_n) = \sum_{n=1}^\infty \widetilde{\Lambda}(B_n)$ *for every countable family of disjoint sets* $B_n \in \pi(\mathcal{B}^*, \mathcal{B}_\Lambda), n \in \mathbb{N}$.

Proof. $\widetilde{\Lambda}$ is well defined if and only if the function $h : T \to \mathbb{R} \cup \{\infty\}$, $h(t) = \#(B \cap (\{t\} \times P))$, is measurable with respect to the σ-algebra $\mathcal{B}_\Lambda$ in T; i.e., if $h^{-1}(\{n\}) \in \mathcal{B}_\Lambda$ for all $n \in \mathbb{N}$. To prove this property, we proceed by induction on n. It is clear that $h^{-1}(\{0\}) = T \setminus \pi(B)$ belongs to $\mathcal{B}_\Lambda$ since $B \in \pi(\mathcal{B}^*, \mathcal{B}_\Lambda)$. Now, suppose $h^{-1}(\{i\}) \in \mathcal{B}_\Lambda$ for $i \in \{0, ..., n-1\}$ and let us check that $h^{-1}(\{n\}) \in \mathcal{B}_\Lambda$. Let

$$C_n = \{ ((t, p_1), (t, p_2), ..., (t, p_{n+1})) \mid t \in T,\ p_1, ..., p_{n+1} \in P \},$$

which is a closed in $(T \times P)^{n+1}$. Observe that C_n is the set of $(n+1)$-uples in $T \times P$ that lie in the same plaque. We remark that C_n is homeomorphic to $\Delta_T \times P^{n+1}$, where Δ_T is the diagonal of the product T^{n+1}, and the projection $\pi_T : \Delta_T \to T$, $(t, ..., t) \mapsto t$ is a homeomorphism. The measure Λ becomes a measure on Δ_T via π_T. The intersection $B^{n+1} \cap C_n$, denoted by D_n, is the set of $(n+1)$-uples in B that lie in the same plaque. Let

$$\Delta_n = \{ ((t, p_1), (t, p_2), ..., (t, p_{n+1})) \in C_n \mid \exists\, i, j \text{ with } i \neq j \text{ and } p_i = p_j \},$$

which is closed in C_n. This set consists of the $(n+1)$-uples in each plaque such that two components are equal. The set $D_n \setminus \Delta_n$ consists of the $(n+1)$-uples of different elements in B that lie in the same plaque. Therefore, $\pi_T \circ \pi_\Delta(D_n \setminus \Delta_n)$ consists of the points $t \in T$ such that the corresponding plaque $\{t\} \times P$ contains more than n points of B, where $\pi_\Delta : C_n \to \Delta_T$ is the natural projection.

Now, let us prove that $\pi_\Delta(D_n \setminus \Delta_n) \in \pi_T^{-1}(\mathcal{B}_\Lambda) = \pi_T^{-1}(\mathcal{B})_\Lambda$. By Lemma 1, $\widehat{\pi}(B^{n+1} \setminus \Delta_n) \in \langle \mathcal{B}_\Lambda^{n+1} \rangle$, where $\widehat{\pi} : (T \times P)^{n+1} \to T^{n+1}$ is the natural projection. Therefore

$$\pi_\Delta(D_n \setminus \Delta_n) = \Delta_T \cap \widehat{\pi}(B^{n+1} \setminus \Delta_n) \in \langle \mathcal{B}_\Lambda^{n+1} \rangle|_{\Delta_T},$$

where $\langle \mathcal{B}_\Lambda^{n+1}\rangle|_{\Delta_T}$ denotes the restriction of the σ-algebra $\langle \mathcal{B}_\Lambda^{n+1}\rangle$ to Δ_T. We only have to prove that $\langle B_\Lambda^{n+1}\rangle|_{\Delta_T} \subset \pi_T^{-1}(\mathcal{B}_\Lambda)$. For that purpose, we have to check that the generators $\prod_{k=1}^{n+1} F_k$, with $F_k \in \mathcal{B}_\Lambda$, satisfy $(\prod_{k=1}^{n+1} F_k) \cap \Delta_T \in \pi_T^{-1}(\mathcal{B}_\Lambda)$. For each k, take $A_k, B_k \in \mathcal{B}$ with $A_k \subset F_k \subset B_k$ and $\Lambda(B_k \setminus A_k) = 0$. Then

$$\left(\prod_{k=1}^{n+1} A_k\right) \cap \Delta_T \subset \left(\prod_{k=1}^{n+1} F_k\right) \cap \Delta_T \subset \left(\prod_{k=1}^{n+1} B_k\right) \cap \Delta_T,$$

and $\left(\prod_{k=1}^{n+1} A_k\right) \cap \Delta_T$ and $\left(\prod_{k=1}^{n+1} B_k\right) \cap \Delta_T$ belong to $\pi_T^{-1}(\mathcal{B})$ because

$$\left(\prod_{k=1}^{n+1} A_k\right) \cap \Delta_T = \bigcap_{k=1}^{n+1} \pi_T^{-1}(A_k), \qquad \left(\prod_{k=1}^{n+1} B_k\right) \cap \Delta_T = \bigcap_{k=1}^{n+1} \pi_T^{-1}(B_k).$$

Moreover

$$\Lambda\left(\left(\left(\prod_{k=1}^{n+1} B_k\right) \cap \Delta_T\right) \setminus \left(\left(\prod_{k=1}^{n+1} A_k\right) \cap \Delta_T\right)\right)$$

$$= \Lambda\left(\bigcap_{k=1}^{n+1}(\pi_T^{-1}(B_k) \setminus \pi_T^{-1}(A_k))\right) = \Lambda\left(\pi_T^{-1}\left(\bigcap_{k=1}^{n+1}(B_k \setminus A_k)\right)\right)$$

$$= \Lambda\left(\bigcap_{k=1}^{n+1}(B_k \setminus A_k)\right) = 0.$$

This shows that $\pi_\Delta(D_n \setminus \Delta_n) \in \pi_T^{-1}(\mathcal{B}_\Lambda) = \pi_T^{-1}(\mathcal{B})_\Lambda$. By induction, we have

$$h^{-1}(n) = T \setminus \left((\pi_T \circ \pi_\Delta(D_n \setminus \Delta_n) \cup h^{-1}(\{0, ..., n-1\}))\right) \in \mathcal{B}_\Lambda.$$

Property (a) is obvious. To show property (b), observe that $\chi_{\bigcup B_n} = \sum \chi_{B_n}$, and then use the monotonous convergence theorem. $\square$

Definition 2. If $B \in \mathcal{B}^* \setminus \pi(\mathcal{B}^*, \mathcal{B}_\Lambda)$, then define $\widetilde{\Lambda}(B) = \infty$.

Proposition 4. $(T \times P, \mathcal{B}^*, \widetilde{\Lambda})$ is a measure space and $\widetilde{\Lambda}$ extends Λ.

Proof. We only have to prove that $\widetilde{\Lambda}(\bigcup_n B_n) = \sum_n \widetilde{\Lambda}(B_n)$ for every countable family of disjoint sets $B_n, n \in \mathbb{N}$, in $\mathcal{B}^*$. By Proposition 3, this holds if $B_n \in \pi(\mathcal{B}^*, \mathcal{B}_\Lambda)$ for all $n \in \mathbb{N}$. If $\bigcup_n B_n \in \mathcal{B}^* \setminus \pi(\mathcal{B}^*, \mathcal{B}_\Lambda)$, then the above equality is obvious. So we only have to consider the case where some $B_j \in \mathcal{B}^* \setminus \pi(\mathcal{B}^*, \mathcal{B}_\Lambda)$ and, however, $\bigcup_n B_n \in \pi(\mathcal{B}^*, \mathcal{B}_\Lambda)$. We can suppose $B_j = B_1$, and let $B = \bigcup_n B_n$. Let

$$B^\infty = \{t \in T \mid \#(B \cap (\{t\} \times P)) = \infty\},$$

which belongs to $\mathcal{B}_\Lambda$ by Proposition 3. The proof will be finished by checking that $\Lambda(B^\infty) > 0$. We have $B_1^\infty \subset B^\infty$, where

$$B_1^\infty = \{\, t \in T \mid \#(B_1 \cap (\{t\} \times P)) = \infty \,\}.$$

Suppose $\Lambda(B^\infty) = 0$. Since $\mathcal{B}^\infty \in \mathcal{B}_\Lambda$, there is some $A \in \mathcal{B}$ such that $B^\infty \subset A$ and $\Lambda(A) = 0$. The Borel set $\pi^{-1}(A)$ satisfies $B_1 \cap \pi^{-1}(A) \in \pi(\mathcal{B}^*, \mathcal{B}_\Lambda)$ since $\emptyset \subset \pi(B_1 \cap \pi^{-1}(A) \cap U) \subset A$ and $\Lambda(A) = 0$ for each open set $U \subset T \times P$. On the other hand, $B_1 \setminus \pi^{-1}(A)$ is a Borel set meeting every plaque in a finite set, which is σ-compact, and therefore projects to a Borel set by Proposition 1. Hence $B_1 \in \pi(\mathcal{B}^*, \mathcal{B}_\Lambda)$ by Proposition 2, which is a contradiction. □

We have constructed an extension of each transverse invariant measure in a product foliated measurable space, but its uniqueness was not proved. This uniqueness is false in general. For instance, take the foliated product $\mathbb{R} \times \{*\}$ and let Λ be the null measure on the singleton $\{*\}$; our extension $\widetilde{\Lambda}$ is the zero measure in the total space. Now, let μ be the measure defined by

(i) $\mu(B) = 0$ for all countable set B and
(ii) $\mu(B) = \infty$ for all uncountable Borel set B.

This measure μ extends Λ too and is quite different from $\widetilde{\Lambda}$. In order to solve this problem, we require some conditions to the extension. These conditions have the spirit of coherency with the concept of transverse invariant measures. We will prove that our extension is the unique coherent extension.

Definition 3. Let μ be an extension of a transverse invariant measure Λ on $T \times P$. The measure μ is called a *coherent extension* of Λ if satisfies the following conditions:

(a) If $B \in \mathcal{B}^*$, $B \not\subset \pi^{-1}(S)$ for any $S \in \mathcal{B}$ with $\Lambda(S) = 0$, and $\#B \cap \{t\} \times P = \infty$ for each plaque $\{t\} \times P$ which meets B, then $\mu(B) = \infty$.
(b) If $\Lambda(S) = 0$ for some $S \in \mathcal{B}$, then $\mu(\pi^{-1}(S)) = 0$.
(c) If $B \in \mathcal{B}^*$ and $\Lambda(S) = \infty$ for all $S \in \mathcal{B}$ with $B \subset \pi^{-1}(S)$, then $\mu(B) = \infty$.

Remark 4. Condition (a) determines μ on Borel sets with infinite points in plaques which are not contained in the saturation of a Λ-null set. Condition (b) means certain coherency between the support of Λ and the support of the extension μ. Condition (c) determines μ on any Borel set so that any Borel set containing its projection has infinity Λ-measure.

Proposition 5. $\widetilde{\Lambda}$ *is the unique coherent extension.*

Proof. We prove that every coherent extension has the same values as $\widetilde{\Lambda}$ on $\mathcal{B}^*$. First, we consider the case $B \in \pi(\mathcal{B}^*, \mathcal{B}_\Lambda)$. Let

$$B^\infty = \{\, t \in T \mid \#(B \cap (\{t\} \times P)) = \infty \,\}.$$

This set belongs to $\mathcal{B}_\Lambda$ by Proposition 3. Therefore there exist Borel sets A, C such that $A \subset B^\infty \subset C$ and $\Lambda(C \setminus A) = 0$. Let $\widetilde{B}^\infty = B \cap \pi^{-1}(C)$. The Borel set $B \setminus \widetilde{B}^\infty$ is a generalized transversal and hence $\mu(B \setminus \widetilde{B}^\infty) = \widetilde{\Lambda}(B \setminus \widetilde{B}^\infty)$. On the other hand, if $\Lambda(\pi(\widetilde{B}^\infty))) = 0$, then $\Lambda(C) = 0$ and $\mu(\widetilde{B}^\infty) \le \mu(\pi^{-1}(C)) = 0$ by (b). If $\Lambda(B^\infty)) > 0$, let $\widehat{B}^\infty = B \cap \pi^{-1}(A)$ and $B'^\infty = B \cap \pi^{-1}(C \setminus A)$. Then $\mu(\widehat{B}^\infty) = \infty$ by (a), and $\mu(B'^\infty) = 0$ by (b). Therefore μ equals $\widetilde{\Lambda}$ on $\pi(\mathcal{B}^*, \mathcal{B}_\Lambda)$.

The case $B \in \mathcal{B}^* \setminus \pi(\mathcal{B}^*, \mathcal{B}_\Lambda)$ is similar. The set B^∞ is not a Borel set in this case, but observe that $B \cap \pi^{-1}(B^\infty) \not\subseteq \pi^{-1}(S)$ with $\Lambda(S) < \infty$ or we obtain $\pi(B \cap \pi^{-1}(S) \cap U) \in \mathcal{B}_\Lambda$ for all open set $U \subset T \times P$ by Remark 1, since $B \cap \pi^{-1}(S) \cap U$ is a Borel set in $S \times P$ and Λ is finite in S. Hence $B \in \pi(\mathcal{B}^*, \mathcal{B}_\Lambda)$ by Propositions 2 and 1. Therefore $\mu(B) = \infty$ by (c). This proves that μ and $\widetilde{\Lambda}$ agree on $\mathcal{B}^*$, as desired. $\qquad\square$

4. The general case

In this section, we prove the following theorem.

Theorem 2. *Let X be a foliated measurable space with a transverse invariant measure Λ. There exists a measure $\widetilde{\Lambda}$ on X that extends Λ.*

Let $\{U_i, \varphi_i\}_{i \in \mathbb{N}}$ be a foliated measurable atlas with $\varphi_i(U_i) = T_i \times \mathbb{R}^n$, where T_i is a standard Borel space. It is clear that $\varphi_i^{-1}(T_i \times \{*\})$ is a generalized transversal and, via φ_i, we obtain a Borel measure Λ_i on T_i. Proposition 4 provides a measure $\widetilde{\Lambda}_i$ on $U_i \approx T_i \times \mathbb{R}^n$ that extends Λ_i. Moreover Proposition 1 gives $\widetilde{\Lambda}_i(T) = \Lambda(T)$ for all generalized transversal $T \subset U_i$. Let $\pi_i : U_i \to \varphi_i^{-1}(T_i \times \{*\})$ denote the natural projections.

We begin with a description of the change of foliated measurable charts.

Theorem 3 (Kunugui, Novikov [7]). *Let $\{V_n\}_{n \in \mathbb{N}}$ be a countable base of open sets for a Polish space P. Let $B \subset T \times P$ be a Borel set such that $B \cap (\{t\} \times P)$ is an open set for all $t \in T$. Then there exists a sequence $\{B_n\}_{n \in \mathbb{N}}$ of Borel subsets of T such that*

$$B = \bigcup_n (B_n \times V_n).$$

We take a countable base $\{V_m\}_{m\in\mathbb{N}}$ of $\mathbb{R}^n$ by connected open sets.

Lemma 2. *For $i, j \in \mathbb{N}$, there exists a sequence of Borel subsets of T_i, $\{B_m\}_{m\in\mathbb{N}}$, and a sequence of open sets $\{W_m\}_{m\in\mathbb{N}}$ such that $\varphi_i(U_i \cap U_j) = \bigcup_m (B_m \times W_m)$ and $\varphi_j \circ \varphi_i^{-1}(t, x) = (f_{ijm}(t), g_{ijm}(t, x))$ for $(t, x) \in B_m \times W_m$, where each f_{ijm} is a Borel isomorphism.*

Proof. We apply Theorem 3 to $\varphi_j(U_i \cap U_j)$ and obtain a family $\{B'_m\}_{m\in\mathbb{N}}$ such that $\varphi_j(U_i \cap U_j) = \bigcup_m (B'_m \times V_m)$. Now we apply Theorem 3 to each set $\varphi_i \circ \varphi_j^{-1}(B'_k \times V_k)$, $k \in \mathbb{N}$. We obtain sequences $B'_{k,n}$ such that $\varphi_i \circ \varphi_j^{-1}(B'_k \times V_k) = \bigcup_m (B'_{k,m} \times V_m)$, $k \in \mathbb{N}$. The set $\varphi_j \circ \varphi_i^{-1}(\{t\} \times V_m)$, $t \in B'_{k,m}$, is contained in only one plaque $\{*\} \times \mathbb{R}^n$ since each V_m is connected. Therefore, $\varphi_j \circ \varphi_i^{-1}(t, x) = (f_{ijkm}(t), g_{ijkm}(t, x))$ for $(t, x) \in B'_{k,m} \times V_m$. We must show that f_{ijkm} is one-to-one. If there exist $t, t' \in T_i$ with $f_{ijkm}(t) = f_{ijkm}(t') = t''$, then $g_{ijkm}(\{t\} \times V_m)$ and $g_{ijkm}(\{t'\} \times V_m)$ are connected open subsets of the plaque $\{t''\} \times V_k$ in $B'_k \times V_k$, but this plaque is the image by $\varphi_j \circ \varphi_i^{-1}$ of a connected open set since φ_j and φ_i are homeomorphisms. Hence this set is contained in a single plaque of U_i. This is a contradiction with the assumption $t \neq t'$.

The functions f_{ijkm} are, obviously, measurable and they have measurable image by Proposition 1, hence the image is a standard Borel space. Since f_{ijkn} is a one-to-one measurable function between standard Borel spaces, it is a Borel isomorphism [7]. The proof is completed by observing that the required sequence $B_m \times W_m$ is the bisequence $B'_{k,m} \times V_m$ $k, m \in \mathbb{N}$. □

Lemma 3. *Let B be a Borel subset of $U_i \cap U_j$, $i, j \in \mathbb{N}$. Then*

$$B \in \pi_i(\mathcal{B}^*, \mathcal{B}_{\Lambda_i}) \iff B \in \pi_j(\mathcal{B}^*, \mathcal{B}_{\Lambda_j}).$$

Proof. By Lemma 2, there exists a countable family of measurable holonomy transformations from $\varphi_i^{-1}(T_i \times \{*\})$ to $\varphi_j^{-1}(T_j \times \{*\})$ whose domains and ranges cover $\pi_i(U_i \cap U_j)$ and $\pi_j(U_i \cap U_j)$, respectively. Therefore, if A is a Borel set contained in $U_i \cap U_j$ and $\pi_i(A)$ is a Borel set, then $\pi_j(A)$ is a Borel set and

$$\Lambda(\pi_i(A)) = 0 \iff \Lambda(\pi_j(A)) = 0. \qquad \square$$

Lemma 4. $\widetilde{\Lambda}_i(B) = \widetilde{\Lambda}_j(B)$ *for all Borel set $B \subset U_i \cap U_j$, $i, j \in \mathbb{N}$.*

Proof. We remark that $\widetilde{\Lambda}_i$ and $\widetilde{\Lambda}_j$ have the same values in generalized transversals of $U_i \cap U_j$. By Lemma 3, we only consider Borel sets in $\pi_i(\mathcal{B}^*, \mathcal{B}_\Lambda)$. Suppose that $\pi_i(B)$ is a Borel set; otherwise, $\pi_i(B)$ is Λ-measurable and we can choose a Borel set $A \subset \pi_i(B)$ with $\Lambda(\pi_i(B) \backslash A) = 0$. We take the Borel set $\widetilde{B} = B \cap \pi_i^{-1}(A)$. This Borel set projects onto the Borel set A and $\widetilde{\Lambda}_i(B \backslash \widetilde{B}) = \widetilde{\Lambda}_j(B \backslash \widetilde{B}) = 0$ by Definition 1 and Lemma 3, hence $\widetilde{\Lambda}_i(B) = \widetilde{\Lambda}_i(\widetilde{B})$ and $\widetilde{\Lambda}_j(B) = \widetilde{\Lambda}_j(\widetilde{B})$. Let

$$B^k = \{\, t \in T_i \mid \#(\varphi_i(B) \cap (\{t\} \times \mathbb{R}^n)) = k \,\}, \qquad k \in \mathbb{N} \cup \{\infty\}.$$

These are Λ-measurable sets by Proposition 3, and we assume that they are Borel sets by the same reason as above. Let $\widetilde{B}^k$ denote $B \cap \pi_i^{-1}(\varphi_i^{-1}(B^k \times \{*\}))$, which is a Borel set. It is obvious that $\bigcup_{i=1}^\infty \widetilde{B}^k$ is a generalized transversal, hence $\widetilde{\Lambda}_i(\bigcup_{k=1}^\infty \widetilde{B}^k) = \widetilde{\Lambda}_j(\bigcup_{k=1}^\infty \widetilde{B}^k)$. Now consider

$$\widetilde{B}_l^\infty = \{\, x \in \widetilde{B}^\infty \mid \#(B \cap P_x^j) = l \,\}, \qquad l \in \mathbb{N} \cup \{\infty\},$$

where P_x^j denotes the plaque of U_j that contains x. The proof is finished in the case $\Lambda(\pi_i(\widetilde{B}_\infty^\infty)) = 0$ (we can restrict to the case of a generalized transversal). If $\Lambda(\pi_i(\widetilde{B}_\infty^\infty)) > 0$, then $\Lambda(\pi_j(\widetilde{B}_\infty^\infty)) > 0$. Therefore we obviously obtain $\widetilde{\Lambda}_i(B) = \infty = \widetilde{\Lambda}_j(B)$. $\qquad\square$

Definition 4. Let B be a measurable set in X, and

$$B_1 = B \cap U_1, \ B_k = (B \cap U_k) \backslash (B_1 \cup ... \cup B_{k-1}),$$

for $k \geq 2$. Define

$$\widetilde{\Lambda}(B) = \sum_{i=1}^\infty \widetilde{\Lambda}_i(B_i).$$

By Lemma 4, it is easy to prove that Definition 4 does not depend neither on the ordering of the charts nor on the choice of the countable foliated measurable atlas. It is also easy to prove that $\widetilde{\Lambda}$ extends Λ since both of them have the same values on generalized transversals contained in each chart and, hence, in every generalized transversal. Theorem 2 is now established.

Definition 5. Let μ be an extension of a transverse invariant measure Λ on a foliated measurable space X. The measure μ is called a *coherent extension* of Λ if it is a coherent extension on each foliated measurable chart with the induced transverse invariant measure.

Corollary 1. *The extension $\widetilde{\Lambda}$ is the unique coherent extension of Λ.*

Theorem 1 gives a new interpretation of transverse invariant measures. It can be also used to introduce the following version of the concept of transversal for foliated measurable spaces with transverse invariant measures.

Definition 6. Let X be a foliated measurable space with a transverse invariant measure Λ. A Borel subset of X with finite $\widetilde{\Lambda}$-measure is called a Λ-*generalized transversal*.

Remark 5. In Section 3, we have only used that the plaques are Polish spaces. We can weaken the conditions of foliated measurable spaces taking charts with the form $T \times P$, where P is any connected and locally connected Polish space. In this way our result can be extended to other interesting cases like *measurable graphs* [1].

References

[1] M. Bermúdez, Laminations Boréliennes, Tesis, Université Claude Bernard-Lyon 1, 2004.

[2] M. Bermúdez and G. Hector, *Laminations hyperfinies et revêtements*, Ergod. Th. & Dynam. Sys., **26** (2006), 305–339.

[3] A. Candel and L. Conlon, Foliations I, Amer. Math. Soc., 1999.

[4] A. Connes, *A survey of foliations and operator algebras*, in Operator algebras and aplications, Part I, Proc. Sympos. Pure Math., **38**, Amer. Math. Soc., Providence, R.I., 1982.

[5] R. Kallman, *Certain quotient spaces are countably separated, III.* J. Funct. Analysis, **22** (3) (1976), 225–241.

[6] A.S. Kechris, Classical Descriptive Set Theory, Graduate Texts in Mathematics, Springer-Verlag, New York, 1994.

[7] S.M. Srivastava, A course on Borel sets, Graduate Texts in Mathematics, Springer, 1998.

[8] M. Takesaki, Theory of Operator Algebras, Springer-Verlag, 1979.

Received November 8, 2012.

FOLIATIONS 2012
ed. by Paweł WALCZAK et al.
World Scientific, Singapore, 2013
pp. 115–137

On a Poincaré lemma for foliations*

Eva Miranda

Departament de Matemàtica Aplicada I, Universitat Politècnica de Catalunya
EPSEB, Avinguda del Doctor Marañón, 44-50, 08028, Barcelona, Spain
e-mail: eva.miranda@upc.edu

Romero Solha

Departament de Matemàtica Aplicada I, Universitat Politècnica de Catalunya
ETSEIB, Avinguda Diagonal 647, 08028, Barcelona, Spain
e-mail: romero.barbieri@upc.edu

1. Introduction

In [9] Vu Ngoc and the first author of this paper proved a *singular Poincaré lemma* for the deformation complex of an integrable system with nondegenerate singularities. This complex is the Chevalley-Eilenberg complex [1] associated to a representation by Hamiltonian vector fields of this integrable system on the set of functions (modulo basic functions). The initial motivation for [9] was to give a complete proof for a crucial lemma used in proving a deformation result for pairs of local integrable systems with compatible

*Both authors have been partially supported by the DGICYT/FEDER project MTM2009-07594: Estructuras Geometricas: Deformaciones, Singularidades y Geometria Integral until December 2012 and by the MINECO project GEOMETRIA ALGEBRAICA, SIMPLECTICA, ARITMETICA Y APLICACIONES with reference: MTM2012-38122-C03-01 starting in January 2013. This research has also been partially supported by ESF network CAST, *Contact and Symplectic Topology*. Romero Solha has been partially supported by Start-Up Erasmus Mundus External Cooperation Window 2009-2010 project.

symplectic forms. This deformation proves a Moser path lemma which is a key point in establishing symplectic normal forms *à la Morse-Bott* for integrable systems with nondegenerate singularities [2, 3, 7]. This normal form proof can be seen as a *"infinitesimal stability theorem implies stability"* result in this context ([8]). So the Poincaré lemma turns out to be an important ingredient in the study of the Symplectic Geometry of integrable systems with singularities.

In this paper we use the Poincaré lemma of the deformation complex to compute some cohomology groups associated to the singular foliation defined by the Hamiltonian vector fields of an integrable system. In particular, we consider the analytic case, in which this computation becomes simpler and can be done in full generality.

A Poincaré lemma exists when the foliation is regular, and an offspring of this is a Poincaré lemma in the context of Geometric Quantization, in which the considered complex is a twisted complex from foliated cohomology: the *Kostant complex*. This Poincaré lemma turns out to be handy because it allows to compute a sheaf cohomology associated to Geometric Quantization. We enclose a sketch of the proof of these two Poincaré lemmas.

If we consider singularities into the picture, the whole scenario changes. As concerns the analytical tools, what makes the difference between the regular and singular case are the solutions of the equation $X(f) = g$ for a given g and a given vector field X. When the vector field is regular, we can solve this equation by simple integration no matter which function g is considered. If the vector field is singular, then this is a nontrivial question. Solutions may exist or not depending on some properties of the function g and the singularity of the vector field X. For instance, solutions of this equation are studied in [4].

The nonexistence of solutions of equations of type $X(f) = g$ are interpreted in this paper as an obstruction for local solvability of the cohomological equation $d_{\mathcal{F}}\beta = \alpha$, for a given foliated closed k-form α. Indeed, the fact that the vector fields defining the foliation commute adds an additional ingredient for the simultaneous solution of several equations of this type, which was already exploited in [9] and is further studied in this paper.

Organization of this paper: In Section 2 we describe the geometry of the singular foliations considered in this paper. We recall in Section 3 the singular Poincaré lemma for a deformation complex contained in [9]. We revisit in Section 4 the proof of the regular Poincaré lemma using homotopy operators provided in [5], indicate how to apply these techniques to prove a Poincaré lemma for regular foliations and show an application to Geometric Quantization. In Section 5 we consider the case when the foliation given by the integrable system has rank 0 singularities and compute the foliated cohomology groups.

2. Singular foliations given by nondegenerate integrable systems

An integrable system on a symplectic manifold (M, ω) of dimension $2n$ is a set of n functions $f_1, \ldots, f_n \in C^\infty(M)$ satisfying $df_1 \wedge \cdots \wedge df_n \neq 0$ over an open dense subset of M and $\{f_i, f_j\} = 0$ for all i, j. The mapping $F = (f_1, \ldots, f_n) : M \to \mathbb{R}^n$ is called a moment map.

The Poisson bracket is defined by $\{f, g\} = X_f(g)$, where X_f is the unique vector field defined by $\imath_{X_f}\omega = -df$: the Hamiltonian vector field of f.

The distribution generated by the Hamiltonian vector fields of the moment map, $\langle X_{f_1}, \ldots, X_{f_n} \rangle$, is involutive because $[X_f, X_g] = X_{\{f,g\}}$. Since $0 = \{f_i, f_j\} = \omega(X_{f_i}, X_{f_j})$, the leaves of the associated (possibly singular) foliation are isotropic submanifolds; they are Lagrangian at points where the functions are functionally independent.

There is a notion of nondegenerate singular points which was initially introduced by Eliasson [2, 3]. We may consider different ranks for the singularity. To define the k-rank case we reduce to the 0-rank case considering a Marsden-Weinstein reduction associated to a natural Hamiltonian $\mathbb{T}^k$-action [17, 10] given by the joint flow of the moment map F.

We denote by $(x_1, y_1, \ldots, x_n, y_n)$ a set of coordinates centred at the origin of $\mathbb{R}^{2n}$ and by $\sum_{i=1}^{n} dx_i \wedge dy_i$ the Darboux symplectic form.

In the rank zero case, since the functions f_i are in involution with respect to the Poisson bracket, their quadratic parts commute, defining in this way an Abelian subalgebra of $Q(2n, \mathbb{R})$ (the set of quadratic forms on $2n$-variables). These singularities are said to be of nondegenerate type if this subalgebra is a Cartan subalgebra.

Cartan subalgebras of $Q(2n, \mathbb{R})$ were classified by Williamson in [16].

Theorem 1 (Williamson). *For any Cartan subalgebra $\mathcal{H}$ of $Q(2n, \mathbb{R})$ there is a symplectic system of coordinates $(x_1, y_1, \ldots, x_n, y_n)$ in $\mathbb{R}^{2n}$ and a basis $h_1, \ldots, h_n$ of $\mathcal{H}$ such that each h_i is one of the following:*

$$h_i = x_i^2 + y_i^2 \qquad\qquad \text{for } 1 \leq i \leq k_e, \qquad\qquad \text{(elliptic)}$$

$$h_i = x_i y_i \qquad\qquad \text{for } k_e + 1 \leq i \leq k_e + k_h, \text{(hyperbolic)} \qquad (2.1)$$

$$\begin{cases} h_i = x_i y_i + x_{i+1} y_{i+1}, & \text{for } i = k_e + k_h + 2j - 1, \\ h_{i+1} = x_i y_{i+1} - x_{i+1} y_i & 1 \leq j \leq k_f \end{cases} \text{(focus-focus pair)}$$

Observe that the number of elliptic components k_e, hyperbolic components k_h and focus-focus components k_f is therefore an invariant of the algebra $\mathcal{H}$. The triple (k_e, k_h, k_f) is an invariant of the singularity and it is called the Williamson type of $\mathcal{H}$. We have that $n = k_e + k_h + 2k_f$. Let $h_1, \ldots, h_n$ be a Williamson basis of this Cartan subalgebra. We denote by X_i the Hamiltonian vector field of h_i with respect to the Darboux form. These vector fields are a basis of the corresponding Cartan subalgebra of $\mathfrak{sp}(2n, \mathbb{R})$. We say that a vector field X_i is hyperbolic (resp. elliptic) if the corresponding function h_i is so. We say that a pair of vector fields X_i, X_{i+1} is a focus-focus pair if X_i and X_{i+1} are the Hamiltonian vector fields associated to functions h_i and h_{i+1} in a focus-focus pair.

In the local coordinates specified above, the vector fields X_i take the following form:

- X_i is an elliptic vector field,

$$X_i = 2\left(-y_i \frac{\partial}{\partial x_i} + x_i \frac{\partial}{\partial y_i}\right); \qquad (2.2)$$

- X_i is a hyperbolic vector field,

$$X_i = -x_i \frac{\partial}{\partial x_i} + y_i \frac{\partial}{\partial y_i}; \qquad (2.3)$$

- X_i, X_{i+1} is a focus-focus pair,

$$X_i = -x_i \frac{\partial}{\partial x_i} + y_i \frac{\partial}{\partial y_i} - x_{i+1} \frac{\partial}{\partial x_{i+1}} + y_{i+1} \frac{\partial}{\partial y_{i+1}} \qquad (2.4)$$

and

$$X_{i+1} = x_{i+1} \frac{\partial}{\partial x_i} + y_{i+1} \frac{\partial}{\partial y_i} - x_i \frac{\partial}{\partial x_{i+1}} - y_i \frac{\partial}{\partial y_{i+1}}. \qquad (2.5)$$

Assume that $\mathcal{F}$ is a linear foliation on $\mathbb{R}^{2n}$ with a rank 0 singularity at the origin. Assume that the Williamson type of the singularity is (k_e, k_h, k_f). The linear model for the foliation is then generated by the vector fields above, it turns out that these type of singularities are symplectically linearizable and we can read off the local symplectic geometry of the foliation from the algebraic data associated to the singularity (Williamson type).

This is the content of the following symplectic linearization result in [2, 3, 7] (smooth category) and [13] (analytic category),

Theorem 2. *Let ω be a smooth (resp. analytic) symplectic form defined in a neighbourhood U of the origin and $\mathcal{F}$ a linear foliation with a rank zero singularity, of prescribed Williamson type, at the origin. Then, there exists a local diffeomorphism (resp. analytic diffeomorphism) $\phi : U \longrightarrow \phi(U) \subset \mathbb{R}^{2n}$ such that ϕ preserves the foliation and $\phi^*(\sum_{i=1}^{n} dx_i \wedge dy_i) = \omega$, with $(x_1, y_1, \ldots, x_n, y_n)$ local coordinates on $\phi(U)$.*

Futhermore, if $\mathcal{F}'$ is a foliation that has $\mathcal{F}$ as a linear foliation model near a point, one can symplectically linearize $\mathcal{F}'$ ([7]).

This is equivalent to Eliasson's theorem [2, 3] when the Williamson type of the singularity is $(k_e, 0, 0)$.

The classification of singularities of integrable system changes in the analytic category. This was already considered by Vey in [13] and it is simpler because the Williamson type of the singularities is $(k_e, k_h, 0)$.

There are normal forms for higher rank which have been obtained by the first author together with Nguyen Tien Zung [7, 10] also in the case of singular nondegenerate compact orbits. When the rank of the singularity is greater than 0, a collection of regular vector fields is also attached to it.

3. A singular Poincaré lemma for a deformation complex

This section revisits the main results contained in [9].

Consider the family X_i of singular vector fields given by Williamson's theorem above which form a basis of a Cartan subalgebra of the Lie algebra $\mathfrak{sp}(2r, \mathbb{R})$ with $r \leq n$.

Theorem 3 (Miranda and Vu Ngoc). *Let $g_1, \ldots, g_r$, be a set of smooth functions on $\mathbb{R}^{2n}$ with $r \leq n$ fulfilling the following commutation relations*

$$X_i(g_j) = X_j(g_i), \quad \forall i, j \in \{1, \ldots, r\}, \tag{3.1}$$

where the X_i's are the vector fields defined above. Then there exists a smooth function G and r smooth functions f_i such that,

$$X_j(f_i) = 0 , \ \forall \ i, j \in \{1, \ldots, r\} \quad and \tag{3.2}$$

$$g_i = f_i + X_i(G) , \ \forall \ i \in \{1, \ldots, r\}. \tag{3.3}$$

It is also included in [9] an interesting reinterpretation of this statement in terms of the deformation complex associated to an integrable system. We think that it is instructive to explain this succinctly here (we refer the reader to [14] and [9] for more details).

Using the same notation of the last section, let $\mathbf{h} = \langle h_1, \ldots, h_n \rangle_{\mathbb{R}}$ and $\mathcal{C}_{\mathbf{h}} = \{f \in C^\infty(\mathbb{R}^{2n}) \; ; \; X_h(f) = 0, \; \forall \, h \in \mathbf{h}\}$. The set $\mathbf{h}$ is an Abelian Lie subalgebra of $(C^\infty(\mathbb{R}^{2n}), \{\cdot, \cdot\})$ and $\mathcal{C}_{\mathbf{h}}$ is its centralizer.

The components of the moment map induce a representation of the commutative Lie algebra $\mathbb{R}^n$ on $C^\infty(\mathbb{R}^{2n})$,

$$\mathbb{R}^n \times C^\infty(\mathbb{R}^{2n}) \ni (v, f) \mapsto \{\mathbf{h}(v), f\} \in C^\infty(\mathbb{R}^{2n}). \tag{3.4}$$

Where, denoting by $(e_1, \ldots, e_n)$ a basis of $\mathbb{R}^n$, $v = v_1 e_1 + \cdots + v_n e_n$ and

$$\{\mathbf{h}(v), f\} = v_1 X_1(f) + \cdots + v_n X_n(f). \tag{3.5}$$

We can consider two Chevalley-Eilenberg complexes with the above action in mind, and the deformation complex is built from them. The first is the Chevalley-Eilenberg complex of $\mathbb{R}^n$ with values in $C^\infty(\mathbb{R}^{2n})$, we denote $\mathrm{Hom}_{\mathbb{R}}(\wedge^k \mathbb{R}^n; C^\infty(\mathbb{R}^{2n}))$ by A^k:

$$0 \longrightarrow C^\infty(\mathbb{R}^{2n}) \longrightarrow A^1 \longrightarrow A^2 \longrightarrow A^3 \longrightarrow \cdots . \tag{3.6}$$

The second is the Chevalley-Eilenberg complex of $\mathbb{R}^n$ with values in $C^\infty(\mathbb{R}^{2n})/\mathcal{C}_{\mathbf{h}}$ (with respect to this action, $\mathbb{R}^n$ acts trivially on $\mathcal{C}_{\mathbf{h}}$), where we denote $\mathrm{Hom}_{\mathbb{R}}(\wedge^k \mathbb{R}^n; C^\infty(\mathbb{R}^{2n})/\mathcal{C}_{\mathbf{h}})$ by B^k:

$$0 \longrightarrow C^\infty(\mathbb{R}^{2n})/\mathcal{C}_{\mathbf{h}} \longrightarrow B^1 \longrightarrow B^2 \longrightarrow B^3 \longrightarrow \cdots . \tag{3.7}$$

Finally we define the *deformation complex* as follows:

$$0 \longrightarrow C^\infty(\mathbb{R}^{2n})/\mathcal{C}_{\mathbf{h}} \xrightarrow{\bar{d}_{\mathbf{h}}} B^1 \xrightarrow{\partial_{\mathbf{h}}} A^2 \xrightarrow{d_{\mathbf{h}}} A^3 \xrightarrow{d_{\mathbf{h}}} \cdots , \tag{3.8}$$

the map $\partial_{\mathbf{h}}$ is defined by the following diagram (where all small triangles are commutative):

$$
\begin{array}{ccccccc}
0 \longrightarrow & C^\infty(\mathbb{R}^{2n}) & \xrightarrow{d_{\mathbf{h}}} & A^1 & \xrightarrow{d_{\mathbf{h}}} & A^2 & \xrightarrow{d_{\mathbf{h}}} \cdots \\
& \downarrow {\scriptstyle \partial_{\mathbf{h}}} & \nearrow & \downarrow {\scriptstyle \partial_{\mathbf{h}}} & \nearrow & \downarrow {\scriptstyle \partial_{\mathbf{h}}} & \nearrow \\
0 \longrightarrow & C^\infty(\mathbb{R}^{2n})/\mathcal{C}_{\mathbf{h}} & \xrightarrow{\bar{d}_{\mathbf{h}}} & B^1 & \xrightarrow{\bar{d}_{\mathbf{h}}} & B^2 & \xrightarrow{\bar{d}_{\mathbf{h}}} \cdots
\end{array}
$$

The cohomology groups associated to this complex are denoted by $H^k(\mathbf{h})$.

If α is a 1-cocycle, then for any smooth function g_i with $\alpha(e_i) = [g_i] \in C^\infty(\mathbb{R}^{2n})/\mathcal{C}_{\mathbf{h}}$ the commutation condition $X_i(g_j) = X_j(g_i)$ is fulfilled. Theorem 3 says that there exists a function G such that $g_i = f_i + X_i(G)$, so $[g_i] = [X_i(G)]$ and this is exactly the coboundary condition.

Theorem 3 combined with Theorem 2 can be, then, reformulated as follows:

Theorem 4 (Miranda and Vu Ngoc). *An integrable system with non-degenerate singularities is C^∞-infinitesimally stable at the singular point, that is*

$$H^1(\mathbf{h}) = 0. \tag{3.9}$$

4. Homotopy operators and a regular Poincaré lemma for foliated cohomology

Let us recall the following construction due to Guillemin and Sternberg [5] which generalizes, in a way [†], the classical proof of Poincaré lemma.

Consider $Y \subset M$ an embedded submanifold and let ϕ_t be a smooth retraction from M to Y. Given any smooth k-form α, the following formula holds:

$$\alpha - \phi_0^*(\alpha) = \int_0^1 \frac{\mathrm{d}}{\mathrm{d}t}\phi_t^*(\alpha) = \int_0^1 \phi_t^*(\iota_{\xi_t}\mathrm{d}\alpha)\mathrm{d}t + \mathrm{d}\int_0^1 \phi_t^*(\iota_{\xi_t}\alpha)\mathrm{d}t, \tag{4.1}$$

where ξ_t is the vector field associated to ϕ_t. Thus, defining $I(\alpha) = \int_0^1 \phi_t^*(\iota_{\xi_t}\alpha)\mathrm{d}t$, we obtain,

$$\alpha - \phi_0^*(\alpha) = I \circ \mathrm{d}(\alpha) + \mathrm{d} \circ I(\alpha). \tag{4.2}$$

Now assume that α is a closed form, formula (4.2) yields $\alpha - \phi_0^*(\alpha) = \mathrm{d} \circ I(\alpha)$, and therefore $I(\alpha)$ is a primitive for the closed k-form $\alpha - \phi_0^*(\alpha)$.

This has been classically applied considering retractions to a point in contractible sets or to retractions to the base of a fiber bundle. In the context of Symplectic and Contact Geometry, this homotopy formula leads to the so-called Moser's path method [11]. As said before, formula (4.2) does not, a priori, give a primitive for α but for the difference $\alpha - \phi_0^*(\alpha)$[‡].

This technique can also be applied for regular foliations. This approach using the general homotopy formula of Guillemin and Sternberg has the advantage that some choices on the retraction can be done in such a way that the vector field ξ_t is tangent to special directions in M, thus, allowing an adaptation to the foliated cohomology case.

[†]The proof contained in [15] makes a particular choice of retraction on star-shaped domains.

[‡]The vector field ξ_t is the radial one when the retraction is $\phi_t(p_1, \ldots, p_n) = (tp_1, \ldots, tp_n)$, and this formula coincides with the one of Warner [15], giving a primitive for α.

4.1. Foliated cohomology

Let $(M, \mathcal{F})$ be a foliated m-dimensional manifold and n the dimension of the leaves. The (regular) foliation can be thought as a subbundle of TM, which is often denoted by $T\mathcal{F}$.

The foliated cohomology is the one associated to the following cochain complex:

$$0 \longrightarrow C_{\mathcal{F}}^{\infty}(M) \hookrightarrow C^{\infty}(M) \xrightarrow{\mathrm{d}_{\mathcal{F}}} \Omega_{\mathcal{F}}^1(M) \xrightarrow{\mathrm{d}_{\mathcal{F}}} \cdots \xrightarrow{\mathrm{d}_{\mathcal{F}}} \Omega_{\mathcal{F}}^n(M) \xrightarrow{\mathrm{d}_{\mathcal{F}}} 0, \quad (4.3)$$

where $\Omega_{\mathcal{F}}^k(M) = \Gamma(\wedge^k T\mathcal{F}^*)$, $C_{\mathcal{F}}^{\infty}(M)$ is the space of smooth functions which are constant along the leaves of the foliation, and $\mathrm{d}_{\mathcal{F}}$ is the restriction of the exterior derivative, d, to $T\mathcal{F}$.

We can prove a Poincaré lemma for foliated cohomology, of a regular foliation, using equation (4.2) by considering local coordinates in which the foliation is given by local equations $\mathrm{d}p_{n+1} = 0, \ldots, \mathrm{d}p_m = 0$, and the retraction is given by $(tp_1, \ldots, tp_n, p_{n+1}, \ldots, p_m)$; the vector field ξ_t is tangent to the relevant foliation.

Theorem 5 (Poincaré lemma for foliated cohomology). *The foliated cohomology groups vanish for* degree ≥ 1.

One could try to mimic similar formulae to prove a singular Poincaré lemma for a foliation given by an integrable system with nondegenerate singularities. The main issue of adapting such a proof is the smoothness of the procedure. Indeed, as we will see later, the adaptation of such a procedure is not possible since the cohomology groups do not vanish if the foliation is singular.

Whilst the de Rham complex is a fine resolution for the constant sheaf $\mathbb{R}$ on M, the foliated cohomology is a fine resolution for the sheaf of smooth functions which are constant along the leaves of the foliation.

4.2. Geometric Quantization à la Kostant

A symplectic manifold (M, ω) such that the de Rham class $[\omega]$ is integral is called prequantizable. A prequantum line bundle of (M, ω) is a Hermitian line bundle over M with connection, compatible with the Hermitian structure, (L, ∇^{ω}) that satisfies $curv(\nabla^{\omega}) = -i\omega$ (the curvature of ∇^{ω} is proportional to the symplectic form). And a real polarization $\mathcal{F}$ is an integrable subbundle of TM (the bundle $T\mathcal{F}$) whose leaves are Lagrangian submanifolds, i.e., $\mathcal{F}$ is a Lagrangian foliation.

The restriction of the connection ∇^ω to the polarization induces an operator

$$\nabla : \Gamma(L) \to \Gamma(T\mathcal{F}^*) \otimes \Gamma(L). \tag{4.4}$$

Let $\mathcal{J}$ denotes the space of local sections s of a prequantum line bundle L such that $\nabla s = 0$. The space $\mathcal{J}$ has the structure of a sheaf and it is called the sheaf of flat sections.

The quantization of $(M, \omega, L, \nabla, \mathcal{F})$ is given by

$$\mathcal{Q}(M) = \bigoplus_{k \geq 0} \check{H}^k(M; \mathcal{J}), \tag{4.5}$$

where $\check{H}^k(M; \mathcal{J})$ are Čech cohomology groups with values in the sheaf $\mathcal{J}$.

If $\mathcal{S}$ denotes the sheaf of sections of the line bundle L, the Kostant complex is

$$0 \longrightarrow \mathcal{J} \hookrightarrow \mathcal{S} \xrightarrow{d^\nabla} \Omega^1_\mathcal{F} \otimes \mathcal{S} \xrightarrow{d^\nabla} \cdots \xrightarrow{d^\nabla} \Omega^n_\mathcal{F} \otimes \mathcal{S} \xrightarrow{d^\nabla} 0, \tag{4.6}$$

where $d^\nabla(\alpha \otimes s) = d_\mathcal{F}(\alpha) \otimes s + (-1)^{\mathrm{degree}(\alpha)} \alpha \wedge \nabla s$ and $d^\nabla \circ d^\nabla = 0$ because the curvature of ∇ vanishes along the leaves.

Lemma 1. *There is always a local unitary flat section on each point of M.*

Proof. Let $U \subset M$ be a trivializing neighbourhood of L with a unitary section $s : U \subset M \to L$. Since $\nabla s \in \Omega^1_{\mathcal{F}|_U}(U) \otimes \Gamma(L|_U)$ there is a $\alpha \in \Omega^1_{\mathcal{F}|_U}(U)$ such that $\nabla s = \alpha \otimes s$. The condition $d^\nabla \circ d^\nabla = 0$ implies $d_\mathcal{F}\alpha = 0$;

$$\begin{aligned}
0 = d^\nabla(\nabla s) = d^\nabla(\alpha \otimes s) &= d_\mathcal{F}\alpha \otimes s - \alpha \wedge \nabla s \\
&= d_\mathcal{F}\alpha \otimes s - (\alpha \wedge \alpha) \otimes s = d_\mathcal{F}\alpha \otimes s.
\end{aligned} \tag{4.7}$$

By the Poincaré lemma for foliations (Theorem 5) there exists a neighbourhood $V \subset U$ and $f \in C^\infty(V)$ such that $d_\mathcal{F}f = \alpha|_V$. Setting $r = e^{-f}s|_V$,

$$\nabla r = e^{-f}\nabla s|_V + d_\mathcal{F}(e^{-f}) \otimes s|_V = e^{-f}(\alpha \otimes s)\big|_V - e^{-f}d_\mathcal{F}f \otimes s|_V = 0, \tag{4.8}$$

so r is a unitary flat section of $L|_V$. □

Wherefore, for each point on M there exists a trivializing neighbourhood $V \subset M$ of L with a unitary flat section $s : V \subset M \to L$, and any element of $\Omega^k_\mathcal{F}(M) \otimes \Gamma(L)$ can be locally written as $\alpha \otimes s$, where $\alpha \in \Omega^k_{\mathcal{F}|_V}(V)$. The condition $d^\nabla(\alpha \otimes s) = 0$ is, then, equivalent to $d_\mathcal{F}\alpha = 0$, because $d^\nabla(\alpha \otimes s) = d_\mathcal{F}\alpha \otimes s + (-1)^k \alpha \wedge \nabla s$, $s \neq 0$ and $\nabla s = 0$.

The Kostant complex is just the foliated complex twisted by the sheaf of sections $\mathcal{S}$, and exactness of the foliated complex implies exactness of the Kostant complex.

Theorem 6. *The Kostant complex is a fine resolution for $\mathcal{J}$. Therefore, its cohomology groups are isomorphic to the cohomology groups with coefficients in the sheaf of flat sections $\check{H}^k(M; \mathcal{J})$ and thus compute Geometric Quantization.*

Rawnsley provided a proof of this fact in [12].

5. The singular case

The main objective of this section is to use the Poincaré lemma of the deformation complex to compute foliated cohomology. We start this section by recalling a definition of foliated cohomology that is going to be used in the singular case. We then introduce some analytical tools that we need to compute these groups. These analytical tools are mainly a series of decomposition results for functions with respect to vector fields. Finally, in the last subsection we enclose explicit computations of the cohomology groups.

Roughly, the elements of these cohomology groups are given by a collection of functions wich are constant along the leaves of the foliation, fulfilling additional constraints.

5.1. *Singular foliated cohomology*

Integrable systems defined on (M, ω) induce Lie subalgebras of $(\Gamma(TM), [\cdot, \cdot])$, namely $(\mathcal{F} = \langle X_1, \ldots, X_n \rangle_{C^\infty(M)}, [\cdot, \cdot]|_{\mathcal{F}})$, where X_i is the Hamiltonian vector field of the ith component of a moment map $F : M \to \mathbb{R}^n$.

Now, considering $C^\infty(M)$ as a $C^\infty(M)$-module, $(\mathcal{F}, [\cdot, \cdot]|_{\mathcal{F}})$ can be represented on $C^\infty(M)$ as vector fields acting on smooth functions.

This is an example of a Lie pseudo algebra representation (see [6] for precise definitions and a nice account for the history and, various, names of this structure) and one can, then, consider the following complex[§]:

$$0 \longrightarrow C^\infty_{\mathcal{F}}(M) \hookrightarrow C^\infty(M) \xrightarrow{\mathrm{d}_{\mathcal{F}}} \Omega^1_{\mathcal{F}}(M) \xrightarrow{\mathrm{d}_{\mathcal{F}}} \cdots \xrightarrow{\mathrm{d}_{\mathcal{F}}} \Omega^n_{\mathcal{F}}(M) \xrightarrow{\mathrm{d}_{\mathcal{F}}} 0, \quad (5.1)$$

[§]The Lie pseudo algebra cohomology with respect to that particular representation.

with the differential defined by

$$
d_{\mathcal{F}}\alpha(Y_1,\ldots,Y_{k+1}) = \sum_{i=1}^{k+1} (-1)^{i+1} Y_i(\alpha(Y_1,\ldots,\hat{Y}_i,\ldots,Y_{k+1}))
$$

$$
+ \sum_{i<j} (-1)^{i+j} \alpha([Y_i,Y_j], Y_1,\ldots,\hat{Y}_i,\ldots,\hat{Y}_j,\ldots,Y_{k+1}),
$$

$$(5.2)$$

with $Y_1,\ldots,Y_{k+1} \in \mathcal{F}$. The cochain spaces are defined by

$$
\Omega_{\mathcal{F}}^k(M) = \mathrm{Hom}_{C^\infty(M)}(\wedge_{C^\infty(M)}^k \mathcal{F}; C^\infty(M)),
\tag{5.3}
$$

and $C_{\mathcal{F}}^\infty(M) = \ker(d_{\mathcal{F}} : C^\infty(M) \to \Omega_{\mathcal{F}}^1(M))$.

The differential is a coboundary operator and the associated cohomology is denoted by $H_{\mathcal{F}}^\bullet(M)$.

Remark 1. This construction is also well defined in the analytic category and that is the notion used in Theorem 9.

From now on (M,ω) will be a symplectic manifold near a rank zero non-degenerate singularity of Williamson type $(k_e, k_h, 0)$. Thus $(\mathbb{R}^{2n}, \sum_{i=1}^n dx_i \wedge dy_i)$ is endowed with a distribution $\mathcal{F}$ generated by a Williamson basis.

Definition 1. The vanishing set of a vector field of a Williamson basis X_i is denoted by $\Sigma_i = \{p \in \mathbb{R}^{2n};\ x_i(p) = y_i(p) = 0\}$.

Proposition 1. If $\alpha \in \Omega_{\mathcal{F}}^k(\mathbb{R}^{2n})$ then $\alpha(X_{j_1},\ldots,X_{j_k})\big|_{\Sigma_{j_1} \cup \cdots \cup \Sigma_{j_k}} = 0$.

Proof. At every point $p \in \mathbb{R}^{2n}$ the map $\alpha \in \Omega_{\mathcal{F}}^k(\mathbb{R}^{2n})$ reduces to an element of $\wedge^k \mathcal{F}\big|_p^*$. Since $X_i = 0$ at Σ_i, for any $p \in \Sigma_i$ and vectors $Y_1(p),\ldots,Y_{k-1}(p) \in \mathcal{F}\big|_p$, the following expression holds:

$$
\alpha_p(X_i(p), Y_1(p),\ldots,Y_{k-1}(p)) = 0.
\tag{5.4}
$$

Therefore $\alpha(X_{j_1},\ldots,X_{j_k})\big|_{\Sigma_i} = 0$ for $i = j_1,\ldots,j_k$.   □

5.2. Analytical tools: special decomposition of smooth functions

Here we present special decompositions for functions with respect to vector fields of a Williamson basis. In order to fix notation, we recall what we mean by a *flat function at a subset*,

Definition 2. Consider $\mathbb{R}^m$ endowed with coordinates $(p_1, \ldots, p_m)$. A smooth function $g \in C^\infty(\mathbb{R}^m)$ is said to be Taylor flat at the subset $\{p_1 = \cdots = p_k = 0\}$ when

$$\frac{\partial^{j_1 + \cdots + j_k} g}{\partial p_1^{j_1} \cdots \partial p_k^{j_k}}\bigg|_{\{p_1 = \cdots = p_k = 0\}} = 0, \tag{5.5}$$

for all $j_1, \ldots, j_k$ and some fixed $k \leq m$.

Remark 2. It is important to point out that if $g \in C^\omega(\mathbb{R}^m)$ is Taylor flat at $\{p_1 = \cdots = p_k = 0\}$, then it is the zero function.

Definition 3. Given integer numbers k_e and k_h, we will call a smooth function $f \in C^\infty(\mathbb{R}^{2n})$ complanate if it can be written as

$$f = \sum_{i = k_e + 1}^{k_e + k_h} T_i, \tag{5.6}$$

where each T_i is a Taylor flat function at Σ_i. A noncomplanate function is one for which such an expression cannot be found.

We can find special decompositions for smooth functions like $f = f_i + X_i(F_i)$. The following result is a summary of results contained in [7] and [9],

Lemma 2. *Assume that the origin is a singularity of Williamson type $(k_e, k_h, 0)$, then for any $f \in C^\infty(\mathbb{R}^{2n})$ there exist $f_i, F_i \in C^\infty(\mathbb{R}^{2n})$ such that, for each vector field X_i in a Williamson basis, $f = f_i + X_i(F_i)$. Moreover,*

(1) $X_i(f_i) = 0$;
(2) f_i *is uniquely defined if* X_i *defines an* S^1-*action, otherwise* f_i *is uniquely defined up to Taylor flat functions at* Σ_i;
(3) *one can choose* f_i *and* F_i *such that* $X_j(f_i) = X_j(F_i) = 0$ *whenever* $X_j(f) = 0$ *for* $j \neq i$;
(4) *if* f *vanishes at the zero set of any vector of a Williamson basis, so does the function* f_i *and one can choose* F_i *vanishing at the zero set, as well;*
(5) $X_i(f) = 0$ *implies that* f *depends on* x_i *and* y_i *via* h_i:

$$f(x_1, y_1, \ldots, x_n, y_n) = \tilde{f}(x_1, y_1, \ldots, x_i^2 + y_i^2, \ldots, x_n, y_n),$$

$$f(x_1, y_1, \ldots, x_n, y_n)\big|_{Q_i^j} = \tilde{f}(x_1, y_1, \ldots, x_i y_i, \ldots, x_n, y_n),$$

where $Q_i^1 = \{x_i > 0, y_i > 0\}$, $Q_i^2 = \{x_i > 0, y_i < 0\}$, $Q_i^3 = \{x_i < 0, y_i > 0\}$ *and* $Q_i^4 = \{x_i < 0, y_i < 0\}$.

The case when X_i is an elliptic vector field was proved in [2, 3, 7]; [7] also has a proof when X_i is a hyperbolic vector field.

Remark 3. The proofs contained in [7] can be adapted for the analytic category: for hyperbolic singularities the formal proof yields the corresponding analytic statement, whilst the integrals defining the elliptic decomposition entail the analyticity of the construction. This is the version of the lemma used in the proof of Theorem 9. Furthermore, the uniqueness of the decomposition holds for both types of singularities, since there are no flat functions (apart from the zero function) in the analytic category.

5.3. *Computation of foliated cohomology groups*

We will distinguish between the smooth and the analytic category. In the smooth case we can completely determine the cohomology groups in degree 1 and n for Williamson type $(k_e, k_h, 0)$, and in all degrees for Williamson type $(k_e, 0, 0)$. In the analytic case the computations are done in all degrees.

Theorem 7 (Degree 1 smooth case). *Consider* $(\mathbb{R}^{2n}, \sum_{i=1}^{n} dx_i \wedge dy_i)$ *endowed with a smooth distribution $\mathcal{F}$ generated by a Williamson basis of type $(k_e, k_h, 0)$, then the following decomposition holds:*

$$\ker(d_{\mathcal{F}} : \Omega^1_{\mathcal{F}}(\mathbb{R}^{2n}) \to \Omega^2_{\mathcal{F}}(\mathbb{R}^{2n})) = W^1_{\mathcal{F}}(\mathbb{R}^{2n}) \oplus d_{\mathcal{F}}(C^\infty(\mathbb{R}^{2n})), \qquad (5.7)$$

where $W^1_{\mathcal{F}}(\mathbb{R}^{2n})$ is the set of 1-forms $\beta \in \Omega^1_{\mathcal{F}}(\mathbb{R}^{2n})$ such that $\pounds_{X_i}(\beta) = 0$ for all i, and if X_i is of hyperbolic type $\beta(X_i)$ is not Taylor flat at Σ_i (when it is nonzero).

Thus, the foliated cohomology group in degree 1 is given by:

$$H^1_{\mathcal{F}}(\mathbb{R}^{2n}) \cong \bigoplus_{i=1}^{k_e} \{f \in C^\infty_{\mathcal{F}}(\mathbb{R}^{2n}); f|_{\Sigma_i} = 0\}$$

$$\bigoplus_{i=k_e+1}^{n} \{f \in C^\infty_{\mathcal{F}}(\mathbb{R}^{2n}); f = 0 \text{ or } f|_{\Sigma_i} = 0 \text{ and not Taylor flat at } \Sigma_i\}.$$

Proof. For any $\alpha \in \Omega^1_{\mathcal{F}}(\mathbb{R}^{2n})$ the condition $d_{\mathcal{F}}\alpha = 0$ implies

$$d_{\mathcal{F}}\alpha(X_i, X_j) = X_i(\alpha(X_j)) - X_j(\alpha(X_i)) = 0, \qquad (5.8)$$

and Theorem 3 says that $\alpha(X_i) = f_i + X_i(F)$, where $F \in C^\infty(\mathbb{R}^{2n})$ and $f_i \in C^\infty_{\mathcal{F}}(\mathbb{R}^{2n})$. Thus any closed foliated 1-form α is cohomologous to a foliated 1-form β satisfying $\pounds_{X_i}(\beta) = 0$ for all i (Proposition 1 and item (4) of Lemma 2 guarantee that the forms are well defined); the condition $\mathcal{L}_{X_i}(\beta) = 0$ automatically implies that β is closed.

There exists $g \in C^\infty(\mathbb{R}^{2n})$ such that $d_{\mathcal{F}}g = \beta$ if and only if $\beta(X_i) = X_i(g)$. Since $\mathcal{L}_{X_i}(\beta) = 0$, this implies $X_i(\beta(X_i)) = 0$ and by uniqueness (up to Taylor flat functions, Lemma 2) $0 = \beta(X_i) + X_i(-g)$ has a solution if and only if $\beta(X_i) = 0$ or $\beta(X_i)$ is Taylor flat at Σ_i (for $i = k_e + 1, \ldots, n$). Wherefore, β is exact if and only if $\beta = 0$ or, if $\beta(X_i) \neq 0$ (for $i = k_e + 1, \ldots, n$), $\beta(X_i)$ is Taylor flat at Σ_i.

The expression $\ker = W_{\mathcal{F}}^1(\mathbb{R}^{2n}) \oplus d_{\mathcal{F}}(C^\infty(\mathbb{R}^{2n}))$ implies $H_{\mathcal{F}}^1(\mathbb{R}^{2n}) = W_{\mathcal{F}}^1(\mathbb{R}^{2n})$, by definition any $\beta \in W_{\mathcal{F}}^1(\mathbb{R}^{2n})$ can be given by n functions vanishing at certain points (Proposition 1) and satisfying some Taylor flat condition, e.g.: $\beta(X_n) = f \in C^\infty(\mathbb{R}^{2n})$, $f|_{\Sigma_i} = 0$ and not Taylor flat at Σ_n, if it is nonzero. The Lie derivative condition yields $f \in C_{\mathcal{F}}^\infty(\mathbb{R}^{2n})$.   □

We now consider the case of top degree forms in the smooth category.

Theorem 8 (Top degree smooth case). *Consider* $(\mathbb{R}^{2n}, \sum_{i=1}^{n} dx_i \wedge dy_i)$ *endowed with a smooth distribution $\mathcal{F}$ generated by a Williamson basis of type $(k_e, k_h, 0)$, then the following decomposition holds:*

$$\Omega_{\mathcal{F}}^n(\mathbb{R}^{2n}) = W_{\mathcal{F}}^n(\mathbb{R}^{2n}) \oplus d_{\mathcal{F}}(\Omega_{\mathcal{F}}^{n-1}(\mathbb{R}^{2n})), \tag{5.9}$$

where $W_{\mathcal{F}}^n(\mathbb{R}^{2n})$ is the set of n-forms $\beta \in \Omega_{\mathcal{F}}^n(\mathbb{R}^{2n})$ such that $\mathcal{L}_{X_i}(\beta) = 0$ for all i, and if $\beta(X_1, \ldots, X_n) \neq 0$, it is noncomplanate.

Thus, the foliated cohomology group in degree n is given by:

$$H_{\mathcal{F}}^n(\mathbb{R}^{2n}) \cong \{f \in C_{\mathcal{F}}^\infty(\mathbb{R}^{2n})$$
$$; f|_{\Sigma_1 \cup \cdots \cup \Sigma_n} = 0 \text{ and } f \text{ is noncomplanate or zero}\}.$$

Proof. For $\alpha \in \Omega_{\mathcal{F}}^n(\mathbb{R}^{2n})$ it holds $d_{\mathcal{F}}\alpha = 0$. Since $\alpha(X_1, \ldots, X_n) \in C^\infty(\mathbb{R}^{2n})$, Lemma 2 asserts that $\alpha(X_1, \ldots, X_n) = f_1 + X_1(F_{2\cdots n})$ with $X_1(f_1) = 0$. Applying again Lemma 2, $f_1 = f_2 + X_2(-F_{13\cdots n})$ with $X_2(f_2) = 0$, but also $X_1(f_2) = 0$ because $X_1(f_1) = 0$. Repeating this process for all X_i, one finally gets

$$\alpha(X_1, \ldots, X_n) = f + \sum_{i=1}^{n} (-1)^{i+1} X_i(F_{1\cdots \hat{i} \cdots n}), \tag{5.10}$$

with $f \in C_{\mathcal{F}}^\infty(\mathbb{R}^{2n})$, i.e.: there exists $\beta \in \Omega_{\mathcal{F}}^n(\mathbb{R}^{2n})$ and $\zeta \in \Omega_{\mathcal{F}}^{n-1}(\mathbb{R}^{2n})$ satisfying $\alpha = \beta + d_{\mathcal{F}}\zeta$ and $\mathcal{L}_{X_i}(\beta) = 0$ for all i (again, Proposition 1 and item 4 of Lemma 2 guarantee that the forms are well defined).

The foliated n-form β is exact if and only if there exists $\sigma \in \Omega_{\mathcal{F}}^{n-1}(\mathbb{R}^{2n})$ such that

$$\beta(X_1, \ldots, X_n) = \sum_{i=1}^{n} (-1)^{i+1} X_i(\sigma(X_1, \ldots, \hat{X}_i, \ldots, X_n)). \quad (5.11)$$

Applying Lemma 2,

$$\sigma(X_1, \ldots, \hat{X}_i, \ldots, X_n) = g^1_{1\ldots\hat{i}\ldots n} + X_1(G^1_{1\ldots\hat{i}\ldots n}), \quad (5.12)$$

with $X_1(g^1_{1\ldots\hat{i}\ldots n}) = 0$. Then, substituting equation 5.12 in 5.11, using $[X_i, X_j] = 0$ and invoking Lemma 2,

$$0 = \beta(X_1, \ldots, X_n) + \sum_{i=2}^{n} (-1)^i X_i(g^1_{1\ldots\hat{i}\ldots n}) + X_1\left(\sum_{i=1}^{n} (-1)^i X_i(G^1_{1\ldots\hat{i}\ldots n}) \right) \quad (5.13)$$

has a solution if and only if

$$T_1 = \beta(X_1, \ldots, X_n) + \sum_{i=2}^{n} (-1)^i X_i(g^1_{1\ldots\hat{i}\ldots n}) , \quad (5.14)$$

where $X_1(T_1) = 0$ and, if $T_1 \neq 0$, it is Taylor flat at Σ_1.

Once more, applying Lemma 2 to T_1 and $g^1_{1\ldots\hat{i}\ldots n}$ with respect to X_2,

$$T_1 = T_{12} + X_2(t_{12}) \quad (5.15)$$

and

$$g^1_{1\ldots\hat{i}\ldots n} = g^{12}_{1\ldots\hat{i}\ldots n} + X_2(G^{12}_{1\ldots\hat{i}\ldots n}), \quad (5.16)$$

where $X_2(T_{12}) = X_2(g^{12}_{1\ldots\hat{i}\ldots n}) = 0$, $X_1(T_{12}) = X_1(g^{12}_{1\ldots\hat{i}\ldots n}) = 0$ and T_{12} is Taylor flat at Σ_1 because $X_1(T_1) = X_1(g^{12}_{1\ldots\hat{i}\ldots n}) = 0$ and T_1 is Taylor flat at Σ_1.

Now, replacing equation (5.15) and (5.16) in (5.14), using $[X_i, X_j] = 0$ and because of Lemma 2,

$$0 = \beta(X_1, \ldots, X_n) - T_{12} + \sum_{i=3}^{n} (-1)^i X_i(g^{12}_{1\ldots\hat{i}\ldots n})$$
$$+ X_2\left(-t_{12} + \sum_{i=2}^{n} (-1)^i X_i(G^{12}_{1\ldots\hat{i}\ldots n}) \right) \quad (5.17)$$

has a solution if and only if

$$T_2 + T_{12} = \beta(X_1, \ldots, X_n) + \sum_{i=3}^{n} (-1)^i X_i(g^{12}_{1\ldots\hat{i}\ldots n}), \quad (5.18)$$

where $X_1(T_2) = X_2(T_2) = 0$ and, if $T_2 \neq 0$, it is Taylor flat at Σ_2.

The next step is to decompose T_2, T_{12} and $g_{1\ldots\hat{i}\ldots n}^{12}$ with respect to X_3 and argue as before. Continuing with this process for all X_i one obtains $\beta(X_1, \ldots, X_n) = T_n + T_{(n-1)n} + \cdots + T_{1\ldots n}$, where $T_n, \ldots, T_{1\ldots n} \in C_{\mathcal{F}}^{\infty}(\mathbb{R}^{2n})$ and, if $T_{i\ldots n} \neq 0$, it is Taylor flat at Σ_i.

We were assuming $k_e = 0$ and $k_h = n$. The case when $k_e \neq 0$ is straightforward: just forget about Taylor flatness for those indices.

From $\Omega_{\mathcal{F}}^n(\mathbb{R}^{2n}) = W_{\mathcal{F}}^n(\mathbb{R}^{2n}) \oplus d_{\mathcal{F}}(\Omega_{\mathcal{F}}^{n-1}(\mathbb{R}^{2n}))$ we obtain $H_{\mathcal{F}}^n(\mathbb{R}^{2n}) = W_{\mathcal{F}}^n(\mathbb{R}^{2n})$, by definition any $\beta \in W_{\mathcal{F}}^n(\mathbb{R}^{2n})$ can be given by a function vanishing at certain points (Proposition 1) and being noncomplanate: $\beta(X_1, \ldots, X_n) = f \in C^{\infty}(\mathbb{R}^{2n})$, $f = 0$ at $\Sigma_1 \cup \cdots \cup \Sigma_n$ and is noncomplanate, if nonzero. The Lie derivative condition further implies that such a function is constant along the leaves. □

Remark 4. The proofs of Theorems 7 and 8 also hold in the analytic category after, obvious and minor, modifications (essentially getting rid of Taylor flat functions).

Before proving Theorem 9, it is worthwhile to look at a particular (smooth) case to illustrate its intricacy.

Proposition 2. *Consider* $(\mathbb{R}^6, \sum_{i=1}^{3} dx_i \wedge dy_i)$ *with* $h_1, h_2, h_3 \in C^{\infty}(\mathbb{R}^6)$ *a Williamson basis. If both* X_1, X_2 *are of hyperbolic type and* X_3 *is of elliptic type, then:*

$$\ker(d_{\mathcal{F}} : \Omega_{\mathcal{F}}^2(\mathbb{R}^6) \to \Omega_{\mathcal{F}}^3(\mathbb{R}^6)) = W_{\mathcal{F}}^2(\mathbb{R}^6) \oplus d_{\mathcal{F}}(\Omega_{\mathcal{F}}^1(\mathbb{R}^6)), \qquad (5.19)$$

where $W_{\mathcal{F}}^2(\mathbb{R}^6)$ *is the set of 2-forms* $\beta \in \Omega_{\mathcal{F}}^2(\mathbb{R}^6)$ *such that* $\pounds_{X_i}(\beta) = 0$ *for* $i = 1, 2, 3$, *and if* $\beta(X_i, X_j) \neq 0$ *it is noncomplanate.*

Proof. The condition $d_{\mathcal{F}}\alpha = 0$ implies, for any $\alpha \in \Omega_{\mathcal{F}}^2(\mathbb{R}^6)$,

$$0 = X_1(\alpha(X_2, X_3)) - X_2(\alpha(X_1, X_3)) + X_3(\alpha(X_1, X_2)). \qquad (5.20)$$

Lemma 2 gives

$$\alpha(X_1, X_3) = f_{13} + X_3(F_{13}) \text{ and } \alpha(X_2, X_3) = f_{23} + X_3(F_{23}), \qquad (5.21)$$

with $X_3(f_{13}) = X_3(f_{23}) = 0$.

Because $[X_i, X_j] = 0$,

$$0 = X_1(f_{23}) - X_2(f_{13}) + X_3\left(\alpha(X_1, X_2) + X_1(F_{23}) - X_2(F_{13})\right), \qquad (5.22)$$

by uniqueness (Lemma 2),

$$\alpha(X_1, X_2) = f_{12} + X_2(F_{13}) - X_1(F_{23}), \tag{5.23}$$

with $X_3(f_{12}) = 0$ and

$$X_1(f_{23}) = X_2(f_{13}). \tag{5.24}$$

Defining $\alpha_3 \in \Omega^1_{\mathcal{F}}(\mathbb{R}^6)$ by

$$\alpha_3(X_1) = f_{13}, \ \alpha_3(X_2) = f_{23} \text{ and } \alpha_3(X_3) = 0, \tag{5.25}$$

it is clear that $d_{\mathcal{F}}\alpha_3 = 0$ (Proposition 1 and item (4) of Lemma 2 guarantee that it is well defined). Theorem 7, then, implies $\alpha_3 = \beta_3 + d_{\mathcal{F}}G_3$, with $\beta_3 \in W^1_{\mathcal{F}}(\mathbb{R}^6)$. In other words:

$$f_{13} = g_{13} + X_1(G_3), \ f_{23} = g_{23} + X_2(G_3) \text{ and } X_3(G_3) = 0. \tag{5.26}$$

Applying repeatedly Lemma 2, for each X_i with $i \neq 3$, to the function f_{12} one gets

$$f_{12} = g_{12} - X_1(G_{23}) + X_2(G_{13}), \tag{5.27}$$

with $X_1(g_{12}) = X_2(g_{12}) = 0$ and $X_3(g_{12}) = X_3(G_{13}) = X_3(G_{23}) = 0$, because $X_3(f_{12}) = 0$.
Summing up, plugging equation (5.26) in (5.21), using $X_3(G_{13}) = X_3(G_{23}) = 0$, and equation (5.27) in (5.23):

$$\alpha(X_1, X_3) = g_{13} + X_1(G_3) + X_3(F_{13} + G_{13}) \tag{5.28}$$
$$\alpha(X_2, X_3) = g_{23} + X_2(G_3) + X_3(F_{23} + G_{23}) \tag{5.29}$$

and

$$\alpha(X_1, X_2) = g_{12} - X_1(F_{23} + G_{23}) + X_2(F_{13} + G_{13}). \tag{5.30}$$

Wherefore $\alpha = \beta + d_{\mathcal{F}}\zeta$ with $\beta \in W^1_{\mathcal{F}}(\mathbb{R}^6)$;

$$\beta(X_1, X_2) = g_{12}, \ \beta(X_1, X_3) = g_{13}, \ \beta(X_2, X_3) = g_{23} \tag{5.31}$$

and

$$\zeta(X_1) = -F_{13} - G_{13}, \ \zeta(X_2) = -F_{23} - G_{23}, \ \zeta(X_3) = G_3, \tag{5.32}$$

(as always, Proposition 1 and item (4) of Lemma 2 guarantee that the forms are well defined).

The condition $\mathcal{L}_{X_i}(\beta) = 0$ for $i = 1, 2, 3$ implies $d_{\mathcal{F}}\beta = 0$, and there exists $\sigma \in \Omega^1_{\mathcal{F}}(\mathbb{R}^6)$ such that $d_{\mathcal{F}}\sigma = \beta$ if and only if

$$\beta(X_i, X_j) = X_i(\sigma(X_j)) - X_j(\sigma(X_i)). \tag{5.33}$$

Applying Lemma 2,

$$\sigma(X_i) = s_{i3} + X_3(S_{i3}), \tag{5.34}$$

with $X_3(s_{i3}) = 0$. Then, plugging equation 5.34 in 5.33, using $[X_i, X_j] = 0$ and using uniqueness (Lemma 2),

$$0 = \beta(X_i, X_j) + X_j(s_{i3}) - X_i(s_{j3}) + X_3(X_j(S_{i3}) - X_i(S_{j3})) \tag{5.35}$$

has a solution if and only if

$$0 = \beta(X_i, X_j) + X_j(s_{i3}) - X_i(s_{j3}). \tag{5.36}$$

Again, applying Lemma 2,

$$s_{i3} = s_{i23} + X_2(S_{i23}), \tag{5.37}$$

with $X_2(s_{i23}) = 0$ and $X_3(s_{i23}) = 0$, because $X_3(s_{i3}) = 0$. Then, replacing equation (5.37) in (5.36), using $[X_i, X_j] = 0$ and because of Lemma 2,

$$0 = \beta(X_i, X_j) + X_j(s_{i23}) - X_i(s_{j23}) + X_2(X_j(S_{i23}) - X_i(S_{j23})) \tag{5.38}$$

has a solution if and only if

$$T_{ij2} = \beta(X_i, X_j) + X_j(s_{i23}) - X_i(s_{j23}), \tag{5.39}$$

where $X_3(T_{ij2}) = X_2(T_{ij2}) = 0$ and T_{ij2} is Taylor flat at Σ_2. Explicitly,

$$T_{122} = \beta(X_1, X_2) - X_1(s_{223}), \tag{5.40}$$

$$T_{132} = \beta(X_1, X_3) - X_1(s_{323}) \tag{5.41}$$

and

$$T_{232} = \beta(X_2, X_3). \tag{5.42}$$

Once more, applying Lemma 2,

$$T_{ij2} = t_{ij} + X_1(T_{ij12}), \tag{5.43}$$

with $X_1(t_{ij}) = 0$, $X_2(t_{ij}) = X_3(t_{ij}) = 0$ and t_{ij} is Taylor flat at Σ_2, because $X_2(T_{ij2}) = X_3(T_{ij2}) = 0$ and T_{ij2} is Taylor flat at Σ_2. Then, substituting equation (5.43) in (5.40), (5.41) and (5.42), using $[X_i, X_j] = 0$ and using Lemma 2,

$$\beta(X_1, X_2) - t_{12} = X_1(s_{223} + T_{1212}), \tag{5.44}$$
$$\beta(X_1, X_3) - t_{13} = X_1(s_{323} + T_{1312}) \tag{5.45}$$

and

$$\beta(X_2, X_3) - t_{23} = X_1(T_{2312}). \tag{5.46}$$

have solution if and only if

$$\beta(X_1, X_2) = t_{12} + T_{12}, \ \beta(X_1, X_3) = t_{13} + T_{13} \text{ and } \beta(X_2, X_3) = t_{23} + T_{23}, \quad (5.47)$$

where each $T_{ij} \in C_{\mathcal{F}}^{\infty}(\mathbb{R}^6)$ and is Taylor flat at Σ_1. $\square$

Theorem 9 (Analytic case). *Consider* $(\mathbb{R}^{2n}, \sum_{i=1}^{n} \mathrm{d}x_i \wedge \mathrm{d}y_i)$ *endowed with a analytic distribution* $\mathcal{F}$ *generated by a Williamson basis of type* $(k_e, k_h, 0)$, *then the following decomposition holds:*

$$\ker(\mathrm{d}_{\mathcal{F}} : \Omega_{\mathcal{F}}^k(\mathbb{R}^{2n}) \to \Omega_{\mathcal{F}}^{k+1}(\mathbb{R}^{2n})) = W_{\mathcal{F}}^k(\mathbb{R}^{2n}) \oplus \mathrm{d}_{\mathcal{F}}(\Omega_{\mathcal{F}}^{k-1}(\mathbb{R}^{2n})), \quad (5.48)$$

where $W_{\mathcal{F}}^k(\mathbb{R}^{2n})$ *is the set of analytic k-forms* $\beta \in \Omega_{\mathcal{F}}^k(\mathbb{R}^{2n})$ *such that* $\pounds_{X_i}(\beta) = 0$ *for all i.*

Proof. It remains to prove when the degree is different from 1 and n, since the proofs of Theorems 7 and 8, as mentioned, work for these particular cases.

If $\alpha \in \Omega_{\mathcal{F}}^k(\mathbb{R}^{2n})$ the condition $\mathrm{d}_{\mathcal{F}}\alpha = 0$ implies

$$0 = X_{j_1}(\alpha(X_{j_2}, \ldots, X_{j_{k+1}})) + \sum_{i=2}^{k+1} (-1)^{i+1} X_{j_i}(\alpha(X_{j_1}, \ldots, \hat{X_{j_i}}, \ldots, X_{j_{k+1}})). \quad (5.49)$$

Applying successively of Lemma 2, with respect to each X_{j_i} $(i \neq 1)$, gives

$$\alpha(X_{j_1}, \ldots, \hat{X_{j_i}}, \ldots, X_{j_{k+1}}) = g_{j_1 \cdots \hat{j_i} \cdots j_{k+1}}^{j_1 \cdots j_{k+1}} + X_{j_1}(F_{j_1 \cdots \hat{j_i} \cdots j_{k+1}}^{j_1 \cdots j_{k+1}})$$
$$+ \sum_{\substack{l=2 \\ l \neq i}}^{k+1} (-1)^{l_i+1} X_{j_l}(G_{j_1 j_2 \cdots \hat{j_i} \cdots \hat{j_l} \cdots j_{k+1}}^{j_1 \cdots j_{k+1}}), \quad (5.50)$$

where $X_{j_m}(g_{j_1 \cdots \hat{j_i} \cdots j_{k+1}}^{j_1 \cdots j_{k+1}}) = 0$ for $m = 1, \ldots, k+1 \neq i$ and

$$l_i = \begin{cases} l & \text{if } l < i \\ l+1 & \text{if } l > i. \end{cases} \quad (5.51)$$

Substituting equation (5.50) in equation (5.49) (using $[X_i, X_j] = 0$),

$$0 = \sum_{i=2}^{k+1} (-1)^{i+1} X_{j_i} \left(g_{j_1 \cdots \hat{j_i} \cdots j_{k+1}}^{j_1 \cdots j_{k+1}} + \sum_{\substack{l=2 \\ l \neq i}}^{k+1} (-1)^{l_i+1} X_{j_l}(G_{j_1 j_2 \cdots \hat{j_i} \cdots \hat{j_l} \cdots j_{k+1}}^{j_1 \cdots j_{k+1}}) \right)$$
$$+ X_{j_1} \left(\alpha(X_{j_2}, \ldots, X_{j_{k+1}}) + \sum_{i=2}^{k+1} (-1)^{i+1} X_{j_i}(F_{j_1 \cdots \hat{j_i} \cdots j_{k+1}}^{j_1 \cdots j_{k+1}}) \right), \quad (5.52)$$

and by uniqueness (Lemma 2):

$$\alpha(X_{j_2}, \ldots, X_{j_{k+1}}) = f_{j_2 \cdots j_{k+1}}^{j_1 \cdots j_{k+1}} + \sum_{i=2}^{k+1} (-1)^i X_{j_i}(F_{j_1 \cdots \hat{j}_i \cdots j_{k+1}}^{j_1 \cdots j_{k+1}}), \tag{5.53}$$

with $X_{j_1}(f_{j_2 \cdots j_{k+1}}^{j_1 \cdots j_{k+1}}) = 0$.

Again, applying repeatedly Lemma 2, for each X_{j_i} with $i \neq 1$, to the function $f_{j_2 \cdots j_{k+1}}^{j_1 \cdots j_{k+1}}$ one gets

$$f_{j_2 \cdots j_{k+1}}^{j_1 \cdots j_{k+1}} = g_{j_2 \cdots j_{k+1}}^{j_1 \cdots j_{k+1}} + \sum_{i=2}^{k+1} (-1)^i X_{j_i}(G_{j_1 \cdots \hat{j}_i \cdots j_{k+1}}^{j_1 \cdots j_{k+1}}), \tag{5.54}$$

with $X_{j_i}(g_{j_2 \cdots j_{k+1}}^{j_1 \cdots j_{k+1}}) = 0$ for $i = 1, \ldots, k+1$ and $X_{j_1}(G_{j_1 \cdots \hat{j}_i \cdots j_{k+1}}^{j_1 \cdots j_{k+1}}) = 0$,

because $X_{j_1}(f_{j_2 \cdots j_{k+1}}^{j_1 \cdots j_{k+1}}) = 0$.

Using $X_{j_1}(G_{j_1 \cdots \hat{j}_i \cdots j_{k+1}}^{j_1 \cdots j_{k+1}}) = 0$, and substituting equation (5.54) in equation (5.53):

$$\begin{aligned}
\alpha(X_{j_1}, \ldots, \hat{X}_{j_i}, \ldots, X_{j_{k+1}}) = {}& g_{j_1 \cdots \hat{j}_i \cdots j_{k+1}}^{j_1 \cdots j_{k+1}} + X_{j_1}(F_{j_1 \cdots \hat{j}_i \cdots j_{k+1}}^{j_1 \cdots j_{k+1}} \\
+ G_{j_1 \cdots \hat{j}_i \cdots j_{k+1}}^{j_1 \cdots j_{k+1}}) {}& + \sum_{\substack{l=2 \\ l \neq i}}^{k+1} (-1)^{l_i+1} X_{j_l}(G_{j_1 j_2 \cdots \hat{j}_i \cdots \hat{j}_l \cdots j_{k+1}}^{j_1 \cdots j_{k+1}}),
\end{aligned} \tag{5.55}$$

for $i \neq 1$, and

$$\begin{aligned}
\alpha(X_{j_2}, \ldots, X_{j_{k+1}}) = {}& g_{j_2 \cdots j_{k+1}}^{j_1 \cdots j_{k+1}} + \sum_{i=2}^{k+1} (-1)^i X_{j_i}(F_{j_1 \cdots \hat{j}_i \cdots j_{k+1}}^{j_1 \cdots j_{k+1}} \\
& + G_{j_1 \cdots \hat{j}_i \cdots j_{k+1}}^{j_1 \cdots j_{k+1}}).
\end{aligned} \tag{5.56}$$

A priori it cannot be guaranteed that the g's belong to $C_{\mathcal{F}}^{\omega}(\mathbb{R}^{2n})$, however, varying j_1 from 1 to n, there is more than one decomposition like equation (5.55) and equation (5.56) for each combinations of vector fields. By the uniqueness of these decompositions (Lemma 2) this yields $\alpha = \beta + d_{\mathcal{F}}\zeta$. There exists a correct number of functions to define β and ζ, the g's and $F + G$'s of equation (5.55) and equation (5.56) (after applying uniqueness and identifying some of them, and using Proposition 1 and item (4) of Lemma 2 to guarantee that the forms are well defined).

The condition $\pounds_{X_i}(\beta) = 0$ for all i implies $d_{\mathcal{F}}\beta = 0$, and there exists $\sigma \in \Omega_{\mathcal{F}}^{k-1}(\mathbb{R}^{2n})$ such that $d_{\mathcal{F}}\sigma = \beta$ if and only if

$$\beta(X_{j_1}, \ldots, X_{j_k}) = \sum_{i=1}^{k} (-1)^{i+1} X_{j_i}(\sigma(X_{j_1}, \ldots, \hat{X}_{j_i}, \ldots, X_{j_k})). \tag{5.57}$$

Applying Lemma 2,

$$\sigma(X_{j_1}, \ldots, \hat{X}_{j_i}, \ldots, X_{j_k}) = s^{j_1}_{j_1 \cdots \hat{j}_i \cdots j_k} + X_{j_1}(S^{j_1}_{j_1 \cdots \hat{j}_i \cdots j_k}), \tag{5.58}$$

with $X_{j_1}(s^{j_1}_{j_1 \cdots \hat{j}_i \cdots j_k}) = 0$. Now plugging equation (5.58) in equation (5.57) and using the commutation of the vector fields ($[X_i, X_j] = 0$) and because of uniqueness (Lemma 2), we obtain,

$$\begin{aligned}
0 = \beta(X_{j_1}, \ldots, X_{j_k}) &+ \sum_{i=2}^{k} (-1)^i X_{j_i}(s^{j_1}_{j_1 \cdots \hat{j}_i \cdots j_k}) \\
&+ X_{j_1} \left(\sum_{i=1}^{k} (-1)^i X_{j_i}(S^{j_1}_{j_1 \cdots \hat{j}_i \cdots j_k}) \right)
\end{aligned} \tag{5.59}$$

has a solution if and only if,

$$0 = \beta(X_{j_1}, \ldots, X_{j_k}) + \sum_{i=2}^{k} (-1)^i X_{j_i}(s^{j_1}_{j_1 \cdots \hat{j}_i \cdots j_k}). \tag{5.60}$$

Again, applying Lemma 2,

$$s^{j_1}_{j_1 \cdots \hat{j}_i \cdots j_k} = s^{j_1 j_2}_{j_1 \cdots \hat{j}_i \cdots j_k} + X_{j_2}(S^{j_1 j_2}_{j_1 \cdots \hat{j}_i \cdots j_k}), \tag{5.61}$$

with $X_{j_2}(s^{j_1 j_2}_{j_1 \cdots \hat{j}_i \cdots j_k}) = 0$. Then, plugging equation (5.61) in equation (5.60), using $[X_i, X_j] = 0$ and invoking uniqueness (Lemma 2),

$$\begin{aligned}
0 = \beta(X_{j_1}, \ldots, X_{j_k}) &+ \sum_{i=3}^{k} (-1)^i X_{j_i}(s^{j_1 j_2}_{j_1 \cdots \hat{j}_i \cdots j_k}) \\
&+ X_{j_2} \left(\sum_{i=2}^{k} (-1)^i X_{j_i}(S^{j_1 j_2}_{j_1 \cdots \hat{j}_i \cdots j_k}) \right)
\end{aligned} \tag{5.62}$$

has a solution if and only if,

$$0 = \beta(X_{j_1}, \ldots, X_{j_k}) + \sum_{i=3}^{k} (-1)^i X_{j_i}(s^{j_1 j_2}_{j_1 \cdots \hat{j}_i \cdots j_k}). \tag{5.63}$$

Following the same procedure for all X_{j_i}, $i = 3, \ldots, k$, this yields $\beta(X_{j_1}, \ldots, X_{j_k}) = 0$. □

This determines all foliated cohomology groups in the analytic case,

Corollary 1. *The foliated cohomology groups in the analytic case are determined for $k = 1, \ldots, n$ by,*

$$H^k_{\mathcal{F}}(\mathbb{R}^{2n}) \cong \bigoplus_{(j_1, \ldots, j_k)} \{f \in C^{\omega}(\mathbb{R}^{2n})$$

$$; f(p) = \tilde{f}(h_1(p), \ldots, h_n(p)) \, and \, f\big|_{\Sigma_{j_1} \cup \cdots \cup \Sigma_{j_k}} = 0\}, \qquad (5.64)$$

where the right hand side has $\binom{n}{k}$ summands.

Proof. Theorem 9 reads $H^k_{\mathcal{F}}(\mathbb{R}^{2n}) = W^k_{\mathcal{F}}(\mathbb{R}^{2n})$, by definition any $\beta \in W^k_{\mathcal{F}}(\mathbb{R}^{2n})$ can be given by $\binom{n}{k}$ functions vanishing at certain points (proposition 1) e.g.: $\beta(X_1, \ldots, X_k) = f_{1 \cdots k} \in C^{\omega}(\mathbb{R}^{2n})$ and $f_{1 \cdots k} = 0$ at $\Sigma_1 \cup \cdots \cup \Sigma_k$. The Lie derivative condition yields (item 5 of Lemma 2) that each such function has a special dependence on its variables, e.g.: $f_{1 \cdots k}(x_1, y_1, \ldots, x_n, y_n) = f(h_1, \ldots, h_n)$. $\qquad \square$

The previous proofs work as well in the smooth category if all the vector fields are of elliptic type. Thus for completely elliptic singularities we can compute all the cohomology groups obtaining the following:

Theorem 10 (Elliptic case). *Consider $(\mathbb{R}^{2n}, \sum\limits_{i=1}^{n} \mathrm{d}x_i \wedge \mathrm{d}y_i)$ with $h_1, \ldots, h_n \in C^{\infty}(\mathbb{R}^{2n})$ a Williamson basis. If all the vector fields, $X_1, \ldots, X_n$ are of elliptic type, for $k = 1, \ldots, n$:*

$$\ker(\mathrm{d}_{\mathcal{F}} : \Omega^k_{\mathcal{F}}(\mathbb{R}^{2n}) \to \Omega^{k+1}_{\mathcal{F}}(\mathbb{R}^{2n})) = W^k_{\mathcal{F}}(\mathbb{R}^{2n}) \oplus \mathrm{d}_{\mathcal{F}}(\Omega^{k-1}_{\mathcal{F}}(\mathbb{R}^{2n})), \quad (5.65)$$

where $W^k_{\mathcal{F}}(\mathbb{R}^{2n})$ is the set of k-forms $\beta \in \Omega^k_{\mathcal{F}}(\mathbb{R}^{2n})$ such that $\pounds_{X_i}(\beta) = 0$ for all i.

Thus, the foliated cohomology groups are given by:

$$H^k_{\mathcal{F}}(\mathbb{R}^{2n}) \cong \bigoplus_{(j_1, \ldots, j_k)} \{f \in C^{\infty}(\mathbb{R}^{2n})$$

$$; f(p) = \tilde{f}(h_1(p), \ldots, h_n(p)) \, and \, f\big|_{\Sigma_{j_1} \cup \cdots \cup \Sigma_{j_k}} = 0\} \qquad (5.66)$$

where the right hand side has $\binom{n}{k}$ summands.

References

[1] C. Chevalley and S. Eilenberg, *Cohomology theory of Lie groups and Lie algebras*, Trans. Amer. Math. Soc. **63** (1948), 85–124.

[2] L.H. Eliasson, *Hamiltonian systems with Poisson commuting integrals*, Ph.D. Thesis, Stockholm University (1984).

[3] L.H. Eliasson, *Normal forms for Hamiltonian systems with Poisson commuting integrals—elliptic case*, Comment. Math. Helv., **65** (1) (1990), 4–35.

[4] V. Guillemin and D. Schaeffer, *On a certain class of Fuchsian partial differential equations*, Duke Math. J., **44** (1) (1977), 157–199.

[5] V. Guillemin and S. Sternberg, *Geometric Asymptotics*, Mathematical Surveys AMS, **14**, 1977.

[6] K. Mackenzie, *Lie algebroids and Lie pseudoalgebras*, Bull. London Math. Soc., **27** (2) (1995), 97–147.

[7] E. Miranda, *On symplectic linearization of singular Lagrangian foliations*, Ph.D. Thesis, Universitat de Barcelona, 2003.

[8] E. Miranda, *Integrable systems and group actions*, to appear in Central European Journal of Mathematics, 2013.

[9] E. Miranda and S. Vũ Ngọc, *A Singular Poincaré Lemma*, IMRN, **1** (2005), 27–46.

[10] E. Miranda and N.T. Zung, *Equivariant normal forms for nondegenerate singular orbits of integrable Hamiltonian systems*, Ann. Sci. Ecole Norm. Sup., **37** (6) (2004), 819–839.

[11] J. Moser, *On the volume elements on a manifold*. Trans. Amer. Math. Soc., **120** (1965), 286–294.

[12] J.H. Rawnsley, *On The Cohomology Groups of a Polarisation and Diagonal Quantisation*, Trans. Amer. Math. Soc., **230** (1977), 235–255.

[13] J. Vey, *Sur certains systèmes dynamiques séparables*. Amer. J. Math., **100** (3) (1978), 591–614.

[14] S. Vũ Ngọc, *Symplectic techniques for semiclassical completely integrable systems*, Topological methods in the theory of integrable systems, Camb. Sci. Publ., Cambridge, 2006, 241–270.

[15] F.W. Warner, *Foundations of Differentiable manifolds and Lie groups*, Graduate Texts in Mathematics, **94**, Springer, 1983.

[16] J. Williamson, *On the algebraic problem concerning the normal forms of linear dynamical systems*, Amer. J. Math., **58** (1) (1936), 141–163.

[17] N.T. Zung, *Symplectic topology of integrable hamiltonian systems, I: Arnold-Liouville with singularities*, Compositio Mathematica, **101** (1996), 179–215.

Received January 16, 2013.

FOLIATIONS 2012
ed. by Paweł WALCZAK *et al.*
World Scientific, Singapore, 2013
pp. 139–162

Compact generation for topological groupoids

Nicolas Raimbaud

Université de Bretagne Sud, UMR 6205
LMBA, F-56000 Vannes, France
e-mail: groupoid.kid@gmail.com

1. Introduction

1.1. Recall that a foliation of an n-manifold M is a partition of M into immersed submanifolds of same dimension p, which are locally 'parallel'. The connected components of the p-dimensional structure on M are called the leaves of the foliation. It is well-known that, even though locally simple, the global behaviour of the leaves may be pretty complicated.

Given a foliated manifold, one can encode the dynamics of the leaves in a pseudogroup structure, the transverse holonomy pseudogroup, which depends on the choice of an exhaustive transversal for the foliation (a submanifold of complementary dimension everywhere transverse to the leaves, and meeting every leaf at least once). This pseudogroup actually depends essentially only on the foliation, in the sense that two different exhaustive transversals always yield two locally equivalent (or Haefliger-equivalent) pseudogroups. When the manifold is compact, André Haefliger remarked that a finite number of compact data was enough to generate all the holonomy; and introduced the corresponding notion of compact generation for pseudogroups, invariant under Haefliger equivalence. In [7], he extended this notion to étale groupoids, using the close relation between pseudogroups and étale groupoids (a pseudogroup is indeed no more than an effective étale groupoid – see Section 4). Given a pseudogroup of local homeomorphisms on a topological space, the set of germs of its elements may be given a natural groupoid structure (using composition of germs) with a compatible (sheaf) topology, making this groupoid a topological groupoid. When translated to germ groupoids, Haefliger equivalence of pseudogroups becomes Morita equivalence of topological groupoids, thus there is already a specific notion of compact generation for germ groupoids, Morita-invariant among germ groupoids (Theorem 3).

In this article, we extend Haefliger's definition of compact generation for germ groupoids (or equivalently pseudogroups) to all locally compact topological groupoids with open source map and Hausdorff base, with a definition which is invariant under the usual Morita equivalence of topological groupoids. We also prove that all s-connected[a] foliation groupoids with compact base are compactly generated, generalizing Haefliger's result about transverse homotopy and holonomy (Section 5).

2. Preliminaries

Recall that a *groupoid* is a small category with only invertible morphisms. Geometrically, we shall see a groupoid G as a set of points G_0, together with a set of arrows G_1 between those points[b]. The *base* G_0 of the groupoid may be seen as a subset of G_1 via the *unit map* $1\!\!1 : G_0 \to G_1$ (sending each point to the associated identity map). Each arrow is assigned a source and a target point in G_0 via the *source and target maps* s and t, and a compatible concatenation operation μ is defined on the arrows: given two arrows $g : x \to y$ and $h : y \to z$ in G_1, we have a *product* arrow $\mu(h, g) = hg : x \to z$. (Note that the product map μ is only defined on the fiber product $G_t\!\times_s G = \{(h, g) \in G^2 / th = sg\}$.) Finally, there is an *inverse map* ι sending each arrow to its inverse: $\iota(g)g = 1\!\!1_{s(g)}$ and $g\iota(g) = 1\!\!1_{t(g)}$.

As algebraic objects, groupoids behave a lot like groups, that's why both the vocabulary of groups and categories is applied to groupoids: composition is called product, identity maps are called units, and so on. In this article, we will particularly be interested in the obvious notions of subgroupoid (a subcategory closed under inversion), and of *subgroupoid* $\langle S \rangle$ *generated by a subset S* (the intersection of all subgroupoids containing the subset S). We shall also call full a subgroupoid which is full as a subcategory.

Notations: Given two subsets X and Y of G_0, the set of arrows of G_1

---

[a]Recall that an s-connected groupoid is a (topological) groupoid with connected s-fibers.
[b]It could be helpful for this article to imagine each arrow going leftwards.

which starts from (some point of) X and heads to Y will be denoted G_X^Y. If X or Y is the whole base G_0, it will be omitted in the notation (G^Y, G_X).

2.1. Given a groupoid G, a (left) *action* of G on a set X is a consistent way of sending the arrows of G over X so that the relation $(hg)x = h(gx)$ holds. But the product in a groupoid is only partially defined, so for the relation to make sense it is necessary to define a map $m : X \to G_0$, the *momentum map*, which tells which arrow is allowed to act on which point. An action of G on X over m is then a map $\cdot : G_{1s} \times_m X \to X$, with the property that $m(gx) = t(g)$ and satisfying the extended associativity.

In particular, G acts on itself over t by left multiplication. We shall call 0-*orbit* of G an orbit of this action; given $x \in G_0$, the 0-orbit of x is the set of points of G_0 which are linked to x by an arrow, $t(s^{-1}(x))$. The product groupoid $G \times G$ (with obvious structure) acts on G_1 over (s, s) by $(k, h) \cdot g = kgh^{-1}$. We shall call 1-*orbit* of G an orbit for this second action; the 1-orbit of an arrow g is the set of arrows in G_1 which are composable with an arrow composable with g, that is $t^{-1}(t(s^{-1}(s(g))))$ or equivalently $s^{-1}st^{-1}tg$.

Given a subset $S \subset G_1$, we define $S_0 = \langle S \rangle_0 = t(S) \cup s(S)$ and call it the base of S. It is an easy exercise to check that S meets every 1-orbit of G if and only if S_0 meets every 0-orbit of G. When this is the case, we shall say that S is *exhaustive* in G.

2.2. This article is intended to be read either in the topological or in the $\mathcal{C}^\infty$ setting. Therefore, you may choose to work in the category of topological spaces and continuous maps, or in the category of smooth $\mathcal{C}^\infty$ manifolds and smooth maps. To allow our constructions to make sense in both cases without specifying each time two sets of hypothesis and conclusions, we will introduce a common vocabulary in the next two paragraphs. The reason for this double discourse is that we wish the definition of compact generation to make sense in the general setting of non-differentiable spaces, whereas our applications take place mainly in the $\mathcal{C}^\infty$ setting.

We wish to emphasize the fact that we don't require our spaces to be Hausdorff, neither in the topological nor in the $\mathcal{C}^\infty$ category. Indeed, we wish to work with groupoids arising from (smooth) foliations, and many of these groupoids are naturally non-Hausdorff[c]. On the other hand, we wish to study properties associated to compactness, so we will assume that all

---

[c]For example, the holonomy groupoid of the celebrated Reeb foliation of the 3-sphere associated to a connected exhaustive transversal is a smooth connected 1-manifold with an infinite number of ends.

our spaces are locally compact (here compactness shall be understood as quasi-compactness in the Bourbaki sense, i.e. we do not assume Hausdorff property).

Remark 1. Also, we will use a non-standard definition for *relative compactness* to fit the non-Hausdorff case: we shall say that a subset $Y \subset X$ is relatively compact in X if it may be included inside a compact subset K of X, $Y \subset K \subset X$.

Note that this definition is strictly weaker than the classical "having a compact closure" (we do not ask the compact set in the definition to be closed in X).

Definition 1 (Regular map). In the topological case, we shall say that a map is regular if it is an open surjection. In the smooth case, we shall say that a map is regular if it is a surjective submersion.

This kind of map will appear in every construction we will make. Those maps have two important properties that should be kept in mind: first, the composition of two regular maps is regular. The second property is about fiber products, which we will extensively use. Given two maps, one of whose is regular, their fiber product always exists (i.e. is a smooth manifold in the smooth case), and the pullback of a regular map is always regular (left as an exercise).

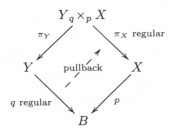

These remarks considerably simplify the proofs we make (especially in the smooth case), reducing a lot of questions to simple diagram chasing. The property about fiber products admits the following extension to triple fiber products: given two pairs of maps $Z \to B' \leftarrow Y$ and $Y \to B \leftarrow X$ such that the two maps issuing from Y are regular, then the two fiber products $Z \times_{B'} Y$ and $Y \times_B X$ exist.

If moreover, the compositions $r \circ \pi_Y$ and $q \circ (\pi_Y)'$ are regular, then the triple product $Z_s \times_r Y_q \times_p X$ exists (i.e. is smooth in the smooth case) and is canonically isomorphic to its two decompositions into nested double fiber products. This property will be used to define pullback groupoids. Note that if the four maps at the beginning are regular, then the two conditions are fulfilled and every map in the following diagram is regular.

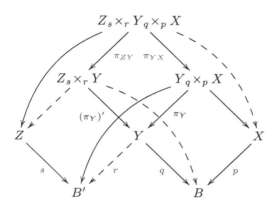

Definition 2 (Regular topological groupoid). A groupoid G will be called a regular topological groupoid if G_0 and G_1 are locally compact topological spaces, such that all structure maps are continuous, and the source map s is regular.

Definition 3 (Lie groupoid). A groupoid G will be called a Lie groupoid if G_0 and G_1 are finite dimensional non-necessarily Haussdorff manifolds, such that the source map is regular, and all structure maps are smooth.

Note that the product in a groupoid is only defined on the fiber product $G_t \times_s G$. Thus in the Lie case the assumption that the source map (or target map, which is equivalent) is regular ensures that the product map is defined on a smooth manifold. We shall use the generic term 'regular' to mean that a groupoid is topological and regular or is Lie, depending on the working category.

Remark 2. Unless explicitly stated, the groupoids we consider will always be regular

There's of course an obvious associated notion of groupoid morphism: a groupoid morphism is just a continuous functor between regular groupoids (smooth in the Lie case).

2.3. Given a groupoid H and a map $f : X \to H_0$, it is always possible to build the fiber product of the source s of H with f (because s is regular).

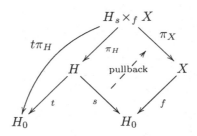

This fiber product may intuitively be seen as a set of arrows from points of X to points of H_0: we have pulled back the sources of the arrows of H_1 from H_0 to X through f. If the projection $t\pi_H$ is also regular, we may again consider its cross product with f to get 'arrows' from X to X:

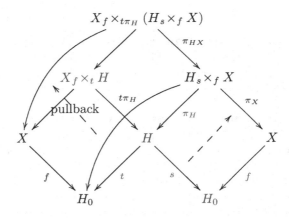

Using the inverse map of H, it is easy to see that $t\pi_H : H_s \times_f X \to H_0$ is regular if and only if $s(\pi_H)' : H_s \times_f X \to H_0$ is. Under this condition the triple product property applies, so that the order of the two pullbacks may be reversed, and the last space we get is in both cases canonically isomorphic to $X_f \times_t H_s \times_f X$. This set has a natural structure of regular groupoid inherited from H, with obvious source and target maps ($f^*(s(\pi_H)')$ and $f^*(t\pi_H)$) and product $(z, h, y)(y, g, x) = (z, hg, x)$. Note that when f is itself regular, the condition on $t\pi_H$ is automatically fulfilled, as $\pi_H = s^*(f)$ is regular in this case.

Definition 4. Given a groupoid H and a map $f : X \to H_0$ such that $t\pi_H : H_s\times_f X \to H_0$ is regular, then the space $X_f\times_t H_s\times_f X$ has a natural structure of regular groupoid with base X. It is called the *pullback groupoid* of H by f, and denoted f^*H.

2.4. In particular, given a groupoid morphism $\varphi : G \to H$, we may consider the induced base map $\varphi_0 : G_0 \to H_0$ and try to carry out a pullback construction. If the condition of Definition 4 is fulfilled by φ_0 ($t\pi_H$ regular), then the pullback groupoid φ_0^*H exists, and comes with a canonical morphism $(t, \varphi, s) : G \to \varphi_0^*H$. In the case when this canonical morphism is an isomorphism (homeomorphism or diffeomorphism), i.e. in the case when G canonically identifies to φ_0^*H, φ is called an *essential equivalence*. In particular, φ induces bijections $G_m^n \to H_{\varphi m}^{\varphi n}$, that is: given an arrow $h : x \to y \in H_1$ between two points x and y in the image of φ_0, and given two points m and n in their repective φ_0-fibers, their exists a unique arrow $g : m \to n$ which is sent to h by φ. This important property is called the *unique lifting property*. Note that the existence of an essential equivalence from G to H is not an equivalence relation; it is indeed not symmetric.

Definition 5 (Morita equivalence). Two groupoids G and H are said to be Morita equivalent if there exists a third groupoid K, and two essential equivalences $\varphi : K \to G$ and $\psi : K \to H$.

It can be proved that this relation is an equivalence relation (see for example section 2 of [10]). Morita equivalence admits a lot of equivalent defintions, we shall especially be interested in the following one:

Theorem 1. *Two groupoids G and H are Morita equivalent if and only if there exists a third groupoid K, and two Morita morphisms $\varphi : K \to G$ and $\psi : K \to H$.*

A *Morita morphism* is a special kind of essential equivalence, where the base map is already regular (so that the pullback by the base automatically exists without further assumption). It is an easy exercise to check that for an essential equivalence φ, the base φ_0 is regular if and only if φ is (use paragraph 1).

3. Compact generation

Definition 6 (Compact generation). A regular groupoid G is said to be *compactly generated* if it contains an exhaustive relatively compact open subset $\mathcal{U}$, which generates a full subgroupoid.

Definitions for relative compactness and exhaustiveness may be found in Sections 2.2 and 2.1 respectively. Note that the product of a regular groupoid is an open map, hence the subgroupoid generated by an open subset is itself a regular open subgroupoid ($\langle U \rangle = \cup_{n \in \mathbb{N}} (\mathcal{U} \cup \iota(\mathcal{U}))^n$). With this remark, it is not hard to see that this definition is equivalent to "there exists a relatively compact open subset $\mathcal{U}$ of G such that the inclusion $\langle \mathcal{U} \rangle \subset G$ is an essential equivalence[d]".

Theorem 2. *Let G and H be two Morita-equivalent object-separated regular groupoids. Then G is compactly generated if and only if H is.*

Recall that a topological groupoid is said to be *object-separated* if its base is a Hausdorff space[e]. Using Theorem 1, it is enough to prove the following:

Lemma 1. *If $\varphi : G \to H$ is a Morita morphism between groupoids with H object-separated, and if G or H is compactly generated, then so is the other one.*

The easy case is when G is compactly generated: let $\mathcal{U}$ be an exhaustive relatively compact open subset of G with full generation, and let $\mathcal{V} := \varphi(\mathcal{U})$. Then $\mathcal{V}$ is immediately relatively compact and open, as φ is continuous and open (see Sections 2.4 and 1). Using the unique lifting property (Section 2.4), one can check that the 1-orbits of G and H are in one-to-one correspondence via φ ($\mathrm{Orb}(g) \mapsto \mathrm{Orb}(\varphi g)$). Thus the H_1-orbits met by $\mathcal{V}$ are precisely the images of the G_1-orbits met by $\mathcal{U}$. But $\mathcal{U}$ meets *every* G_1-orbit, thus $\mathcal{V}$ is exhaustive. Finally, as φ is a functor, the subgroupoid generated by $\mathcal{V}$, the image of $\mathcal{U}$, is the image of the subgroupoid generated by $\mathcal{U}$, which is full. Thus $\langle \mathcal{V} \rangle$ is also full, the unique lifting property ensuring it can't miss any arrow.

The other case is a bit tricky, because we have to climb up φ preserving both openness and relative compactness. Assume we have a $\mathcal{V} \subset H$ giving compact generation for H. Replacing $\mathcal{V}$ by $\mathcal{V} \cup \iota(\mathcal{V})$ if necessary, we may suppose that $\mathcal{V}$ is symmetric. Let $\mathcal{K}$ be some compact set containing $\mathcal{V}$ (relative compactness), and write $K := \mathcal{K}_0 = t(\mathcal{K}) \cup s(\mathcal{K})$.

For each point $x \in K$, we choose the following data: a point $m \in G_0$ in the φ_0-fiber over x, an open neighborhood W_m of m in G_0, a compact subneighborhood $C_m \subset W_m$, and an open subneighborhood $U_m \subset C_m$.

---

[e]This assumption is technical: we need fiber products over base spaces to be closed in their respective cartesian products.

The collection of the $\varphi(U_m)$ is an open cover of the compact set K, thus it admits a finite subcover associated to a finite sequence $m_1, m_2, \ldots, m_N$. We forget the points m and just reindex the associated data W_i, C_i, U_i. Intersecting with $\varphi_0^{-1}(V)$ if necessary, we may assume that $\varphi(U_i) \subset V$ for each i. Let U be the union of the U_i's, and consider the set of all arrows between points of U that are sent in V by φ:

$$\mathcal{U} := (t, \varphi, s)^{-1}\big(U \times V \times U\big).$$

This is an open subset of G, which is included in the compact set

$$\mathcal{C} := (t, \varphi, s)^{-1}\big(C \times K \times C\big)$$

where C is the union of the C_i's. The set $\mathcal{C}$ is indeed compact, for it is the direct image under the diffeomorphism $(t, \varphi, s)^{-1} : \varphi_0^* H \to G$ of the intersection of the compact cartesian product $C \times K \times C$, and the *closed* submanifold $\varphi_0^* H \subset G_0 \times H \times G_0$ (here we use that $\varphi_0^* H = G_0\,{}_{\varphi_0}\!\times_t H\,{}_s\!\times_{\varphi_0} G_0$ is a fiber product over twice the diagonal of G_0, which is *closed*). Thus $\mathcal{U}$ is relatively compact.

It is easy to see that $\mathcal{U}$ is exhaustive: by unique lifting every G_1-orbit is the preimage by φ of an H_1-orbit, which necessarily crosses V (exhaustiveness), and thus its preimage crosses $\mathcal{U}$ ($\varphi(\mathcal{U}) = V$ by construction).

It only remains to check the full generation property. Take any $g \in G_U^U$. We have $\varphi g \in H_V^V = \langle V \rangle$, thus we can write φg as a finite product of elements of V

$$\varphi g = v_l \cdots v_1.$$

Choose indices $\alpha(k)$ such that $x_0 := sv_1 \in V_{\alpha(0)}$, and $x_k := tv_k \in V_{\alpha(k)}$ for all $k > 0$. We then choose points $m_k \in U_{\alpha(k)}$ over the x_k's by φ_0, with two special choices $m_0 = sg$ and $m_l = tg$, and define

$$g_k := (t, \varphi, s)^{-1}\big(m_k\,,\, v_k\,,\, m_{k-1}\big) \quad \text{for } 0 < k < l.$$

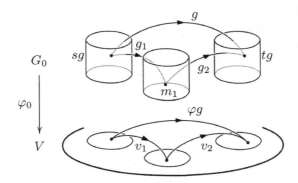

By definition of $\mathcal{U}$, $g_k \in \mathcal{U}$ for all j, so that the (well-defined) product $g_l \cdots g_1$ is in $\langle \mathcal{U} \rangle$. But

$$\varphi(g_l \cdots g_1) = \varphi g_l \cdots \varphi g_1 = v_l \cdots v_1 = \varphi g$$

and unicity of the pullback between $m_0 = sg$ and $m_l = tg$ implies $g = g_l \cdots g_1 \in \langle \mathcal{U} \rangle$. Thus $\mathcal{U}$ has full generation, and G is compactly generated.

3.1. As in the case of pseudogroups, it is important to require $\mathcal{U}$ to be open in Definition 6. We construct here an example of a non-compactly generated groupoid, which admits an exhaustive relatively compact non-open subset $\mathcal{U}$ with full generation. This example is inspired from an exercise in [7].

Consider the Klein bottle, seen as a cylinder $S^1 \times [-1, 1]$ ($S^1 \subset \mathbb{C}$) with its ends identified $(z, 1) = (\bar{z}, -1)$, and foliated by the directrices $\{z\} \times [-1, 1]$.

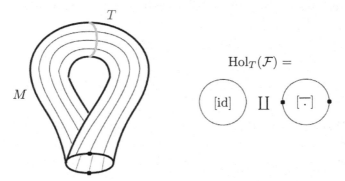

$$\mathrm{Hol}_T(\mathcal{F}) =$$

Take a circular transversal $T = S^1 \times \{0\}$, then the transverse holonomy groupoid $H = \mathrm{Hol}_T(\mathcal{F})$ associated to T may be seen as two copies of S^1, one for trivial holonomy germs and one for the germs of the holonomy diffeomorphism obtained by turning once along the directrices (which induces the complex conjugation on S^1).

Now make two holes in the Klein bottle at $(\pm 1, 1)$, so that we cannot turn around the associated leaves. In this case, the points $[\overline{\cdot}]_{\pm 1}$ disappear from H, and it is no more compactly generated. Indeed, if we had some $\mathcal{U}$ satisfying Definition 6, $\mathcal{U}_0$ would have to contain $(1, 0)$ (alone in its orbit), and thus would contain some small neighborhood V of 1, which we may assume stable under conjugation. The full generation property would then force $\{[\overline{\cdot}]_z; z \in V \setminus \{1\}\} \subset \mathcal{U}$, and thus $\mathcal{U}$ could not be relatively compact in H, absurd. However, if we set

$$\mathcal{U} = \{[\mathrm{id}]_z; \Im m(z) \geqslant 0\}$$

then $\mathcal{U}$ is relatively compact, exhaustive, and generates a full subgroupoid of H. It statisfies all conditions for compact generation, *but* being open.

3.2. In the classical definition of compact generation for pseudogroups (Proposition 5), the existence of an exhaustive relatively compact open subset which admits some suitable extra data satisfying the definition of compact generation implies that any other exhaustive relatively compact open subset may be equipped with some analogous data. This choice on U may be translated as follows for Lie groupoids:

Proposition 1. *Let G be a compactly generated Lie groupoid. For any exhaustive relatively compact open subset U in G_0, there exists a relatively compact open subset $\mathcal{U}$ in G_1 which generates a full subgroupoid and satisfies $\mathcal{U}_0 = U$.*

The proof of this proposition is given in Section 3.2. The idea of the proof is basically to use local bisections (Definition 7) between an open set $\mathcal{V}$ satisfying Definition 6, and use these bisections to 'send' $\mathcal{V}$ over U, inheriting in this way a full, relatively compact set of generators. Let us for now give a criterion for compact generation, whose proof is essentially the same as that of Proposition 1.

Lemma 2. *Let G be a groupoid, and assume there exists a subset $S \subset G$ satisfying the following conditions:*

- *S is exhaustive in G*
- *S is relatively compact in G*
- *S generates a full (set-theoric) subgroupoid of G*
- *S_0 is locally G-reachable in G_0, that is: every point $x \in G_0$ admits a neighborhood V_x, equipped with a section $\sigma_x : V_x \to G^{S_0}$ of s.*

Then G is compactly generated.

Proof. We simply use the local sections σ_x to spread the full generation property of S to one of its (relatively compact open) neighborhoods.

Let $\mathcal{U}_S$ be some relatively compact open neighborhood of S in G_1, and let K be some compact set containing $U := (\mathcal{U}_S)_0$. For each point x of K, choose the following: a neighborhood V_x equipped with a reaching map σ_x, a compact subneighborhood $K_x \subset V_x$, and an open subneighborhood $U_x \subset K_x$. The open sets U_x cover K, thus a finite number of them $U_1, \ldots, U_N$ cover K. We intersect these open sets with U to get an open cover of U, again denoted $(U_1, \ldots, U_N)$. For each i, let $\mathcal{U}_i$ be a relatively compact open neighborhood of $\sigma_i(U_i)$ (which is included in the compact set $\sigma_i(K_i)$, and thus relatively compact). Intersecting with $s^{-1}(U_i) \cap t^{-1}(U)$ if necessary, we assume that $s(\mathcal{U}_i) = U_i$ and $t(\mathcal{U}_i) \subset U$.

Let $\mathcal{U}$ be the union of the $\mathcal{U}_i$'s and $\mathcal{U}_S$. We claim that $\mathcal{U}$ satisfies Definition 6. It is obviously relatively compact and open, as finite union of relatively compact and open subsets. It is also clearly exhaustive, as it contains S which already is. It only remains to check that it generates a full subgroupoid.

Let $g : x \to y$ be an arrow between points of U, with $x \in U_i$ and $y \in U_j$. Then $h := \sigma_j(y) g \iota(\sigma_i(x))$ has its source $t\sigma_i x$ and its target $t\sigma_j y$ both in S_0, and thus is a product of elements of $S \cup \iota(S)$: $h = h_l \cdots h_1$. But $\sigma_i x \in \mathcal{U}_i$, $\sigma_j y \in \mathcal{U}_j$, and $S \subset \mathcal{U}_S$, hence $g = \iota(\sigma_j(y)) h_l \cdots h_1 \sigma_i(x)$ is a product of elements of $\mathcal{U} \cup \iota(\mathcal{U})$, and therefore is in the subgroupoid generated by $\mathcal{U}$. $\qquad\square$

Definition 7 (bisection[f]). Let G be a Lie groupoid. A local section β of the source map s of G such that the composition $t\beta$ is a diffeomorphism of G_0 is called a (local) bisection of G.

Note that this notion is symmetric in s and t: when we identify any s-section to its image in G, bisections are those submanifolds of G for which the restrictions of both s and t are embeddings into G_0. We won't use any global bisection, so we shall for convenience refer to local bisections simply as bisections. Bisections are very efficient in Lie groupoid theory, due to the following fact:

Proposition 2. *For any element g of a Lie groupoid G, $sg \mapsto g$ may be extended to a local bisection $\beta : U \to G$ (where U is a neighborhood of sg).*

---

[f]See [9] for details on all definitions given here.

The proof of the proposition is straightforward with the submanifold point of view: any small enough smooth submanifold containing g, simultaneously transverse to the s- and t-fibers through g fits (see [9]). As we have plenty of them, we shall for convenience refer to local bisections as bisections.

Given a bisection $\beta : U \to G$, one can easily define a (local) *left-translation* $L_\beta : G^U \to G^{t\beta(U)}$ by letting any element of the image of β act by left-translation on the t-fiber of its source: $L_\beta(h) = \beta(th) \cdot h$. Such a translation is a diffeomorphism (compose β with the inversion and build an inverse mapping). There is an analogous definition of right-translation $G_{t\beta(U)} \to G_U$ associated to β, defined by $R_\beta(h) = h \cdot \beta((t\beta)^{-1}sh)$.

Proof of Proposition 1. Let $\mathcal{V}$ be an open set satisfying Definition 6, chosen symmetric, and write $V := \mathcal{V}_0$. As U is exhaustive, every point $y \in G_0$ may be linked to a point $x_y \in U$ by an element $g_y : x_y \to y$ of G_1. We extend g_y to a bisection α_y, and shrink it if necessary to have its domain included in U. We then choose a relatively compact shrinking of its domain around x_y, that is: a compact subset K_y and an open subset V_y of its domain satisfying $x_y \in V_y \subset K_y$. Now V is relatively compact, thus it may be covered by only a finite number of open sets of the form $t\alpha_y(V_y)$. Shrinking if necessary, we may assume that $t\alpha_y(V_y) \subset V$. Symmetrically, V is exhaustive, thus we can find for every point $x \in G_0$ a point $y_x \in V$ and an arrow $h_x : x \to y_x$. We again extend this arrow to a bisection β_x, consider that the image of $t\beta_x$ is in V, and choose a relatively compact shrinking of its domain $x \in U_x \subset C_x$. By relative compactness of U, it is covered by only a finite number of sets U_x, which we assume included in U.

Finally, we get the following data: a finite number of bisections, say $\beta_i : U_i \to G^V$ (each being the restriction of a larger bisection $\beta_i : W_i \to G^V$), with $U_i \subset U$, such that U is covered by the U_i's and V is covered by the $t\beta(U_i)$'s. This is essentially the situation encountered in Lemma 2: we have a subset $S = \mathcal{V}$ with full generation and a finite number of 'reaching maps' β_i which link every point of U to a point of S_0. The difference here is that U and S_0 may be disjoint, so we have to pull both the tails *and heads* of the arrows of $\mathcal{V}$ with our reaching functions.

For each pair of indices, let $\mathcal{U}_{ji}$ be the set of arrows from U_i to U_j which are conjugated to an arrow of $\mathcal{V}$ by σ_i at their source and σ_j at their target (precisely: $L_{\iota \circ \beta_j (t\beta_j)^{-1}} \circ R_{\beta_i}(\mathcal{V})$). Each such set $\mathcal{U}_{ji}$ is obviously open, and is relatively compact as it is included in $\iota(\beta_j(K_j)) \cdot \mathcal{K} \cdot \beta_i(K_i)$, where $\mathcal{K}$ is some compact set containing $\mathcal{V}$. Let $\mathcal{U}$ be the (finite) union of all the $\mathcal{U}_{ji}$'s. The set $\mathcal{U}$ is relatively compact and open, and satisfies $\mathcal{U}_0 = U$. It remains to check that it generates a full subgroupoid.

Let $g : m \to n$ be an element of G_1 with m in some U_i and n in some U_j. We send this element over V as $h := \beta_j(n)g\iota(\beta_i(m))$, and decompose it as a product of elements of V: $h = h_l \cdots h_1$. Now each point $y_k = th_k$, $k < l$, is in some $t\beta_{i(k)}(U_{i(k)})$. We set $x_k = (t\beta_{i(k)})^{-1}(y_k)$, and pull the decomposition of h as a decomposition of g:

$$g = (\iota\beta_j(n)h_l\beta_{i(l-1)}(x_{l-1})) \cdots (\iota\beta_{i(k)}(x_k)h_k\beta_{i(k-1)}(x_{k-1}))$$
$$\cdots (\iota\beta_{i(1)}(x_1)h_1\beta_i(m)).$$

But each $\iota\beta_{i(k)}(x_k)h_k\beta_{i(k-1)}(x_{k-1})$ is in $\mathcal{U}_{i(k)i(k-1)}$ (with $i(l) = j$ and $i(0) = i$), thus g is in the subgroupoid generated by $\mathcal{U}$. $\qquad\square$

4. Relation to compactly generated pseudogroups

4.1. Recall that a *pseudogroup* of diffeomorphisms on a manifold T is a set of diffeomophisms between open sets of T, containing id_T, which is closed under restriction, composition, inversion and gluing. To any pseudogroup $(\mathcal{H}, T)$, one can associate the set H of all germs of elements of $\mathcal{H}$, which is a groupoid for the composition of germs $[g]_y \cdot [f]_x = [g \circ f]_x$. This groupoid has a natural (sheaf) topology: given an element $h \in \mathcal{H}$, the collection of all germs of h at all points of its domain represents a basic open set for this topology. Each such base open set may be identified to an open set of T, therefore the changes of coordinates in H identify to changes of coordinates in T, and hence are smooth. Moreover, it is obvious in these particular coordinates that the source and target maps of H are étale (i.e. local diffeomorphisms), and that its other structure maps are smooth. We shall call H the *germ groupoid* of $\mathcal{H}$, and denote it $[\mathcal{H}]$. It is an *étale groupoid*, a Lie groupoid with étale source and target.

Conversely, given an étale Lie groupoid H, one can consider its pseudogroup $\mathcal{T}H$ of H-transitions, that is the set of all homemorphisms between open sets of H_0 which can be locally written $t\beta$ for some bisection β – or equivalently which are locally the restriction to H_0 of the left-translation associated to β (see Definition 7). An étale groupoid H is called *effective* whenever it is naturally isomorphic to its effect[g] $[\mathcal{T}H]$. It is not hard to see that the germ groupoid of a pseudogroup $\mathcal{H}$ is always effective, with $\mathcal{T}[\mathcal{H}] = \mathcal{H}$. Therefore there is a one-to-one correspondence between pseudogroups and their germ groupoids, given by $[.]$ and $\mathcal{T}(.)$. It actually extends to a one-to-one correspondence between isomorphism classes of pseu-

---

[g]See [11] for more information about the 'effect-functor' $[\mathcal{T}\cdot]$.

dogroups[h] and those of effective étale groupoids. The goal of this section is to prove the following

Theorem 3. *Germification yields an identification of pseudogroups as germ groupoids. This identification is compatible with equivalence, and induces a one-to-one correspondence between Haefliger classes of pseudogroups and Morita classes restricted to germ groupoids. Through this correspondence, the pseudogroup classes which are compactly generated in the sense of Haefliger correspond to the groupoid classes which are compactly generated in the sense of Definition 6.*

Let us recall here the classical definiton of Haefliger equivalence of pseudogroups: two pseudogroups $(\mathcal{H}, T)$ and $(\mathcal{H}', T')$ are said to be Haefliger-equivalent if there exists a set Φ of diffeomorphisms from open sets covering T to open sets covering T' such that:

$$\Phi \mathcal{H} \Phi^{-1} \subset \mathcal{H}' \quad \text{and} \quad \Phi^{-1} \mathcal{H}' \Phi \subset \mathcal{H}.$$

The definition of compact generation for pseudogroups is recalled in Proposition 5.

Proposition 3. *If two pseudogroups $(\mathcal{H}, T)$ and $(\mathcal{H}', T')$ are Haefliger-equivalent, then their germ groupoids H and H' are Morita-equivalent.*

Proof. Let Φ be a set of diffeomorphisms giving an Haefliger equivalence from $\mathcal{H}$ to $\mathcal{H}'$, and let us write $Z_0 = H'[\Phi]H$ the collection of all germs coming from compositions of maps of $\mathcal{H}$, Φ and $\mathcal{H}'$ (whenever defined).

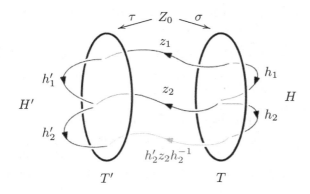

Write σ and τ the maps sending each element of Z_0 repectively to its source (in H) and its target (in H'), and give the manifold $Z = H'_s \times_\tau Z_0 {}_\sigma \times_s H$ a source map $s = \pi_{Z_0}$, a target map $t(h', z, h) = h'zh^{-1}$, and a product

$$(h'_2, z_2, h_2)(h'_1, z_1, h_1) = (h'_2 h'_1, z_1, h_2 h_1)$$

Using the diagrams given in Definition 1 (all maps being étale here), it is easy to check that Z is an étale groupoid, and that the two obvious smooth functors $\mathfrak{S} = \pi_H : Z \to H$ and $\mathfrak{T} = \pi_{H'} : Z \to H'$ are actually Morita morphisms. Indeed, their base maps $\sigma = \mathfrak{S}_0$ and $\tau = \mathfrak{T}_0$ are étale (easily seen on the diagrams), and we can use the global composition of germs over $T \cup T'$ to build smooth inverses for the canonical functors (for example with $\mathfrak{S}$):

$$
\begin{array}{ccc}
H'_s \times_\tau Z_0 {}_\sigma \times_s H & \xrightarrow{\ \cong\ } & Z_0 {}_\sigma \times_t H_s \times_\sigma Z_0 \\
(h', z_1, h) & \xmapsto{\ (t, \mathfrak{S}, s)\ } & (h'z_1 h^{-1}, h, z_1) \\
(z_2 h z_1^{-1}, z_1, h) & \longleftarrow\!\!\!| & (z_2, h, z_1)
\end{array}
$$

This implies that H and H' are Morita-equivalent. $\qquad\square$

Proposition 4. *If H and H' are two Morita-equivalent germ groupoids with $\dim H_0 = \dim H'_0$, then their associated pseudogroups are Haefliger-equivalent.*

Of course, if $\dim H_0 \neq \dim H'_0$, there's no way their groupoids can be Haefliger-equivalent, as their base manifolds do not even have the same dimension.

The proof of this proposition involves an adapted reformulation of Morita equivalence, whose proof is given in Section 4.2:

Lemma 3. *If H and H' are Morita-equivalent étale groupoids with $\dim H_0 = \dim H'_0 = d$, then there exists a third étale groupoid Z with $\dim Z_0 = d$ and two Morita morphisms $\varphi, \psi : Z \to H, H'$.*

Proof of Proposition 4. Considering Lemma 3, it is sufficient to prove the statement when there exists a Morita morphism $\varphi : H \to H'$. In this case, $\varphi_0 : H_0 \to H'_0$ is a surjective submersion between manifolds of the same dimension, and thus it is surjective and étale. Cover H_0 with open sets U_i such that the restrictions $\varphi_i = \varphi_{0|U_i} : U_i \to V_i$ are diffeomorphisms, and set $\Phi = \{\varphi_i\}$. We claim that Φ gives a Haefliger equivalence from $\mathcal{T}H$ to $\mathcal{T}H'$.

Due to the stability of pseudogroups under gluing, it is enough to prove the following: for any bisection $\beta : U \to H_{U_i}^{U_j}$ with domain contained in a single U_i and $t\beta(U)$ contained in a single U_j (resp $\beta' : V \to (H')_{V_i}^{V_j}$), the composition $\varphi_j \circ t\beta \circ \varphi_i^{-1}$ is an element of $\mathcal{T}H'$ (resp $\varphi_j^{-1} \circ t\beta' \circ \varphi_i \in \mathcal{T}H$). But all such bisections β and β' are in one-to-one correspondence through:

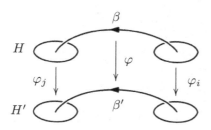

$$\beta \longmapsto \qquad\qquad \beta' = \varphi\beta\varphi_i^{-1}$$
$$\beta' \longmapsto \qquad \beta = (t,\varphi,s)^{-1}(\varphi_j^{-1}t\beta'\varphi_i, \beta'\varphi_i, \mathrm{id})$$

Thus every $\varphi_j t\beta\varphi_i^{-1} = t\varphi\beta\varphi_i^{-1}$ is some $t\beta' \in \mathcal{T}H'$, and every $\varphi_j^{-1}t\beta'\varphi_i$ is some $t\beta$, with β given by the correspondence. $\square$

Proposition 5. *Let $(\mathcal{H}, T)$ be a pseudogroup and H its germ groupoid. Then $\mathcal{H}$ is compactly generated if and only if H is.*

Let us first recall the definition of compact generation for pseudogroups:

Definition 8. A pseudogroup $(\mathcal{H}, T)$ is said to be *compactly generated* if the following holds:

(1) T admits an open subset U, relatively compact, and exhaustive (meeting every $\mathcal{H}$-orbit).
(2) There exist finitely many $h_i \in \mathcal{H}$ and open sets $U_i \subset U$, with $U_i \subset Dom(h_i)$ relatively compact and $h_i(U_i) \subset U$, such that the induced pseudogroup $\mathcal{H}_{|U}$ (elements of $\mathcal{H}$ with domain and image included in U) is generated by the restrictions $(h_i)_{|U_i}$.

Proof of Proposition 5. Assume that $\mathcal{H}$ is compactly generated, and consider the set of germs

$$\mathcal{U} = \cup_i \{[h_i]_x \; ; \; x \in U_i\}$$

This is an open subset of $[\mathcal{H}] = H$, relatively compact as finite union of relatively compact sets (check it in the charts provided by the domains of the h_i's), and exhaustive because U is. It also generates a full subgroupoid because every arrow of $[\mathcal{H}]$ between points of $\mathcal{U}_0 = U$ is a germ of some element in $\mathcal{H}_{|U}$, so is a product of germs of the $(h_i)_{|U_i}$ at suitably chosen points. Therefore $[\mathcal{H}]$ is compactly generated.

Conversely, assume that H is compactly generated, and let $\mathcal{U}$ be a symmetric exhaustive relatively compact open subset of $[\mathcal{H}]$ with full generation (we may suppose $\mathcal{U}$ symmetric, for $\mathcal{U} \cup \mathcal{U}^{-1}$ has the same properties as $\mathcal{U}$ regarding compact generation). Decompose $\mathcal{U}$ into a finite union of open subsets $\mathcal{U}_i$ such that each $\mathcal{U}_i$ is included in a compact set $\mathcal{K}_i$, itself included in an open set $\mathcal{V}_i$ where s and t are both diffeomorphisms onto open sets of T (the union is finite because $\mathcal{U}$ is included in a compact). Write $s_i = s_{|\mathcal{V}_i}$, and set $h_i = t \circ s_i^{-1} \in \mathcal{T}[\mathcal{H}] = \mathcal{H}$, $U_i = s(\mathcal{U}_i)$ and $U = \mathcal{U}_0 = \cup_i U_i$. We claim that this data satisfies the definition of compact generation for $\mathcal{H}$. It is not hard to see that U is an exhaustive relatively compact open subset of T: exhaustiveness is inherited from that of $\mathcal{U}$; openness and relative compactness are consequences of s and t being étale (and $\mathcal{U}$ relatively compact). Now take some $h \in \mathcal{H}_{|U}$, and any x in the domain of h. The germ $[h]_x$ has source and target in $\mathcal{U}_0$, hence by full generation it may be written as a product of elements of $\mathcal{U}$:

$$[h]_x = u_l \cdots u_1.$$

Each u_k is in some $\mathcal{U}_{\alpha(k)}$, and the definition of $[\mathcal{H}]$ implies $u_k = [h_{\alpha(k)}]_{su_k}$. Thus

$$[h]_x = [h_{\alpha(l)}]_{su_l} \cdots [h_{\alpha(1)}]_{su_1} = [h_{\alpha(l)} \cdots h_{\alpha(1)}]_x$$

and h is locally at x a product of the $h_{i|U_i}$'s. As it is true for every x in its domain, by gluing, h is in the pseudogroup generated by the $h_{i|U_i}$'s.   □

4.2. Let us give now the final details for the proof of Theorem 3. By Proposition 3, we know that Heafliger equivalence implies Morita equivalence. Conversely, if two germ groupoids are Morita-equivalent, we need to prove that their bases have the same dimension in order to apply Proposition 4. This may be done easily using the notion of groupoid dimension[i].

Definition 9 (Groupoid dimension). The groupoid dimension of a Lie groupoid H is the relative integer $\operatorname{gdim} H = 2 \dim H_0 - \dim H_1$.

---

[i]Appears as the dimension of the stack canonically associated to a groupoid in [1].

The groupoid dimension is obviously invariant under pullbacks ($\dim f^* H = 2 \dim X + \dim H_1 - 2 \dim H_0$, see Definition 4), hence it is invariant under Morita equivalence. But in the case of germ (or more generally étale) groupoids, the groupoid dimension is the same as the dimension of the base space. Thus Proposition 4 always applies to Morita-equivalent germ groupoids, and Morita classes coincide with Haefliger classes. We end this section with the proof of the technical Lemma 3, which had been left apart.

Proof of Lemma 3. Take K a Lie groupoid yielding a Morita equivalence between H and H', with Morita morphisms $\eta, \vartheta : K \to H, H'$ (Theorem 1). Take any $m \in K_0$, and let $x = \eta_0 m$, $x' = \vartheta_0 m$. The subspaces $T_m(\eta_0^{-1} x)$ and $T_m(\vartheta_0^{-1} x')$ have the same codimension d in $T_m K_0$, so they admit a common complement F_m. Take some coordinates system around m, and consider a small d-disc D_m containing m, and contained in the subspace F_m in those coordinates.

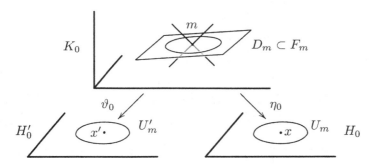

The disc D_m is transverse to the η_0- and ϑ_0-fibers at m, thus we may assume (shrinking if necessary) that the restrictions of η_0 and ϑ_0 to D_m are diffeomorphisms onto open sets $U_m \subset H_0$ and $U'_m \subset H'_0$. We may also assume that there exists a chart around m, containing D_m in its domain, in which the submersion η_0 is locally the projection of a product $H_0 \times F \to H_0$.

Set Z_0 the disjoint union of such discs D_m for all $m \in K_0$. We claim that the canonical map $j : Z_0 \to K_0$ induces an essential equivalence. We first check that $t\pi_K : K_s \times_j Z_0 \to K_0$ is a surjective submersion by constructing local sections of this map (see next figure).

Take any $n \in K_0$, and any $(k, m) \in K \times Z_0$ in the $t\pi_K$-fiber over n, that is: $k : m \to n$ in K. Let D_{m^*} denote the reference disk for m in Z_0. There exists a local bisection β extending k^{-1} in a neighborhood V of n (Proposition 2), we shrink it if necessary to have $\eta t \beta(V) \subset U_{m^*}$. We then crush $W = t\beta(V)$

into D_{m^*} following the arrows given by

$$p : \begin{array}{ccc} W & \longrightarrow & K_W^{D_{m^*}} = s^{-1}(W) \cap t^{-1}(D_{m^*}) \\ m' & \longmapsto & (t, \eta, s)^{-1}\big(\sigma\eta(m'),\, 1_{\eta m'},\, m' \big) \end{array}$$

where σ is the inverse of $\eta_0 : D_{m^*} \to U_{m^*}$. Following β and then p, we get a section of $t\pi_K$, which may be written precisely $(\iota\beta \cdot \iota p \circ (t\beta)^{-1}, j^{-1}tpt\beta)$ (where j is restricted to D_{m^*} for the inverse).

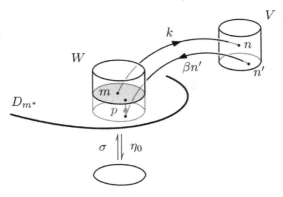

It follows that j induces a groupoid pullback, with an essential equivalence for the canonical map $J : Z \to K$ (see Definition 4 and Section 2.4). Let us denote $\varphi = \eta J$ and $\psi = \vartheta J$. We claim that $\varphi, \psi : Z \to H, H'$ fit the problem.

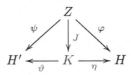

By definition of Z_0, we know that it has dimension d, and that φ_0 and ψ_0 are surjective and étale. Thus it only remains to check whether the canonical functors $Z \to \varphi_0^* H$ and $Z \to \psi_0^* H'$ are diffeomorphisms. This can be achieved by writing:

$$\begin{aligned} Z &= Z_0 \,_j\times_t K \,_s\times_j Z_0 \\ &= Z_0 \,_j\times_{\mathrm{id}} K_0 \,_{\eta_0}\times_t H \,_s\times_{\eta_0} K_0 \,_{\mathrm{id}}\times_j Z_0 \\ &= Z_0 \,_{(\eta_0 j)}\times_t H \,_s\times_{(\eta_0 j)} Z_0 \end{aligned}$$

and same with H'. $\square$

5. The case of foliated manifolds

5.1. Given a foliation $\mathcal{F}$ on a manifold M, a classical groupoid with base M associated to $\mathcal{F}$ is the holonomy groupoid of the foliation, $\mathrm{Hol}(\mathcal{F})$. Recall that, given a path γ in a leaf L of $\mathcal{F}$, and given two transversals T, T' passing respectively through the origin and the end of γ, a holonomy element associated to (γ, T, T') is a diffeomorphism between open neighborhoods of the origin and the end of γ in T and T' which roughly follows the coordinates of the leaves surrounding L along γ. The holonomy groupoid of $\mathcal{F}$ is the set of all germs of holonomy elements of the foliation up to the choice of the local transversals T and T'. (There is indeed a canonical way of changing transversals at a given point: choosing a foliated chart and going in straight line from one transversal to another, i.e. using the holonomy element of the constant path at the point.) There are obvious smooth charts for this groupoid, induced by pairs of foliated charts of M: given a path γ, a neighborhood of the holonomy germ of γ is described by the holonomy germs of paths which start in a fixed foliated chart around the origin of γ, follow the direction of γ and end in another fixed foliated chart around the extremity of γ (in a plaque prescribed by the plaque which they started in).

This groupoid is a modern evolution of the classical holonomy pseudogroup $\mathcal{H}_T$ associated to an exhaustive transversal T [4]. A classical result asserts that the germ groupoid (see section 3) of the pseudogroup $\mathcal{H}_T$ identifies to the pullback of $\mathrm{Hol}(\mathcal{F})$ along the inclusion $T \to M$ – the holonomy groupoid reduced to T, and that the inclusion morphism $\mathrm{Hol}_T(\mathcal{F}) \to \mathrm{Hol}(\mathcal{F})$ is an essential equivalence (see Section 2.4). Using this result, it is easy to get the germ version of Haefliger's result about compact generation:

Proposition 6. *If $(M, \mathcal{F})$ is a compact foliated manifold, then for any choice of exhaustive transversal T, the reduced holonomy groupoid $\mathrm{Hol}_T(\mathcal{F})$ associated to T is compactly generated.*

Indeed, if we cover M by a finite number of open domains U_i of foliated charts, and take a shrinking V_i of this open cover, the set $\mathcal{U}$ of all tangent paths which are entirely included in some V_i obviously satisfies Definition 6: it is clearly open and relatively compact in the charts induced by the pairs (U_i, U_i), and it is obviously exhaustive with full generation, as every tangent path may be decomposed into short paths each contained in a single V_i (hence $\mathcal{U}$ generates the entire holonomy group). This shows that $\mathrm{Hol}(\mathcal{F})$ is compactly generated, thus by the invariance Theorem 2, every reduced holonomy groupoid is also compactly generated (the inclusion morphism naturally yielding a Morita equivalence between $\mathrm{Hol}_T(\mathcal{F})$ and $\mathrm{Hol}(\mathcal{F})$.

5.2. The previous statement that the holonomy groupoid of a compact foliated manifold is compactly generated actually generalizes to all s-connected foliation groupoids with compact base (an s-connected groupoid is a groupoid with connected s-fibers). Let us first recall that a general foliation groupoid is a special kind of groupoid which modelizes foliations in a good way. It is a classical result that for any Lie groupoid G, the restriction of the target map t to a fixed s-fiber has constant rank (see [9]), and that any 0-orbit is an immersion.

Definition 10. A foliation groupoid is a Lie groupoid where the restriction of the tangent map dt to the kernel of the tangent map ds is everywhere zero.

This implies in particular that the restrictions of t to the s-fibers are not only subimmersions but immersions, whose rank is independent of the s-fiber. It can be proved that in this case, the connected components of the 0-orbits are in fact the leaves of a foliation on G_0 [3]. With this definition, the statement on the holonomy groupoids admits the following extension:

Theorem 4. *Every s-connected foliation groupoid with compact base is compactly generated.*

Note that this theorem applies in particular to the holonomy and monodromy[j] groupoids of a foliation, as the s-fibers of a holonomy (resp. monodromy) groupoid are the holonomy (resp. universal) coverings of the leaves, and thus are connected.

The theorem fails to be true if we don't require the groupoid to be s-connected: choose any non-finitely generated group A and any compact manifold M. Give A the discrete topology and consider the trivial groupoid on M with group A, $M \times A \times M$ (with product $(z, b, y)(y, a, x) = (z, ba, x)$). It is a foliation groupoid: it is indeed étale, hence has discrete s-fibers and therefore obviously satisfies Definition 10. It has a compact base, but cannot be compactly generated. Indeed, if it was, there would exist a relatively compact subset $\mathcal{U} \subset M \times A \times M$ generating a full subgroupoid. The group A is not finitely generated, so $\mathcal{U}$ would have to cross an infinite number of $M \times \{a\} \times M$ to have full generation. But it is impossible, because the sets

---

[j]This is the groupoid of tangent paths up to tangent homotopy, also called homotopy groupoid of the foliation.

$M \times \{a\} \times M$ are open and pairwise disjoint, so that $\mathcal{U}$, which is relatively compact, can only meet a finite number of them.

Let G denote an s-connected foliation groupoid, and $\mathcal{F}$ the foliation it induces on its base, which we assume to be compact. The proof of Theorem 4 mimics the case of the holonomy groupoid treated for Proposition 6, but uses both the local structure of foliation groupoids and the construction of the natural factorisation morphsim $h_G : \mathrm{Mon}(\mathcal{F}) \to G$ given in [3], which we recall here.[k]

Lemma 4 ([3]). *A trivializing submersion $\pi : U \to T$ of $\mathcal{F}$ is a submersion on an open set of G_0 with contractible fibers, which are precisely the leaves of $\mathcal{F}_{|U}$.*
Denote $G(U)$ the s-connected component of G_U^U. Then the map (t,s) : $G(U) \to U \times_T U$ is a natural isomorphism of Lie groupoids, where $U \times_T U$ is given the pair product $(z,y)(y,x) = (z,x)$.

Proposition 7 ([3]). *With the preceding notations, there is a natural factorisation of the canonical map $\mathrm{hol} : \mathrm{Mon}(\mathcal{F}) \to \mathrm{Hol}(\mathcal{F})$ into two surjective (functorial) local diffeomorphisms*

$$\mathrm{Mon}(\mathcal{F}) \xrightarrow{\ h_G\ } G \xrightarrow{\ hol_G\ } \mathrm{Hol}(\mathcal{F})$$

Proof of Theorem 4. Let $(V_i)_i$ be a finite open cover of G_0 by domains of trivializing submersions, and $(U_i)_i$ a shrinking of $(V_i)_i$. Let $\mathcal{U}$ be the union of the restrictions $G(U_i) = G(V_i)_{U_i}^{U_i}$ of the local groupoids $G(V_i)$. The set $\mathcal{U}$ is open in G, and clearly exhaustive because $\mathcal{U}_0 = G_0$. Using the isomorphisms $(t,s)_i : G(V_i) \to V_i \times_{T_i} V_i$, we can include each $G(U_i)$ in a compact set (namely $(t,s)_i^{-1}(\overline{U_i} \times_{T_i} \overline{U_i})$), which implies that $\mathcal{U}$ is relatively compact. Finally, as $(U_i)_i$ covers G_0, every tangent path α of G_0 may be decomposed as a product of tangent paths α_k contained each in a single $U_{i(k)}$. Due to the contractibility of the $\pi_{i(k)}$-fibers, each path α_k is entirely defined by its ends. Those ends in turn give an element $g_k = (t,s)_{i(k)}^{-1}(\alpha_k) \in G(U_{i(k)})$. Then the product of the g_k's lies in the subgroupoid generated by $\mathcal{U}$, and is precisely $h_G(\alpha)$ by construction of h_G. It follows that $\mathcal{U}$ generates the entire image of h_G. But G is s-connected, so that h_G is surjective onto G. Thus $\mathcal{U}$ generates G, which is without any doubt a full subgroupoid of itself. $\square$

---

[k]Note that the minimality of Hol and maximality of Mon were already known in various forms long before [3], see for example [14] or [13] (whose main theorem is very close to the proposition we recall here). I'd like to thank the referee for pointing out the existence of these remarkable references.

References

[1] K. Behrend and P. Xu, *Differentiable Stacks and Gerbes*, J. Symplectic Geom., **9** (2011), 286–341.

[2] P. Cartier, *Groupoïdes de Lie et leurs Algébroïdes*, Sém. Bourbaki 2007/2008, Astérisque **326** (2009), Exp. 987, 165–196.

[3] M. Crainic and I. Moerdijk, *Foliation groupoids and their cyclic homology*, Adv. Math., **157** (2001), 177–197.

[4] C. Godbillon, Feuilletages, Études géométriques, Birkhäuser, 1991.

[5] A. Haefliger, *Homotopy and Integrability*, in Manifolds, Lecture Notes in Math., Springer, 1970, 133–163.

[6] A. Haefliger, *Structure transverse des feuilletagesGroupoïdes d'holonomie et classifiants, in Structure transverse des feuilletages*, Astérisque **116** (1984), 70–97.

[7] A. Haefliger, *Foliations and compactly generated pseudogroups*, in Foliations : Geometry and Dynamics, World Scientific, 2000, 275–295.

[8] K. Mackenzie, Lie Groupoids and Lie Algebroids in Differential Geometry, Cambridge University Press, 1987.

[9] K. Mackenzie, General Theory of Lie Groupoids and Lie Algebroids, Cambridge University Press, 2005.

[10] I. Moerdijk, *Orbifolds as Groupoids: an Introduction*, in Orbifolds in Mathematics and Physics, Contemp. Math., **310**, AMS, Providence, 2002, 205–222.

[11] J. Mrčun, *Stability and invariants of Hilsum-Skandalis maps*, PhD thesis, Universiteit Utrecht, 1996, http://arxiv.org/abs/math/0506484.

[12] J. Mrčun, *Functoriality of the Bimodule Associated to a Hilsum-Skandalis Map*, K-Theory, **18** (1999), 235–253.

[13] J. Phillips, *The holonomic imperative and the homotopy groupoid of a foliated manifold*, Rocky Mountain J. Math., **17** (1987), 151–165.

[14] J. Pradines, *Théorie de Lie pour les groupoïdes différentiables. Relations entre propriétés locales et globales*, C.R.A.S. Paris, **263** (1966), 907–910.

Received December 29, 2012.

FOLIATIONS 2012
ed. by Paweł WALCZAK et al.
World Scientific, Singapore, 2013
pp. 163–203

Prescribing the positive mixed scalar curvature of totally geodesic foliations

Vladimir Rovenski*

Mathematical Department, University of Haifa, Israel
e-mail: rovenski@math.haifa.ac.il

Leonid Zelenko

Mathematical Department, University of Haifa, Israel
e-mail: zelenko@math.haifa.ac.il

1. Introduction

1.1. *Extrinsic geometry of foliations*

A Riemannian manifold (M, g) may admit many kinds of geometrically interesting foliations (i.e., partitions into submanifolds of the same dimension). Totally geodesic foliations and foliations with bundle-like metric (called Riemannian) are among these kinds that enjoyed a lot of interest and investigation of many geometers (see [5, 11, 12, 19]).

A foliation $\mathcal{F}$ is called *totally geodesic* if the leaves are totally geodesic submanifolds (i.e., the second fundamental form of the leaves $b_{\mathcal{F}} = 0$), such metric g will be called *totally geodesic*.

The simple examples of totally geodesic foliations are parallel circles or winding lines on a flat torus, and a Hopf field of great circles on the sphere S^3. Totally geodesic foliations appear naturally as null-distributions

---
*Supported by the Marie-Curie actions grant EU-FP7-P-2010-RG, No. 276919.

(or kernels) in the study of manifolds with degenerate curvature-like tensors and differential forms, [9]. In the codimension-one case, totally geodesic foliations on closed non-negatively curved space forms are completely understood: they are given by parallel hyperplanes in the case of a flat torus T^n and they do not exist for spheres S^n. If the codimension is greater than one, examples of geometrically distinct totally geodesic foliations are abundant, see survey in [19]. Riemannian foliations (e.g., submersions) with totally geodesic fibers are a central tool for many geometric constructions, several works are devoted to their classification, see [11] etc.

Let M^{n+p} be a connected closed manifold (i.e., compact without a boundary), endowed with a p-dimensional foliation $\mathcal{F}$. We have the g-orthogonal decomposition $TM = D_{\mathcal{F}} \oplus D$, where the distribution $D_{\mathcal{F}}$ (dim $D_{\mathcal{F}} = p$) is tangent to $\mathcal{F}$. Denote by $(\cdot)^{\mathcal{F}}$ and $(\cdot)^{\perp}$ projections onto $D_{\mathcal{F}}$ and D, respectively. The integrability tensor and the second fundamental tensor of D are given by

$$T(X, Y) := \frac{1}{2}[X, Y]^{\mathcal{F}} = \frac{1}{2}(\nabla_X Y - \nabla_Y X)^{\mathcal{F}},$$
$$b(X, Y) := \frac{1}{2}(\nabla_X Y + \nabla_Y X)^{\mathcal{F}}, \qquad (X, Y \in D),$$

where ∇ is the Levi-Civita connection of g. The mean curvature vector of D is given by $H = \mathrm{Tr}_g b$.

A foliation $\mathcal{F}$ is said to be *conformal, transversely harmonic,* or *Riemannian,* if $b = \frac{1}{n}H\hat{g}$, $H = 0$ or $b = 0$, respectively. In these cases, the distribution D is called totally umbilical, harmonic or totally geodesic, respectively.

Conformal foliations were introduced by Vaisman [24] as foliations admitting a transversal conformal structure. Such foliations extend the class of Riemannian foliations.

Extrinsic geometry of a foliation [7, 21] on a Riemannian manifold describes the properties which are expressed in terms of $b_{\mathcal{F}}$ and its invariants (e.g., mean curvature vector and higher mean curvatures). Extending the definition above, we shall say that *extrinsic geometry of a pair* of complementary orthogonal distributions $(D_{\mathcal{F}}, D)$ on (M, g) describes the properties which are expressed in terms of the integrability tensors and second fundamental tensors of the distributions. One of the principal problems of extrinsic geometry of foliations reads as follows:

Given a foliation $\mathcal{F}$ on a manifold M and an extrinsic geometric property (P), does there exist a Riemannian metric g on M such that $\mathcal{F}$ enjoys (P) w.r.t. g?

Such problems of the existence and classification of metrics or foliations (first posed explicitly by H. Gluck for geodesic foliations) have been studied intensively by many geometers in the 1970's. D. Sullivan provided a topological condition (called *topological tautness*) for a foliation, equivalent to the existence of a Riemannian metric making all the leaves minimal submanifolds. By the Novikov Theorem (see [6]) and Sullivan's results, the sphere S^3 has no 2-dimensional taut foliations. In recent decades, several tools for proving results of this sort have been developed. Among them, one may find Sullivan's *foliated cycles* and new *integral formulae*, see [21, Part 1], [25] etc.

1.2. The mixed scalar curvature

There are three kinds of Riemannian curvature for a foliation: tangential, transversal, and mixed (a plane that contains a tangent vector to the foliation and a vector orthogonal to it is said to be mixed). The geometrical sense of the mixed curvature follows from the fact that for a totally geodesic foliation, certain components of the curvature tensor (i.e., contained in the Riccati equation for the conullity tensor, see Section 4.1), regulate the deviation of the leaf geodesics. In general relativity, the *geodesic deviation equation* is an equation involving the Riemann curvature tensor, which measures the change in separation of neighboring geodesics or, equivalently, the tidal force experienced by a rigid body moving along a geodesic. In the language of mechanics it measures the rate of relative acceleration of two particles moving forward on neighboring geodesics.

Let $\{e_i, \varepsilon_\alpha\}_{i \le n, \alpha \le p}$ be a local orthonormal frame on TM adapted to D and $D_{\mathcal{F}}$. Tracing the Riccati equation yields the equality with the *mixed scalar curvature* that is the following function:

$$\mathrm{Sc}_{\mathrm{mix}} = \sum_{i=1}^{n} \sum_{\alpha=1}^{p} K(e_i, \varepsilon_\alpha),$$

see [19, 21, 25]. Here $K(e_i, \varepsilon_\alpha)$ is the sectional curvature of the mixed plane spanned by the vectors e_i and ε_α. For example, $\mathrm{Sc}_{\mathrm{mix}}$ of a foliated surface (M^2, g) is the gaussian curvature K.

For a foliation $\mathcal{F}$ on a Riemannian manifold (M, g) we have the formula, see [18, 25]:

$$\mathrm{Sc}_{\mathrm{mix}} = \mathrm{div}\, H + |H|^2 + \|T\|^2 - \|b\|^2 + \mathrm{div}\, H_{\mathcal{F}} + |H_{\mathcal{F}}|^2 - \|b_{\mathcal{F}}\|^2, \qquad (1)$$

it shows that $\mathrm{Sc}_{\mathrm{mix}}$ belongs to the extrinsic geometry of a pair $(D_{\mathcal{F}}, D)$ on (M, g). By the Divergence Theorem on a closed manifold, (1) yields the integral formula with the total $\mathrm{Sc}_{\mathrm{mix}}$.

Formula (1) yields decomposition criteria for foliated manifolds with an integrable orthogonal distribution D under the constraints on the sign of $\mathrm{Sc}_{\mathrm{mix}}$ (see [25] and a survey in [19]): Let $\mathcal{F}$ and $\mathcal{F}^{\perp}$ be complementary orthogonal totally geodesic and totally umbilical foliations on a closed oriented Riemannian manifold M with $\mathrm{Sc}_{\mathrm{mix}} \geq 0$, then M splits along the foliations.

One may ask: *Which foliations admit a totally geodesic metric of positive mixed scalar curvature?*

Example 1. (a) Let (M, g) and (B, g_B) be closed Riemannian manifolds, $\pi : M \to B$ a submersion with compact totally geodesic fibers, and the foliation $\mathcal{F}$ composed of fibers is conformal. Denote by D the distribution (i.e., a subbundle in TM) orthogonal to fibers. Suppose that there exists a smooth section $\xi : B \to M$. Relying on $\mathcal{F}$-Jacobi vector fields in D with unit lenght on $\xi(B)$, one may D-conformally deform g to obtain a bundle-like metric $\bar{g}$. The deformation is D-conformal and preserves the property "the fibres are totally geodesic", see Lemma 4. Thus, π becomes a Riemannian submersion from $(M, \bar{g})$ with totally geodesic fibers. By the formula for the mixed sectional curvature (see, for example, [11], we have $\mathrm{Sc}_{\mathrm{mix}}(\bar{g}) \geq 0$, moreover, $\mathrm{Sc}_{\mathrm{mix}}(\bar{g}) > 0$ when D is non-integrable at any point.

(b) For any $n \geq 2$ and $p \geq 1$ there exists a fibre bundle with a closed $(n + p)$-dimensional total space and a compact p-dimensional fiber, having a totally geodesic metric of positive mixed scalar curvature. To show this, consider the Hopf fibration $\tilde{\pi} : S^3 \to S^2$ of a unit sphere S^3 by great circles (closed geodesics). Let $(\tilde{F}, g_1)$ and $(\tilde{B}, g_2)$ be closed Riemannian manifolds with dimensions, respectively, $p - 1$ and $n - 2$. Let $M = \tilde{F} \times S^3 \times \tilde{B}$ be the metric product, and $B = S^2 \times \tilde{B}$. Then $\pi : M \to B$ is a fibration with a totally geodesic fiber $F = \tilde{F} \times S^1$. Certainly, $\mathrm{Sc}_{\mathrm{mix}} = 2 > 0$.

Based on Example 1, we ask the following.

Question 1. Given a foliated manifold $(M, \mathcal{F})$ with a totally geodesic metric g, does there exist a D-conformal to g metric $\bar{g}$ such that $\mathrm{Sc}_{\mathrm{mix}}(\bar{g})$ is positive?

The question has two aspects: the orthogonal distribution D may be fixed or not. In what follows we replace the question by another constructive question.

1.3. *Flows of metrics on foliations*

We shall attack Question 1 using approach of evolution equations which are important tool to study many physical and natural phenomena. A *geometric flow* (GF) of metrics on a manifold is a solution g_t of an evolution equation $\partial_t g = S(g)$, where a geometric functional $S(g)$ (i.e., a symmetric $(0,2)$-tensor) is usually related to some kind of curvature. The theory of GFs is a new subject, of common interest in mathematics and physics. This corresponds to a dynamical system in the infinite-dimensional space of all appropriate geometric structures on a given manifold. The most popular GFs in mathematics are the *heat flow*, the *Ricci flow* ([8] etc) and the *mean curvature flow*. Although the short time existence of solutions is guaranteed by the parabolic or hyperbolic nature of the equations, their long-time convergence to canonical geometric structures is analyzed under various conditions.

The definition of the *D-truncated* (r, k)-tensor field $\hat{S}$ (where $r = 0, 1$) on M will be helpful: $\hat{S}(X_1, \ldots, X_k) = S(X_1^{\perp}, \ldots, X_k^{\perp})$. Rovenski and Walczak [21] introduced extrinsic GFs of metrics on codimension-one foliations (i.e., depending on the extrinsic geometry of the leaves) and posed the question: *Given a geometric property* (P), *can one find an* $\mathcal{F}$-truncated flow $\partial_t g = \hat{S}(g)$ *on a foliation* $(M, \mathcal{F})$ *such that the solution metrics* g_t $(t > 0)$ *converge to a metric for which* $\mathcal{F}$ *enjoys* (P)? Rovenski [20] extended some results of [21] for GFs related to parabolic PDEs, studied the problem of prescribing the mean curvature function of a codimension-one foliation and gave applications to harmonic and totally umbilical foliations. Recently, Rovenski and Wolak [22] studied the *D-conformal flow* of metrics on a foliation of any codimension with the speed proportional to the $\mathcal{F}$-divergence of H. Based on known long-time existence results for the heat flow they showed convergence of a solution to a metric for which $H = 0$; actually under some topological assumptions they prescribe the mean curvature vector H. The conditions in their results are rather different and seem to be stronger of known hypotheses to guarantee tautness in non-constructive results. This makes sense because it is much harder to provide the good GF.

For *D-conformal* flows of metrics on foliations, we have $\hat{S}(g) = s(g)\,\hat{g}$, where $s(g)$ is a smooth function on the space of metrics on M. Here the *D*-truncated metric tensor $\hat{g}$ is given by $\hat{g}(N, \cdot) = 0$ and $\hat{g}(X_1, X_2) = g(X_1, X_2)$ for all $X_i \in D$, $N \in D_{\mathcal{F}}$. For short we shall write

$$\partial_t g = s_t\,\hat{g} \tag{2}$$

for a certain function s_t on M. The paper continues studying *D-conformal*

GFs on a foliated manifold $(M, \mathcal{F})$. In aim to prescribe the positive sign of $\mathrm{Sc}_{\mathrm{mix}}(g)$, we introduce the following flow of totally geodesic metrics:

$$\partial_t g = -2\big(\mathrm{Sc}_{\mathrm{mix}}(g) - \|T\|_g^2 - \Phi\big)\hat{g}. \tag{3}$$

Here $\Phi : M \to \mathbb{R}$ is an arbitrary function constant on the leaves, it is used for normalizing the flow.

The definition is correct because D-conformal adapted variations preserve totally geodesic metrics (see Lemma 4). If $\mathcal{F}$ is conformal and D is integrable (e.g., M is a twisted product, see Corollary 2) then we take $\Phi = 0$, hence the speed of (3) is $-2\,\mathrm{Sc}_{\mathrm{mix}}(g)$, i.e., $\partial_t g = -2\,\mathrm{Sc}_{\mathrm{mix}}(g)\,\hat{g}$.

Example 2. Let (M^2, g_0) be a surface (i.e., $n = p = 1$) foliated by geodesics, and K its Gaussian curvature. Since D is one-dimensional, we have $T = 0$. Taking $\Phi = 0$, we reduce (3) to the form:

$$\partial_t g = -2\,K(g)\,\hat{g}. \tag{4}$$

This looks like the Ricci flow PDE on surfaces, but uses the truncated metric $\hat{g}$ instead of g. The equation (4) on PC surfaces (see definition in Section 4.5) and on surfaces of revolution yields fruitful geometrical interpretation of the classical relation between the Burgers and heat equations.

The basic question that we want to address in the paper is the following (see Question 1):

Question 2. Given a totally geodesic metric g_0 on a foliated manifold $(M, \mathcal{F})$, when do the solution metrics g_t of (3) converge as $t \to \infty$ to the limit metric $\bar{g}$ with $\mathrm{Sc}_{\mathrm{mix}}(\bar{g}) > 0$?

Throughout the paper everything (manifolds, submanifolds, foliations, etc.) is assumed to be smooth (i.e., C^∞-differentiable) and oriented.

1.4. The structure of the paper

Introduction poses the questions on prescribing the positive mixed scalar curvature, surveys flows of metrics on foliations and introduces the D-conformal flow of metrics depending on the mixed scalar curvature. Section 2 collects main results and applications. Section 3 contains results for parabolic PDEs on a closed leaf. Section 4 represents D-conformal adapted variations of geometric quantities, proves main results and gives applications to foliations by curves. Section 5 studies linear operators with parameter in Banach spaces, it is used in Section 3.

2. Main results

An important step in the study of evolutionary PDEs is to show short-time existence/uniqueness.

Proposition 1. *Let g_0 be a totally geodesic metric on a closed foliated manifold $(M, \mathcal{F})$. Then (3) has a unique solution g_t defined on a positive time interval $[0, \varepsilon)$ and smooth on the leaves.*

We denote $\nabla^{\mathcal{F}} f := (\nabla f)^{\mathcal{F}}$. Given a vector field X and a function φ on M, define the functions using the leaf-wise derivatives: the divergence $\operatorname{div}_{\mathcal{F}} X = \sum_{\alpha=1}^{p} g(\nabla_\alpha X, \varepsilon_\alpha)$ and the Laplacian $\Delta_{\mathcal{F}} \varphi = \operatorname{div}_{\mathcal{F}}(\nabla^{\mathcal{F}} \varphi)$. Notice that $\nabla^{\mathcal{F}}, \operatorname{div}_{\mathcal{F}}$ and $\Delta_{\mathcal{F}}$ (i.e., along the leaves) are t-independent.

Based on the "linear algebra" inequality $n \|b\|^2 \geq |H|^2$ with the equality when D is totally umbilical, we introduce the following non-negative measure of "non-umbilicity" of D:

$$\beta_D := n^{-2}\big(n \|b\|^2 - |H|^2\big).$$

For $p = 1$, we have $\beta_D = n^{-2} \sum_{i<j}(k_i - k_j)^2$, where k_i are the principal curvatures of D, (see [26]).

The Schrödinger operator is central to all of quantum mechanics. By Proposition 6 in Section 4, the flow of metrics (3) preserves the leaf-wise Schrödinger operator,

$$\mathcal{H}(u) = -n \Delta_{\mathcal{F}} u - n \beta_D \cdot u. \tag{1}$$

Certainly, β_D and the operator $\mathcal{H}$ depend on the D-*conformal class* $[g_0]$ of the metric g_0 only (the metrics in $[g_0]$ coincide with g_0 on $D_{\mathcal{F}}$). The spectrum of $\mathcal{H}$ is an infinite sequence of isolated real eigenvalues $\lambda_0 \leq \lambda_1 \leq \ldots \lambda_n \leq \ldots$ (constant on the leaves) counting their multiplicities, and $\lim_{n \to \infty} \lambda_n = \infty$. If the leaf $F(x)$ through $x \in M$ is compact then $\lambda_1 > \lambda_0$ and the eigenfunction e_0 of unit L_2-norm (called the *ground state*) for the least eigenvalue λ_0 is positive, see Proposition 3. (The *ground state* of the quantum mechanical system corresponds to the smallest possible energy.)

One has $\lambda_0(x) \geq -n \bar{\beta}(x)$ where $\bar{\beta}(x) = \max_{F(x)} \beta_D$. To show this, define the operator $\tilde{\mathcal{H}}(u) = -n \Delta_{\mathcal{F}} u - n \bar{\beta} \cdot u$. From $\mathcal{H} \geq \tilde{\mathcal{H}} \geq -n\bar{\beta} \operatorname{id}$ (id is the identity function) the estimate of λ_0 follows. Beyond studying the least eigenvalue, the most interesting object is the gap $\lambda_1 - \lambda_0$ between the first two eigenvalues of $\mathcal{H}$ (called the *fundamental gap*). The fundamental gap has interesting physical implications, as well as mathematical ones (e.g., in refinements of Poincaré inequality and a priori estimates); in addition, from numerical point of view, the gap can be used to control the rate of convergeness of numerical methods of computation, see survey in [3].

We shall say that a smooth function $f(t, x)$ on $(0, \infty) \times F$ converges to $\bar{f}(x)$ as $t \to \infty$ in C^∞, if it converges in C^k-norm for any $k \geq 0$. It converges exponentially fast if there exists $\omega > 0$ (called the *exponential rate*) such that $\lim_{t \to \infty} e^{\omega t} \|f(t, \cdot) - \bar{f}\|_{C^k} = 0$ for any $k \geq 0$.

A vector field H_0 in $D_{\mathcal{F}}$ is *leaf-wise conservative*, if it has a scalar potential: $H_0 = -n\nabla^{\mathcal{F}}(\log u_0)$ for a positive function u_0 on M. Note that the condition "H_0 is leaf-wise conservative" is satisfied for the mean curvature H of twisted products (see Corollary 2).

The next proposition shows that the mean curvature vector field of D obeys the leaf-wise Burgers PDE, and has a single-point global exponential attractor on compact leaves. Notice that [13] proved the polynomial convergence of a solution to the Burgers PDE on $\mathbb{R}^n$.

Proposition 2. *Let g_0 be a totally geodesic metric on a smooth foliated manifold $(M, \mathcal{F})$, and the mean curvature vector H_0 of D is leaf-wise conservative. If the metrics g_t $(t \geq 0)$ obey (3) then*

(i) the mean curvature vector field H of D is a unique solution of the forced Burgers PDE

$$\partial_t H + \nabla^{\mathcal{F}} g(H, H) = n\nabla^{\mathcal{F}}(\mathrm{div}_{\mathcal{F}} H) - n^2 \nabla^{\mathcal{F}} \beta_D, \qquad (2)$$

(ii) one has $H = -n\nabla^{\mathcal{F}}(\log u)$, where the function $u > 0$ obeys the heat Schrödinger equation

$$\partial_t u = -\mathcal{H}(u), \qquad (3)$$

(iii) on compact leaves, as $t \to \infty$, H converges in C^∞ with the exponential rate $\lambda_1 - \lambda_0$ to the vector field $-n\nabla^{\mathcal{F}}(\log e_0)$.

The central result of the work is Theorem 1, presenting a constructive method to provide totally geodesic metrics with $\mathrm{Sc}_{\mathrm{mix}} > 0$. It shows that (3) has a unique global solution on a compact foliation (a foliation is *compact* if all leaves are compact), and the special choice of $\Phi = n\lambda_0$ yields the convergence of a solution g_t as $t \to \infty$. The proof of Theorem 1 is based on Proposition 2.

Denote by $[g]_c = \{g_0 \in [g] : \hat{g}(X, X) \leq c\,\hat{g}_0(X, X)$ for all $x \in M$ and $X \in D_x\}$ where $c > 0$.

Theorem 1. *Let g be a smooth totally geodesic metric on a compact foliation $(M, \mathcal{F})$, and the mean curvature vector field H (of D) is leaf-wise conservative.*

(i) Then (3) with $g_0 \in [g]$ admits a unique global solution g_t $(t \geq 0)$ smooth on the leaves.

(ii) Let $\Phi = n\lambda_0$. Then, as $t \to \infty$, the metrics g_t converge in C^∞ with the exponential rate $\lambda_1 - \lambda_0$ to the limit metric $\bar{g}$. Moreover, there exists real $c > 0$ such that $\mathrm{Sc}_{\mathrm{mix}}(\bar{g}) > 0$ when $g_0 \in [g]_c$ and

$$\|T(x)\|_{g_0}^2 > n^2 c^4 \max_{F(x)} \beta_D \quad (x \in M). \tag{4}$$

(iii) Let the leaves have finite holonomy group. Then g_t $(t > 0)$ and $\bar{g}$ are smooth on M.

There are many examples when (4) is satisfied with $\beta_D \not\equiv 0$, e.g., Example 3(a). Since (4) is automatically satisfied when $\beta_D = 0$ and $\|T(x)\|_{g_0} \neq 0$ for all $x \in M$, we have the following.

Corollary 1. *Let $\mathcal{F}$ be a conformal totally geodesic foliation on a smooth Riemannian manifold (M, g_0) with nowhere integrable orthogonal distribution, and H_0 is leaf-wise conservative. Suppose that the leaves are compact with finite holonomy group. Then (3) with $\Phi = n\lambda_0$ has a unique smooth global solution g_t $(t \geq 0)$, converging as $t \to \infty$ in C^∞ with the exponential rate $\lambda_1 - \lambda_0$ to the limit metric $\bar{g}$ for which*

(i) $\mathrm{Sc}_{\mathrm{mix}}(\bar{g})$ is positive and constant on the leaves; and

(ii) $\mathcal{F}$ is Riemannian (i.e., $\bar{b} = 0$).

Remark 1. Using Lemma 3 (in Section 4.1), the identity $\mathrm{div}\, H = \mathrm{div}_{\mathcal{F}}\, H - g(H, H)$ and the Divergence Theorem, we obtain

$$\int_M \mathrm{Sc}_{\mathrm{mix}}(g)\, d\,\mathrm{vol} = \int_M \left(\left(1 - \frac{1}{n}\right) g(H, H) + \|T\|_g^2 - n\beta_D \right) d\,\mathrm{vol}$$

$$\geq \int_M \left(\|T\|_g^2 - n\beta_D \right) d\,\mathrm{vol}$$

on a closed manifold M. Hence, the inequality $\|T\|_g^2 > n\beta_D$ yields $\int_M \mathrm{Sc}_{\mathrm{mix}}(g)\, d\,\mathrm{vol} > 0$.

In conditions of Theorem 1 with $\Phi = n\lambda_0$, we have the following conclusions: (a) if β_D is constant on the leaves then also $e_0 = \mathrm{const}$ on the leaves, hence $\mathcal{F}$ is $\bar{g}$-transversely harmonic (i.e., $\bar{H} = 0$); (b) if $\beta_D = 0$ (see Corollary 1) and D is integrable, then $\mathrm{Sc}_{\mathrm{mix}}(\bar{g}) = 0$. Thus M is locally the product w.r.t. $\bar{g}$ (the product globally when M is simply connected).

We apply Proposition 2 and Theorem 1 to twisted products.

Definition 1. (see [17]) Let (M_1, g_1) and (M_2, g_2) be Riemannian manifolds, and $f \in C^\infty(M_1 \times M_2)$ a positive function. The *twisted product*

$M_1 \times_f M_2$ is the manifold $M = M_1 \times M_2$ with the metric $g = (f^2 g_1) \oplus g_2$. If the warping function f depends on M_2 only, then we have a *warped product*.

The fibers $M_1 \times \{y\}$ of a twisted product are totally umbilical with the mean curvature vector $H = -n \nabla^{\mathcal{F}}(\log f)$ while submanifolds $\{x\} \times M_2$ are totally geodesic. The fibers $M_1 \times \{y\}$ have

(a) constant mean curvature if and only if $\|\nabla^{\mathcal{F}}(\log f)\|$ is a function of M_1, and

(b) parallel mean curvature vector if and only if $f = f_1 f_2$ for some $f_i : M_i \to \mathbb{R}_+$ $(i = 1, 2)$.

If on a simply connected complete Riemannian manifold (M, g) two orthogonal foliations with the properties (a) and (b) are given, then M is a twisted product, see [17].

Corollary 2. *Let* $(M, g_t) = M_1 \times_{f_t} M_2$ *be a family of twisted products of Riemannian manifolds* (M_1, g_1) *and* (M_2, g_2). *Then the following properties are equivalent:*

(i) The metrics g_t *satisfy* $\partial_t g = -2 \operatorname{Sc}_{\mathrm{mix}}(g) \hat{g}$ *(i.e., (3) with* $\Phi = 0$*).*

(ii) The mean curvature vector H *of fibers* $M_1 \times \{y\}$ *satisfies the Burgers equation*

$$\partial_t H + \nabla^{\mathcal{F}} g(H, H) = n \nabla^{\mathcal{F}}(\operatorname{div}_{\mathcal{F}} H). \tag{5}$$

(iii) The warping function f *satisfies the heat equation* $\partial_t f = n \Delta_{\mathcal{F}} f$. *Moreover, if* M_1 *and* M_2 *are closed manifolds then (3) admits a unique smooth solution* g_t $(t \geq 0)$ *consisting of twisted product metrics on* $M_1 \times_{f_t} M_2$; *the metrics* g_t *converge as* $t \to \infty$ *in* C^∞ *with the exponential rate* λ_1 *to the metric* $\bar{g}$ *of the product* $(M_1, \bar{f}^2 g_1) \times (M_2, g_2)$, *where* $\bar{f}(x) = \int_{M_2} f(0, x, y) \, dy_g$.

Example 3. (a) For the Hopf fibration $\pi : (S^{2m+1}, g_{\mathrm{can}}) \to \mathbb{C}P^m$ of a unit sphere with fiber S^1, the orthogonal distribution D is non-integrable while $b = 0$ and $\beta_D = 0$. We have $\lambda_0^{\mathrm{can}} = 0$, $\lambda_1^{\mathrm{can}} = 1$. By (1), $\operatorname{Sc}_{\mathrm{mix}} = \|T\|^2 = 2m$ holds. Thus, g_{can} on S^{2m+1} is a fixed point of (3) with $\Phi = 0$.

Consider the perturbation of the canonical metric: $g = g_{\mathrm{can}} + q \tilde{g}$ (for small $q \in \mathbb{R}$), where $\tilde{g} \notin [g_{\mathrm{can}}]$ is a D-truncated symmetric $(0, 2)$-tensor on S^{2m+1}. Generally, D is not totally umbilical w.r.t. g, hence $\beta_D(g)$ is not identically zero. The fundamental gap of elliptic operator (1) w.r.t. g is continuous in q, i.e., $\lambda_1 - \lambda_0 \approx 1$ for small q. As in the proof of Theorem 1(*i*), we conclude that there is $c > 0$ such that if $g_0 \in [g]_c$ and (4) holds then $Sc_{mix}(\bar{g}) > 0$ for the limit metric of (3).

(b) Let (M^2, g_0) be a two-dimensional torus with Gaussian curvature K, foliated by closed geodesics. Suppose that the geodesic curvature k of orthogonal (to the leaves) curves obeys $k = N(\psi_0)$ for a smooth function ψ_0 on M^2. The operator (1) coincides with the leaf-wise Laplacian, hence $\lambda_0 = 0$ and $e_0 = \text{const}$. Certainly, we have $T = 0$ and $\beta_D = 0$. By Theorem 1, the flow of metrics (4) on M^2 admits a unique solution g_t ($t \geq 0$) converging as $t \to \infty$ to a flat metric, and the leaves compose a rational linear foliation on the torus.

(c) Let $M_t^2 \subset \mathbb{R}^3$ ($t \geq 0$) be foliated surfaces of revolution about the Z-axis,

$$r = [\rho(x,t)\cos\vartheta, \ \rho(x,t)\sin\vartheta, \ h(x,t)] \ (0 \leq x \leq l, \ -\pi \leq \vartheta \leq \pi \quad \rho > 0).$$

The leaves (geodesics) $\vartheta = \text{const}$ are profile curves. Let k be the geodesic curvature of parallels (i.e., circles orthogonal to the leaves). Based on the known formulae $k = -(\log\rho)_{,x}$ and $K = -\rho_{,xx}/\rho$, we observe that the following properties of surfaces M_t^2 are equivalent (see Corollary 2):

- the induced metrics g_t (on M_t^2) satisfy (4);
- the geodesic curvature k satisfies the Burgers equation $k_{,t} + (k^2)_{,x} = k_{,xx}$;
- the distance from the profile curve to the axis, $\rho > 0$, obeys the heat equation $\rho_{,t} = \rho_{,xx}$.

A special case of twisted product metrics (i.e., foliated surfaces) is considered in Section 4.5.

In fluid and gas mechanics the viscid *Burgers equation* $k_{,t} + (k^2)_{,x} = \nu k_{,xx}$ (a constant $\nu > 0$ is the kinematic viscosity) is the prototype for non-linear advection-diffusion processes. It serves as the simplest model equation for solitary waves and describes wave processes [23].

The paper first introduces the use of the Burgers equation in the geometry of foliations.

3. Auxiliary results: PDEs on a closed Riemannian manifold

The section (see also Section 5) plays an important role in proofs of main results (see Section 4).

Let (F^p, g) be a smooth closed Riemannian manifold, e.g., a leaf of a foliation $\mathcal{F}$. The functional spaces over F will be denoted without writing (F), for example, L_2 instead of $L_2(F)$.

Denote by $(\cdot, \cdot)_E$ the inner product in a Hilbert space E. Let H^l be the Hilbert space of differentiable by Sobolev real functions on F, with the inner product $(\cdot, \cdot)_l$ and the norm $\| \cdot \|_l$. In particular, $H^0 = L_2$ with the product $(\cdot, \cdot)_0$ and the norm $\| \cdot \|_0$.

If E is a Banach space, we denote by $\| \cdot \|_E$ the norm of vectors in this space, by $S_1 := \{x \in E : \|x\|_E = 1\}$ the unit sphere in E, and by $\mathrm{Im}(\mathcal{P})$ the image of a linear operator $\mathcal{P}$ acting in E.

If B and C are real Banach spaces, we denote by $\mathcal{B}^r(B, C)$ the Banach space of all bounded r-linear operators $A : \prod_{i=1}^r B \to C$ with the norm $\|A\|_{\mathcal{B}^r(B,C)} = \sup_{v_1, \ldots v_r \in B \backslash 0} \frac{\|A(v_1, \ldots, v_r)\|_C}{\|v_1\|_B \cdots \|v_r\|_B}$. If $r = 1$, we shall write $\mathcal{B}(B, C)$ and $A(\cdot)$, and if $B = C$ we shall write $\mathcal{B}^r(B)$ and $\mathcal{B}(B)$, respectively.

If M is a k-regular manifold or an open neighborhood of the origin in a real Banach space, and N is a real Banach space, we denote by $C^k(M, N)$ ($k \geq 1$) the Banach space of all C^k-regular functions $f : M \to N$, for which the following norm is finite:

$$\|f\|_{C^k(M,N)} = \sup_{x \in M} \max\{\|f(x)\|_N, \max_{1 \leq j \leq k} \|d^j f(x)\|_{\mathcal{B}^j(T_x M, N)}\}.$$

If a Banach space E_1 is continuously embedded into a Banach space E_2, we shall write $E_1 \hookrightarrow E_2$. If $\mathcal{P}$ is a bounded linear operator acting from a Banach space E_1 to a Banach space E_2, we shall write $\mathcal{P} : E_1 \to E_2$. Denote by $\| \cdot \|_{C^k}$, where $0 \leq k < \infty$, the norm in the Banach space C^k; certainly, $\| \cdot \|_C$ when $k = 0$. In coordinates $(x_1, \ldots x_p)$ on F, we have $\|f\|_{C^k} = \max_{x \in F} \max_{|\alpha| \leq k} |d^\alpha f(x)|$, where $\alpha \geq 0$ is the multi-index of order $|\alpha| = \sum_{i=1}^p \alpha_i$ and d^α is the partial derivative.

The *resolvent set* of $\mathcal{P} : E \to E$, is defined by $\rho(\mathcal{P}) = \{\lambda \in \mathbb{C} : \mathcal{P} - \lambda \, \mathrm{id}$ is invertible and $(\mathcal{P} - \lambda \, \mathrm{id})^{-1}$ is bounded$\}$. The *resolvent* of $\mathcal{P}$ is the operator $R_\lambda(\mathcal{P}) = (\mathcal{P} - \lambda \, \mathrm{id})^{-1}$ for $\lambda \in \rho(\mathcal{P})$, and the *spectrum* of $\mathcal{P}$ is the set $\sigma(\mathcal{P}) := \mathbb{C} \backslash \rho(\mathcal{P})$, see [10, Ch. VII, Sect. 9]. Certainly, the spectrum of a self-adjoint operator is real.

3.1. *The ground state of the Schrödinger operator*

For a smooth function f on (F, g), the following operator:

$$\mathcal{H}(u) := -\Delta u - f \cdot u \tag{1}$$

is self-adjoint, bounded from below (but it is unbounded). The domain of definition of $\mathcal{H}$ is H^2. One can add a constant to f such that $\mathcal{H}$ becomes invertible in L_2 (e.g., $\lambda_0 > 0$) and $\mathcal{H}^{-1}$ is bounded in L_2. The resolvent of $\mathcal{H}$ is compact, i.e., for some $\lambda \in \rho(\mathcal{H})$ the operator $R_\lambda(\mathcal{H})$ maps any bounded in L_2 set onto a set, whose closure is compact in L_2. Let λ_0 be the least eigenvalue (with a unit-norm eigenfunction e_0) of $\mathcal{H}$.

The spectrum $\sigma(\mathcal{H})$ consists of an infinite sequence of isolated real eigenvalues $\lambda_0 \leq \lambda_1 \leq \ldots \lambda_n \leq \ldots$ counting their multiplicities, and $\lambda_n \to \infty$ as $n \to \infty$. If we fix in L_2 an orthonormal basis of corresponding eigenfunctions $\{e_n\}$ (i.e., $\mathcal{H}(e_n) = \lambda_n e_n$) then any function $u \in L_2$ is expanded into the series (converging to u in the L_2-norm)

$$u(x) = \sum_{n=0}^{\infty} c_n\, e_n(x), \qquad c_n = (u, e_n)_0 = \int_F u(x)\, e_n(x)\, dx. \qquad (2)$$

The proof of (2) is based on the following facts. Since by Elliptic regularity Theorem, we have $\mathcal{H}^{-1} : L_2 \to H^2$ when $0 \notin \sigma(\mathcal{H})$, and the embedding of H^2 into L_2 is continuous and compact, see [4], then the operator $\mathcal{H}^{-1} : L_2 \to L_2$ is compact. This means that the spectrum $\sigma(\mathcal{H})$ is discrete, hence by the spectral expansion theorem for compact self-adjoint operators, $\{e_n\}_{n \geq 0}$ form an orthonormal basis in L_2, (see [10, I, Chap. VII, Sect. 4; and II, Chap. XII, Sect. 2]).

The Cauchy's problem for the heat equation with a linear reaction term has a form

$$\partial_t u = -\mathcal{H}(u), \qquad u(\cdot, 0) = u_0. \qquad (3)$$

Definition 2. Denote by $\mathcal{K}$ the cone in $C(F)$ consisting of non-negative continuous functions on F, and let y be some fixed element of $\mathcal{K}$. The cone $\mathcal{K}$ is called *reproducing*, i.e., every element $x \in \mathcal{K}$ can be represented in the form $x = x_1 - x_2$ ($x_1, x_2 \in \mathcal{K}$). If for every non-zero $x \in \mathcal{K}$ a natural n and a positive real α can be found such that $\alpha\, y \leq S^n x$, then the linear operator S is called *y-bounded below*. Similarly, if for every non-zero $x \in \mathcal{K}$ a natural m and a positive real β can be found such that $S^m x \leq \beta\, y$, then the linear operator S is called *y-bounded above*. If the operator is *y*-bounded above and below, then it is *y-positive*, see [14, Definition 2.1.1 and Theorem 2.2].

Proposition 3. *Let f be a smooth function on a closed Riemannian manifold (F, g). Then the eigenspace of operator (1), corresponding to the least eigenvalue, is one-dimensional, and it contains a positive smooth eigenfunction.*

Proof. Let $u_0 \geq 0$ be not identically zero continuous function, i.e., $u_0 \in \mathcal{K} \setminus \{0\}$. By Lemma 2, a solution of the Cauchy's problem (3) belongs to

C^∞ for any $t > 0$ and permits the representation

$$u(\cdot, t) = e^{-\lambda_{j_0} t} \sum_{j:\, \lambda_j = \lambda_{j_0}} (u_0, e_j)_0 \, e_j + \vartheta_N(\cdot, t)$$

with $N = j_0 + n_0$, where $j_0 = \min\{j : \sum_{i:\, \lambda_i = \lambda_j} (u_0, e_i)_0^2 \neq 0\}$. Here n_0 is the multiplicity of the eigenvalue λ_{j_0} of $\mathcal{H}$, and for any integer $k \geq 0$ there exists $M_k > 0$ such that (9) of Lemma 2 holds with $l = \left[\frac{p}{4} + \frac{k}{2}\right] + 1$. The sum $e^{-\lambda_{j_0} t} \sum_{j:\, \lambda_j = \lambda_{j_0}} (u_0, e_j)_0 \, e_j$ can be represented in one-term form as $\tilde{c}\, e^{-\lambda_{j_0} t}\, \tilde{e}$, where $\tilde{c} > 0$ and the eigenfunction $\tilde{e}$ (for λ_{j_0}) has unit L_2-norm. By Lemma 1(ii), we conclude that $u > 0$ for all $t > 0$, hence the eigenvalue $\tilde{e}$ is a smooth non-negative function.

By Lemma 2 and the above arguments, the semi-group operator $S(t) = e^{-t\mathcal{H}}$ ($t > 0$) acts in $C(F)$, is bounded in $C(F)$ and is 1_F-bounded below, where 1_F is a constant function of value one. The eigenvalues of self-adjoint operators $S(t)$ are positive: $0 < \mu_j(t) = e^{-t\lambda_j} \leq \mu_0(t) = e^{-t\lambda_0}$, and the eigenspaces are the same as for $\mathcal{H}$. Since $S(t)$ is bounded in $C(F)$, i.e., there exists $c(t) > 0$ such that $\|S(t)u_0\|_C \leq c(t)\|u_0\|_C$ for any $u_0 \in C(F)$, we conclude that the operator $S(t)$ is 1_F-bounded above: $S(t)u_0 \leq c(t)\|u_0\|_C$. Thus, $S(t)$ ($t > 0$) is 1_F-positive.

In view of $\tilde{e} = \mu_0^{-1}(t) S(t)\tilde{e} > 0$ for sufficiently large $t > 0$, we find that $\tilde{e} > 0$ on F. Since μ_{j_0}-eigenspace contains a positive function (eigenvector) $\tilde{e}$, by [14, Theorem 2.13] we have $j_0 = 0$, hence $\lambda_{j_0} = \lambda_0$. Applying [14, Theorem 2.10] to $S(t)$ ($t > 0$), we conclude that the eigenspace of $\mathcal{H}$, corresponding to $\mu_0(t)$, is one-dimensional – with the eigenfunction $e_0 = \tilde{e} > 0$. $\qquad\square$

The rest of the section contains facts which were used in the proof of Proposition 3.

Sobolev embedding Theorem (see [4]). *If a nonnegative $k \in \mathbb{Z}$ and $l \in \mathbb{N}$ are such that $2\,l > p + 2\,k$, then H^l is continuously embedded into C^k.*

Elliptic regularity Theorem (see [4]). *If $\mathcal{H}$ is given by (1) and $0 \notin \sigma(\mathcal{H})$, then for any nonnegative $k \in \mathbb{Z}$ we have $\mathcal{H}^{-1} : H^k \to H^{k+2}$.*

Remark 2. We will briefly discuss the inhomogeneous *heat equation* on (F, g),

$$\partial_t u = \Delta\, u + \tilde{f}(\cdot, t). \tag{4}$$

The fundamental solution of (4) with $\tilde{f} = 0$ (i.e., of $\partial_t u = \Delta\, u$, satisfying $u(x, 0) = \delta_y(x)$, where $\delta_y(x)$ is the Dirac delta-function concentrated at

$y \in F$), $G_0(t, x, y) = \sum_j e^{-\lambda_j t} e_j(x) e_j(y)$, is called the *heat kernel*, see [6, II, Sect. B.6]. A solution of (4), satisfying $u(\cdot, 0) = u_0 \in L_2$, is given by

$$u(x, t) = \int_F G_0(t, x, y) u_0(y) \, dy_g + \int_0^t \int_F G_0(t - \tau, x, y) \tilde{f}(y, \tau) \, dy_g \, d\tau. \quad (5)$$

The solution of the homogeneous heat equation (i.e., $\tilde{f} = 0$) has the property $u(\cdot, t) \in C^\infty$ for $t > 0$. Moreover, $\lim_{t \to \infty} u(\cdot, t) = \bar{u}_0 = \int_F u_0(x) \, dx_g / \operatorname{vol}(F, g)$ and $\|u(\cdot, t) - \bar{u}_0\|_C \leq e^{-\lambda_1 t} \|u_0 - \bar{u}_0\|_C$ for $t > 0$. The least eigenvalue of Δ is $\lambda_0 = 0$ whose normalized eigenfunction is constant $e_0 = \operatorname{vol}(F, g)^{-1/2}$.

For a circle S^1 of radius one, we have $\lambda_j = j^2$ and $e_j(x) = \frac{1}{\sqrt{2\pi}} e^{-ijx}$, where $j \geq 0$.

Lemma 1. *Let $u_0 \in C(F)$ be a non-negative and not identically zero function.*

(i) If a continuous function $u(x, t)$ ($x \in F$, $t \geq 0$) with $u(\cdot, 0) = u_0$ satisfies the inequality

$$\partial_t u \geq \Delta u + \alpha u \quad (t > 0) \quad (6)$$

for some $\alpha \in \mathbb{R}$, then $u(\cdot, t) > 0$ for all $t > 0$.

(ii) If $u(x, t)$ ($x \in F$, $t \geq 0$) obeys the Cauchy's problem (3), then $u(\cdot, t) > 0$ for $t > 0$.

Proof. (i) It is known that $G_0(t, x, y)$ is continuous and positive in the product manifold $(0, \infty) \times F \times F$, [6, II, Proposition B.6.4]. Let $\tilde{f}_\alpha(x, t) \geq 0$ be the difference between l.h.s. and r.h.s. of (6). Notice that $\partial_t u - \Delta u - \alpha u = e^{\alpha t}(\partial_t \tilde{u} - \Delta \tilde{u})$ for $u = e^{\alpha t} \tilde{u}$. Using (5), we obtain the equality

$$e^{-\alpha t} u(x, t) = \int_F G_0(t, x, y) u_0(y) \, dy_g + \int_0^t \int_F G_0(t - \tau, x, y) \tilde{f}_\alpha(y, \tau) \, dy_g \, d\tau.$$

Certainly, the r.h.s. of the equality above is positive for $t > 0$. Hence $u(\cdot, t) > 0$ for all $t > 0$.

(ii) By the maximum principle for scalars [8, Proposition 4.3], we have $u(\cdot, t) \geq 0$ for all $t > 0$. Since $u(\cdot, t)$ obeys the inequality (6) with $\alpha = -\max_{x \in F} |f(x)|$, the claim follows from (i). $\quad \square$

One may consider $(3)_1$ as ODE $\frac{d}{dt} u = -\mathcal{H}(u)$ in the Hilbert space L_2, and assume that $u_0 \in L_2$. A function u is the *weak solution* of (3) if (see [15, Ch. I, Sect. 3, Definition 3.1])

(i) $u(\cdot, t) \in \mathcal{D} = H^2$ for any $t > 0$, and $u \in C([0, \infty), L_2) \cap C^1((0, \infty), L_2)$,

(ii) u satisfies the above ODE in $(0, \infty)$.

Spectral resolution of identity, corresponding to $\mathcal{H}$ is a family E_λ ($\lambda \in \mathbb{R}$) of orthogonal projection operators in L_2 with the properties (see [15, Sect. 5 in Introduction]):

1) E_λ is strongly left-continuous in λ;
2) $E_\lambda E_\mu = E_\mu E_\lambda = E_\lambda$ for $\lambda < \mu$;
3) $E_{-\infty} = \lim\limits_{\lambda \to -\infty} E_\lambda = 0$, $\quad E_\infty = \lim\limits_{\lambda \to \infty} E_\lambda = \mathrm{id}$;
4) $\mathcal{H}(u) = \int_{-\infty}^{\infty} \lambda \, dE_\lambda(u)$ for all $u \in \mathcal{D}$.

Here all limits are given in the sense of convergence in the *strong operator topology* (i.e., the topology, generated on the set $\mathcal{B}(L_2)$ of linear bounded operators acting in L_2 by the sub-basis $\mathcal{S} := \{U_{A,x,r}\}_{A \in \mathcal{B}, x \in L_2, r > 0}$, where $U_{A,x,r} := \{B \in \mathcal{B}(L_2) : \|Bx - Ax\|_0 < r\}$), and all the integrals are considered as the strong limits of corresponding Stieltjes integral sums.

The condition 2) is equivalent to $\mathrm{Im}\, E_\lambda \subseteq \mathrm{Im}\, E_\mu$ if $\lambda < \mu$, and in the case of discrete spectrum, we obtain $E_\lambda = \sum_{n: \lambda_n < \lambda} (\cdot, e_n)_0 \, e_n$. Furthermore, if we denote $E(I) = E_b - E_a$ for $I = [a, b)$, then condition 2) implies the following orthogonality property for $E(I)$:

$$E(I_1)E(I_2) = E(I_1 \cap I_2) \quad \text{where} \quad I_1 = [a_1, b_1) \text{ and } I_2 = [a_2, b_2). \quad (7)$$

Lemma 2. *Suppose that $f \in C^\infty$. Then*

(i) any eigenfunction e_n of the operator (1) belongs to class C^∞;

(ii) for any $N \in \mathbb{N}$, $t > 0$ and $u_0 \in L_2$ the weak solution of (3) has the form

$$u(\cdot, t) = \sum_{j=0}^{N-1} e^{-\lambda_j t}(u_0, e_j)_0 \, e_j + \vartheta_N(\cdot, t), \quad (8)$$

where $\vartheta_N(\cdot, t) \in C^\infty$, and for any integer $k \geq 0$ there exists $M_k > 0$ such that

$$\|\vartheta_N(\cdot, t)\|_{C^k} \leq M_k (1 - e^{-t})^{-\left[\frac{p}{4} + \frac{k}{2}\right] - 2} (|\lambda_N| + 1)^{\left[\frac{p}{4} + \frac{k}{2}\right] + 1} e^{-\lambda_N t} \|u_0\|_0$$
$$(t > 0). \quad (9)$$

Proof. (i) By Elliptic regularity Theorem, $\mathcal{H}^{-1} : H^k \to H^{k+2}$ ($k \geq 0$) holds. For any $l \in \mathbb{N}$ we have

$$\mathcal{H}^{-l} : L_2 \to H^{2l}. \quad (10)$$

One may check that $e_n = \lambda_n^l \mathcal{H}^{-l}(e_n)$ for any n. In view of (10), $e_n \in H^{2l}$ holds. On the other hand, by Sobolev embedding Theorem, if $k \geq 0$ and $4l > p + 2k$ then the space H^{2l} is continuously embedded into C^k. Hence $e_n \in C^k$. Since the integer $k \geq 0$ is arbitrary, we conclude that $e_n \in C^\infty$.

(ii) The Cauchy's problem (3) with $u_0 \in L_2$ admits a unique weak solution, and it has the form $u = \int_{\lambda_0}^\infty e^{-\lambda t} dE_\lambda(u_0)$, see [15, Theorem 4.1]. Thus, (8) holds with $\vartheta_N = \int_{\lambda_N}^\infty e^{-\lambda t} dE_\lambda(u_0)$.

In order to prove (9), let us write $\vartheta_N = \sum_{n=0}^\infty \int_{I_n} e^{-\lambda t} dE_\lambda(u_0)$ where $I_n = [\lambda_N + n, \lambda_N + n + 1)$. Using the property (7) and the representation $\mathcal{H}^{-l} = \int_{-\infty}^\infty \lambda^{-l} dE_\lambda$, we find

$$
\begin{aligned}
\int_{I_n} e^{-\lambda t} dE_\lambda u_0 &= \int_{\mathbb{R}} \int_{I_n} \frac{\lambda^l}{\mu^l} e^{-\lambda t} dE_\mu \, dE_\lambda u_0 \\
&= \int_{\mathbb{R}} \frac{1}{\mu^l} dE_\mu \left(\int_{I_n} e^{-\lambda t} \lambda^l \, dE_\lambda u_0 \right) = \mathcal{H}^{-l} \int_{I_n} e^{-\lambda t} \lambda^l \, dE_\lambda u_0.
\end{aligned}
\tag{11}
$$

For $k \geq 0$ and $l = \left[\frac{p}{4} + \frac{k}{2} \right] + 1$, using (10) and that H^{2l} is continuously embedded into C^k (see Sobolev embedding Theorem), we find that there exists $\delta_k > 0$ such that $\mathcal{H}^{-l}(f) \in C^k$ and $\|\mathcal{H}^{-l}(f)\|_{C^k} \leq \delta_k \|f\|_0$ for any $f \in L_2$. Using (11), we obtain $\int_{I_n} e^{-\lambda t} dE_\lambda(u_0) \in C^k$ and

$$
\left\| \int_{I_n} e^{-\lambda t} dE_\lambda(u_0) \right\|_{C^k} \leq \delta_k e^{-(\lambda_N + n)t} (|\lambda_N| + n + 1)^l \|u_0\|_0 \qquad (t > 0).
$$

Thus, the series $\sum_{n=0}^\infty \int_{I_n} e^{-\lambda t} dE_\lambda u_0$ converges in C^k, therefore, $\vartheta_N(\cdot, t) \in C^k$ for $t > 0$. We have $|\lambda_N| + n + 1 \leq (n+1)(|\lambda_N| + 1)$ when $n \geq 0$, hence

$$
\|\vartheta_N(\cdot, t)\|_{C^k} \leq \delta_k e^{-\lambda_N t} (|\lambda_N| + 1)^l \Psi_l(e^{-t}) \|u_0\|_0,
\tag{12}
$$

where $\Psi_l(z) := \sum_{n=0}^\infty (n+1)^l z^n$ for $0 < z < 1$. Notice that $\Psi_0(z) = (1-z)^{-1}$ and the recurrent relation $\Psi_l(z) = \frac{d}{dz}(z\Psi_{l-1}(z))$ holds. By induction, we find that $\Psi_l(z) = \sum_{i=2}^{l+1} A_{il}(1-z)^{-i}$ for $l \geq 1$. Hence $\Psi_l(z) \leq l\tilde{A}_l (1-z)^{-l-1}$, where $\tilde{A}_l = \max\{|A_{il}| : 2 \leq i \leq l+1\}$. From (12) we get:

$$
\|\vartheta_N(\cdot, t)\|_{C^k} \leq \delta_k \, l\tilde{A}_l (|\lambda_N| + 1)^l \, e^{-\lambda_N t} (1 - e^{-t})^{-l-1} \|u_0\|_0 \qquad (t > 0),
$$

where $l = \left[\frac{p}{4} + \frac{k}{2} \right] + 1$. From the above (9) with $M_k = \delta_k \, l\tilde{A}_l$ follows. $\square$

3.2. The forced Burgers equation

Consider the Burgers equation for a vector field H on (F, g), $\partial_\tau H + a\nabla g(H, H) = \nu \nabla(\text{div } H)$ where $a \in \mathbb{R}$ and $\nu > 0$. Using the scaling of variables $x = z \frac{a}{\nu}$ and $t = \tau \frac{a^2}{\nu}$ in coordinates (x, t) on $F \times \mathbb{R}$, we compute

$\partial_\tau H = (a^2/\nu)\, \partial_t H, \ \nabla_z H = (a/\nu)\nabla_x H, \ (\nabla \operatorname{div})_z H = (a^2/\nu^2)\,(\nabla \operatorname{div})_x H.$
Hence the above equation reduces to the *normalized Burgers equation*

$$\partial_t H + \frac{1}{2}\nabla g(H,H) = \nabla(\operatorname{div} H). \tag{13}$$

Equation (13) can be linearized by the Cole-Hopf transformation $H = -2\nabla(\log u)$, which turns it into the homogeneous heat equation $\partial_t u = \Delta u$ on (F,g).

The *forced* (or *inhomogeneous*) *Burgers equation*, see [13, 23], has attached some attention as an analogue of the Navier-Stokes equations. It can be viewed as the following PDE on (F,g):

$$\partial_t H + \frac{1}{2}\nabla g(H,H) = (\nabla \operatorname{div})\, H - 2\nabla f. \tag{14}$$

Here a smooth function $f : F \to \mathbb{R}$ is defined modulo a constant.

Proposition 4. *Suppose that $f \in C^\infty$.*

(a) If $u(x,t)$ is any positive solution of $\partial_t u = -\mathcal{H}(u)$ on (F,g) then the vector field $H = -2\nabla(\log u)$ solves (14). Every solution of (14), which is conservative at $t = 0$, comes by this way.

(b) Let $u(\cdot,t)$ $(t \geq 0)$ be a solution of (3) with $u_0 > 0$ and $f \geq 0$. Then $u(\cdot,t) > 0$ $(t > 0)$, and the vector-field $H = -2\nabla(\log u)$ obeys (14) and, as $t \to \infty$, approaches with the exponential rate $\lambda_1 - \lambda_0$ to a smooth vector-field $\bar{H} = -2\nabla(\log e_0)$ – a unique conservative solution of

$$\operatorname{div} \bar{H} = g(\bar{H},\bar{H})/2 + 2\,(f + \lambda_0). \tag{15}$$

Proof. (a) Using $\partial_t \nabla = \nabla \partial_t$, we rewrite (14) as the conservation law $\partial_t H = \nabla\big(\operatorname{div} H - \frac{1}{2}g(H,H) - 2f\big)$. This can be regarded as the compatibility condition for a function ψ to exist, such that

$$\nabla \psi = H, \qquad \partial_t \psi = \operatorname{div} H - \frac{1}{2}g(H,H) - 2f. \tag{16}$$

Substituting H from $(16)_1$ into $(16)_2$, and using the definition $\Delta \psi = \operatorname{div} \nabla \psi$, we obtain the PDE $\partial_t \psi + g(\nabla\psi, \nabla\psi)/2 = \Delta\psi - 2f$ for ψ. Introducing a positive function u by $\psi = -2\log u$ so that

$$\partial_t \psi = -\frac{2}{u}\,\partial_t u, \quad \nabla\psi = -\frac{2}{u}\nabla u, \quad \Delta\psi = -\frac{2}{u}\,\Delta u + \frac{2}{u^2}\, g(\nabla u, \nabla u),$$

we find that $\partial_t \psi - \Delta\psi + \frac{1}{2}g(\nabla\psi,\nabla\psi) + 2f = -\frac{2}{u}\big(\partial_t u - \Delta u - f\, u\big).$

(b) In view of Lemma 2 (in Section 3.1), there exists the limit vector field $\lim\limits_{t \to \infty} H(\cdot, t) = \bar{H}$,

$$\bar{H}(x) = -2 \lim_{t \to \infty} \frac{\nabla u(x, t)}{u(x, t)} = -2 \lim_{t \to \infty} \frac{(u_0, e_0)_0 \nabla e_0(x) + e^{\lambda_0 t} \nabla \vartheta_1(x, t)}{(u_0, e_0)_0 e_0(x) + e^{\lambda_0 t} \vartheta_1(x, t)}$$

$$= -2 \frac{\nabla e_0(x)}{e_0(x)}.$$

Denote by $w(x, t, \tau) = \tau e^{\lambda_0 t} u(x, t) + (1 - \tau)(u_0, e_0)_0 e_0(x)$. From the representation

$$\frac{\nabla u(x, t)}{u(x, t)} - \frac{\nabla e_0(x)}{e_0(x)} = \int_0^1 \frac{\partial}{\partial \tau} \left(\frac{\nabla w(x, t, \tau)}{w(x, t, \tau)} \right) d\tau$$

$$= \int_0^1 \left(\frac{e^{\lambda_0 t} \nabla \vartheta_1(x, t)}{w(x, t, \tau)} - \frac{e^{\lambda_0 t} \vartheta_1(x, t) \nabla w(x, t, \tau)}{w^2(x, t, \tau)} \right) d\tau,$$

using the fact that $\inf\{w(x, t, \tau) : x \in F, t \in [t_0, \infty), \tau \in [0, 1]\} > 0$ for a large enough $t_0 > 0$ (see Lemma 2) and the obvious property that there exists $M_k > 0$ such that $\|h_1 \cdot h_2\|_{C^k} \leq M_k \|h_1\|_{C^k} \|h_2\|_{C^k}$ for all $h_1, h_2 \in C^k$, it is not difficult to show that, as $t \to \infty$, $\frac{\nabla u(x,t)}{u(x,t)}$ converges to $\frac{\nabla e_0(x)}{e_0(x)}$ in C^∞ with the exponential rate $\lambda_1 - \lambda_0$. By the above we find

$$\operatorname{div} \bar{H} = -2 \Delta(\log e_0) = -2 (\Delta e_0)/e_0 + 2 g(\nabla e_0, \nabla e_0)/g(e_0, e_0).$$

Hence $\operatorname{div} \bar{H} - g(\bar{H}, \bar{H})/2 = -2 (\Delta e_0)/e_0 = 2 (f + \lambda_0)$, that proves (15).

To show uniqueness, assume the contrary, that (15) has another conservative solution $\tilde{H} = -2\nabla(\log \tilde{e}_0)$, where $\tilde{e}_0$ is a positive function of unit L_2-norm. Next, we calculate (15):

$$\operatorname{div} \tilde{H} - \frac{1}{2} g(\tilde{H}, \tilde{H}) - 2 (f(x) + \lambda_0) = -\frac{2}{\tilde{e}_0} \left[\Delta \tilde{e}_0 + (f + \lambda_0) \tilde{e}_0 \right]$$

and find that $\Delta \tilde{e}_0 + (f + \lambda_0) \tilde{e}_0 = 0$. Since λ_0 is a simple eigenvalue of $\mathcal{H}$ (see Proposition 3) we have $\tilde{e}_0 = e_0$, hence $\tilde{H} = \bar{H}$. □

3.3. The Schrödinger operator with parameter

Let $f : F \times \mathcal{O} \to \mathbb{R}$ be a smooth function, and $\{\mathcal{H}(q)\}_{q \in \mathcal{O}}$ a family of Schrödinger operators

$$\mathcal{H}(q)(u) := -\Delta u - f(\cdot, q) u \tag{17}$$

acting in the complex Hilbert space L_2, where $\mathcal{O}$ is an open neighborhood of origin in a real Banach space Q. Recall that the domain of definition of each operator $\mathcal{H}(q)$ is H^2. Notice that the spectrum of elliptic operator (17) is continuous in q. Since $f(\cdot, q) \in C^\infty$ for any fixed $q \in \mathcal{O}$, by Proposition 3, the spectrum of each operator $\mathcal{H}(q)$ is discrete and the least eigenvalue $\lambda_0(q)$ of $\mathcal{H}(q)$ is simple. Denote by $S_1 = \{u \in L_2 : \|u\|_0 = 1\}$ the unit sphere of L_2. Let $e_0(\cdot, q) \in S_1$ be the positive eigenfunction of $\mathcal{H}(q)$ corresponding to its least eigenvalue $\lambda_0(q)$, see Proposition 3. Denote by $e_0 \in S_1$ a real-valued positive eigenfunction of $\mathcal{H}(0)$, corresponding to its least eigenvalue λ_0. One may add a constant to the potential function in (17) such that $\mathcal{H}(0)$ becomes invertible in L_2.

In next proposition we prove the smooth dependence of e_0 on the parameter q. We shall also show the smooth dependence on the parameter q of a solution to the Cauchy's problem

$$\partial_t u = -\mathcal{H}(q)(u), \qquad u(\cdot, 0, q) = u_0(\cdot, q), \tag{18}$$

where $\mathcal{H}(q)$ is defined in (17) and u_0 is a smooth function on $F \times \mathcal{O}$. Assume that the functions from L_2 may take complex values, that is L_2 is a complex Hilbert space. As above, we add a constant to the potential function such that the least eigenvalue of $\mathcal{H}(0)$ is positive, hence this operator is continuously invertible in $\mathcal{O}$.

Proposition 5. *Let $\{\mathcal{H}(q)\}_{q \in \mathcal{O}}$ be given by (17) with a smooth function $f : F \times \mathcal{O} \to \mathbb{R}$. Then*

(i) there exist open neighborhoods $U \subseteq \mathbb{R}$ of λ_0, $\mathcal{W}_1 \subseteq \mathcal{O}$ of origin and a smooth function $\lambda : \mathcal{W}_1 \to U$ such that $\lambda(0) = \lambda_0$ and $\lambda(q) = \sigma(\mathcal{H}(q)) \cap U$ $(q \in \mathcal{W}_1)$ is a simple eigenvalue of $\mathcal{H}(q)$;

(ii) there exists an open neighborhood $\mathcal{W}_2 \subseteq \mathcal{W}_1$ of origin and a smooth function $e : F \times \mathcal{W}_2 \to S_1$ such that $e(\cdot, 0) = e_0$ and $e(\cdot, q)$ is an eigenfunction of $\mathcal{H}(q)$ related to $\lambda(q)$.

(iii) there exists a neighborhood $\mathcal{W} \subseteq \mathcal{O}$ of origin such that the mapping $u : F \times (0, \infty) \times \mathcal{W} \to \mathbb{R}$ is smooth, where u is the solution to (18).

Proof. Since $f : F \times \mathcal{O}$ is smooth, then for arbitrary $k, j \in \mathbb{N}$ the following conditions are satisfied:

(a) $f(\cdot, q) \in C^{2l-2}$ for $q \in \mathcal{O}$ with $l = \left[\frac{p}{4} + \frac{k}{2}\right] + 1$,

(b) the mapping $q \to f(\cdot, q)$ belongs to $C^j(\mathcal{O}, C^{2l-2})$,

First, we'll show that *there exists an open neighborhood $\mathcal{W} \subseteq \mathcal{O}$ of origin such that for any $k, j \in \mathbb{N}$ and $q \in \mathcal{W}$ the operator $\mathcal{H}(q)^{-l}$ (with l defined in (a)) acts from L_2 into C^k and the mapping $q \to \mathcal{H}(q)^{-l}$ belongs to $C^j(\mathcal{W}, \mathcal{B}(L_2, C^k))$*. To prove the claim, note that by the Elliptic Regularity Theorem we have

$$\mathcal{H}(0)^{-1}_{|H^{2i}} \in \mathcal{B}(H^{2i}, H^{2i+2}), \quad i = 0, \ldots, l-1. \tag{19}$$

Since $4l > p + 2k$, by the Sobolev embedding Theorem, we obtain $H^{2l} \hookrightarrow C^k$. Define the function $\varphi(\cdot, q) := f(\cdot, q) - f(\cdot, 0)$ and the operator function $\mathcal{P}(q) := \mathcal{H}(q) - \mathcal{H}(0)$. By Lemma 11 (see Appendix) with $\mathcal{M}_{\varphi(\cdot, q)} = \mathcal{P}(q)$ and in view of conditions (a)–(b), we conclude that

$$\mathcal{P}(\cdot)_{|H^i} \in C^j(\mathcal{O}, \mathcal{B}(H^i)), \qquad 0 \le i \le 2l - 2. \tag{20}$$

In view of (19) and by Lemma 9, there exists a neighborhood $\mathcal{W} \subseteq \mathcal{O}$ of origin such that $(q \to \mathcal{H}(q)_{|H^{2i}}^{-1}) \in C^j(\mathcal{W}, \mathcal{B}(H^{2i}, H^{2i+2}))$ for any $k, j \in \mathbb{N}$ and $i < l$. The last fact and $H^{2l} \hookrightarrow C^k$ imply the desired claim.

(i) The inclusion (20) with $i = 0$ and the simplicity of the least eigenvalue λ_0 of $\mathcal{H}(0)$ imply that for some open neighborhood $\mathcal{W}_1 \subseteq \mathcal{O}$ of origin there exist functions $(q \to \tilde{e}(\cdot, q)) \in C^j(\mathcal{W}_1, L_2)$ and $(q \to \lambda(q)) \in C^j(\mathcal{W}_1, \mathbb{R})$, for which the claim of Proposition 8 is valid with $E = L_2$. Since $j \in \mathbb{N}$ is arbitrary, claim (i) is proven.

(ii) Since the eigenfunction $e_0 \in S_1$ of $\mathcal{H}(0)$ related to λ_0 is real-valued, then $\|\operatorname{Re} \tilde{e}(\cdot, q)\|_0 > 0$ for any $q \in \mathcal{W}_2$ and for some restriction $\mathcal{W}_2$ of the neighborhood $\mathcal{W}_1$. One may check that $e(\cdot, q) = \frac{\operatorname{Re} \tilde{e}(\cdot, q)}{\|\operatorname{Re} \tilde{e}(\cdot, q)\|_0}$ is a real-valued eigenfunction of $\mathcal{H}(q)$ related to $\lambda(q)$ (as was shown in (i), the mapping $q \to e(\cdot, q)$ belongs to $C^j(\mathcal{W}_2, L_2)$ for each $q \in \mathcal{W}_2$). Hence (ii) follows from the equality $e(\cdot, q) = (\lambda(q))^l \mathcal{H}(q)^{-l} e(\cdot, q)$ $(q \in \mathcal{W}_2)$ with $l = \left[\frac{p}{4} + \frac{k}{2}\right] + 1$, claim (i) and the claim above.

(iii) Define the curve $\Gamma := \{\mu = \beta^2 + i\beta : \beta \in \mathbb{R}\}$ in a complex plane $\mathbb{C}$. Due to [16], the weak solution of the Cauchy's problem (18) with $u_0(\cdot, q) \in H^2$ has the representation

$$u(\cdot, t, q) = -\frac{1}{2\pi i} \int_\Gamma e^{-\mu t} (\mathcal{H}(q) - \mu \operatorname{id})^{-1} u_0(\cdot, q) \, d\mu \quad (q \in \mathcal{O}). \tag{21}$$

By (i), one can take a neighborhood $\mathcal{W}_1 \subset \mathcal{O}$ such that for any $q \in \mathcal{W}_1$ the least eigenvalue $\lambda_0(q)$ of $\mathcal{H}(q)$ is positive, hence this operator is continuously invertible. This means that for any $q \in \mathcal{W}_1$ the parabola Γ doesn't intersect $\sigma(\mathcal{H}(q))$, hence the r.h.s. of (21) is well-defined. Moreover, as was shown in [16], for any fixed $q \in \mathcal{W}_1$ the mapping $t \to u(x, t, q) \in L_2$ is C^∞-smooth on $(0, \infty)$ for all $t > 0$, and $u(\cdot, t, q)$ belongs to the domain of definition of $\mathcal{H}(q)^m$ for any $m \in \mathbb{N}$, and we have

$$\mathcal{H}(q)^m u(\cdot, t, q) = (-1)^m \frac{\partial^m}{\partial t^m} u(\cdot, t, q)$$
$$= -\frac{1}{2\pi i} \int_\Gamma \mu^m e^{-\mu t} (\mathcal{H}(q) - \mu \operatorname{id})^{-1} u_0(\cdot, q) \, d\mu.$$

In particular, for $l = \left[\frac{p}{4} + \frac{k}{2}\right] + 1$ $(k \in \mathbb{N})$ we have

$$u(\cdot, t, q) = -\frac{1}{2\pi i} \mathcal{H}(q)^{-l} \int_\Gamma \mu^l e^{-\mu t} (\mathcal{H}(q) - \mu \, \mathrm{id})^{-1} u_0(\cdot, q) \, \mathrm{d}\mu. \qquad (22)$$

By the claim in the beginning of the proof, there exists an open neighborhood $\mathcal{W}_2 \subseteq \mathcal{W}_1$ of origin such that $(q \to \mathcal{H}(q)^{-l}) \in C^j(\mathcal{W}_2, \mathcal{B}(L_2, C^k))$. We shall show that there exists an open neighborhood of origin, $\mathcal{W} \subseteq \mathcal{W}_2$ such that

$$\int_\Gamma \mu^l e^{-\mu t} (\mathcal{H}(\cdot) - \mu \, \mathrm{id})^{-1} u_0 \, \mathrm{d}\mu \in C^j(\mathcal{W}, L_2) \qquad (23)$$

for any $t > 0$. By Lemma 11 with $\varphi(\cdot, q) = f(\cdot, q) - f(\cdot, 0)$ and $k = 0$, the conditions of Lemma 8 are satisfied (with $E = L_2$). Thus, the function $\mu \to (\mathcal{H}(\cdot) - \mu \, \mathrm{id})^{-1} \in C^j(\mathcal{W}, \mathcal{B}(L_2))$ is continuous and bounded on Γ for a neighborhood $\mathcal{W} \subseteq \mathcal{W}_2$ of origin. Since the function $u_0(x, q)$ is smooth in $F \times \mathcal{O}$, we can choose $\mathcal{W}$ such that the function $q \to u_0(\cdot, q)$ belongs to $C^j(\mathcal{W}, L_2)$. Then the function $(\mu \to (\mathcal{H}(\cdot) - \mu \, \mathrm{id})^{-1} u_0(\cdot, \cdot)) \in C(\Gamma, C^j(\mathcal{W}, L_2))$ is continuous and bounded on Γ. Hence, the function on Γ

$$\alpha + i\sqrt{\alpha} \to (\alpha^2 + \alpha)^{\frac{l}{2}} e^{-\alpha t} \sqrt{1 + (4\alpha)^{-1}} \, \|(\mathcal{H}(\cdot)$$
$$- (\alpha + i\sqrt{\alpha}) \, \mathrm{id})^{-1} u_0\|_{C(\Gamma, \, C^j(\mathcal{W}, L_2))} \qquad (\alpha \geq 0)$$

is a majorant of the $C^j(\mathcal{W}, L_2)$-norm of the integrand in (23). Since the scalar improper integral $\int_0^\infty (\alpha^2 + \alpha)^{\frac{l}{2}} e^{-\alpha t} \sqrt{1 + (4\alpha)^{-1}} \, d\alpha$ converges, also the integral in (23) converges in the $C^j(\mathcal{W}, L_2)$-norm. This property implies the desired inclusion (23). Then by (22), $u(\cdot, t, q) \in C^j(\mathcal{W}, C^k)$. Since k and j are arbitrary naturals, this completes the proof. $\qquad \square$

4. Proof of the main results

4.1. *Preliminaries*

The *Levi-Civita connection* ∇ of a metric g on M is given by the formula

$$2\,g(\nabla_X Y, Z) = X(g(Y, Z)) + Y(g(X, Z)) - Z(g(X, Y))$$
$$+ g([X, Y], Z) - g([X, Z], Y) - g([Y, Z], X), \quad (X, Y, Z \in TM). \qquad (1)$$

The covariant derivative of the $(1, j)$-tensor S is the $(1, j+1)$-tensor given by

$$(\nabla S)(X, Y_1, \ldots, Y_j) = (\nabla_X S)(Y_1, \ldots, Y_j)$$
$$= \nabla_X(S(Y_1, \ldots, Y_j)) - \sum\nolimits_{i \leq j} S(Y_1, \ldots, \nabla_X Y_i, \ldots Y_j).$$

The *co-nullity operator* $C : D_{\mathcal{F}} \times D \to D$ of a totally geodesic foliation $\mathcal{F}$ is defined by $C_N(X) = -(\nabla_X N)^{\perp}$ where $X \in D$, $N \in D_{\mathcal{F}}$. It satisfies the Riccati equation along leaf-geodesics

$$\nabla_N C_N = C_N^2 + R_N \qquad (N \in D_{\mathcal{F}}) \tag{2}$$

where $R_N = R(N, \cdot)N$ is the self-adjoint $(1,1)$-tensor on D, called the *Jacobi operator*. The equation (2) serves as a main tool in proving the splitting of foliations with $\mathrm{Sc}_{\mathrm{mix}} \geq 0$. Namely, if $R_N \geq 0$ and C_N is symmetric for any $N \in D_{\mathcal{F}}$ then $C = 0$ on any complete leaf, see [19]. The equality $C \equiv 0$ means that D is integrable and tangent to a totally geodesic foliation, hence M splits.

We decompose C into symmetric and skew-symmetric parts as $C_N = A_N + T_N^{\sharp}$. The Weingarten operator A_N of D and the operator $T_N^{\sharp}$ are related with tensors b and T by

$$g(A_N(X), Y) = g(b(X,Y), N), \quad g(T_N^{\sharp}(X), Y) = g(T(X,Y), N),$$
$$(X, Y \in D).$$

Next lemma represents $\mathrm{Sc}_{\mathrm{mix}}(g)$ in a more convenient than (1) form and shows that the flow (3) can be reduced to the PDE on the vector bundle $TM_{|F}$ that is the restriction of TM on a leaf F.

Lemma 3. *Let $\mathcal{F}$ be a totally geodesic foliation on (M, g). Then*

$$Sc_{\mathrm{mix}}(g) = \|T\|_g^2 + \mathrm{div}_{\mathcal{F}} H - \frac{1}{n} g(H, H) - n\,\beta_D. \tag{3}$$

Proof. Substituting $C_N = A_N + T_N^{\sharp}$ into (2) and taking the (skew-) symmetric part yield

$$\nabla_N A_N = A_N^2 + (T_N^{\sharp})^2 + R_N, \qquad \nabla_N T_N^{\sharp} = A_N T_N^{\sharp} + T_N^{\sharp} A_N \tag{4}$$

for any unit vector $N \in D_{\mathcal{F}}$. By $(4)_2$, we have

$$-N(\|T\|^2) = 2\mathrm{Tr}\,(T_N^{\sharp}(\nabla_N T_N^{\sharp})) = 4\mathrm{Tr}\,(A_N(T_N^{\sharp})^2). \tag{5}$$

Let e_i be a local orthonormal frame of D. Since the leaves are totally geodesic, we have $\nabla_N e_i \in D$ for any vector $N \in D_{\mathcal{F}}$. Then, using $g(A_N(e_i), \nabla_N e_i) = 0$, we find

$$N(\mathrm{Tr} A_N) = \sum_i [g((\nabla_N A_N)(e_i), e_i) + g(A_N(\nabla_N e_i), e_i) + g(A_N(e_i), \nabla_N e_i)]$$
$$= \mathrm{Tr}\,(\nabla_N A_N).$$

Thus, the contraction of $(4)_1$ over D yields the formula

$$N(\mathrm{Tr}A_N) = \mathrm{Tr}\,(A_N^2) + \mathrm{Tr}\,((T_N^\sharp)^2) + \sum_{j=1}^{n} K(e_j, N) \qquad (6)$$

for any unit vector $N \in D_{\mathcal{F}}$. Note that $\mathrm{Tr}A_N = g(H, N)$. We have

$$\|T\|^2 = -\sum_{\alpha=1}^{p} \mathrm{Tr}\,((T_{\varepsilon_\alpha}^\sharp)^2), \quad \|b\|^2 = \sum_{\alpha=1}^{p} \mathrm{Tr}\,(A_{\varepsilon_\alpha}^2) = n\,\beta_D + \frac{1}{n}\,g(H, H),$$

and $\sum_{\alpha=1}^{p} \varepsilon_\alpha(\mathrm{Tr}A_{\varepsilon_\alpha}) = \mathrm{div}_{\mathcal{F}}\,H$. Hence, the contraction of (6) over $D_{\mathcal{F}}$
yields (3).   □

Definition 3. The notion of *holonomy* uses that of a germ of a locally defined diffeomorphism (i.e. an equivalence class of certain maps). The germs
of diffeomorphisms $(\mathbb{R}^n, 0) \to (\mathbb{R}^n, 0)$ with fixed origin form a group, denoted by $\mathrm{Diff}_0(\mathbb{R}^n)$. We denote by $\mathrm{Diff}_0^+(\mathbb{R}^n)$ the subgroup of germs of
diffeomorphisms which preserve orientation of $\mathbb{R}^n$. Let $(M, \mathcal{F})$ be a foliated
manifold, x, y be two points on a leaf F and Q_x, Q_y be transversal sections (diffeomorphic to $\mathbb{R}^n$). To any path α from x to y in F we associate
a germ of a diffeomorphism $h(\alpha) : (Q_x, x) \to (Q_y, y)$ called the holonomy of the path, for $x = y$, this is a generalization of a first return map.
We obtain a homomorphism of groups $h : \pi_1(F) \to \mathrm{Diff}_0(\mathbb{R}^n)$. The image
$\mathrm{Hol}(F) := h(\pi_1(F))$ is called the *holonomy group* of F. If $\mathcal{F}$ is transversally
orientable then $\mathrm{Hol}(F)$ is a subgroup of $\mathrm{Diff}_0^+(\mathbb{R}^n)$.

Certainly, $\mathrm{Hol}(F)$ is finite when the first fundamental group, $\pi_1(F)$, is
finite. Note that a foliation whose leaves are the fibers of a fibre bundle has
trivial holonomy group.

Definition 4. A *foliated bundle* is a fibre bundle $p : E \to F$ admitting a
foliation $\mathcal{F}$ whose leaves meet transversely all the fibers $E_x = p^{-1}(x)$ ($x \in
F$), and the bundle projection restricted to each leaf $F' \in \mathcal{F}$ is a covering
map $p : F' \to F$. There is a representation $h_x : \pi_1(F, x) \to \mathrm{Diff}(E_x)$ of the
first fundamental group $\pi_1(F, x)$ in the group of diffeomorphims of a fiber,
called the *total holonomy homomorphism* for the foliated bundle.

Local Reeb stability Theorem (see [6, Vol. I]). *Let F be a compact
leaf of a foliated manifold $(M, \mathcal{F})$ and $\mathrm{Hol}(F)$ is finite. Then there is a
normal neighborhood* $\mathrm{pr} : V \to F$ *of F in M such that $(V, \mathcal{F}_{|V}, \mathrm{pr})$ is
a foliated bundle with all leaves compact (and transversal to the fibers).
Furthermore, each leaf $F' \subset V$ has finite holonomy group of order at most*

the order of $\mathrm{Hol}(F)$ *and the covering* $\mathrm{pr}_{|F'} : F' \to F$ *has* k *sheets, where* $k \leq$ *order of* $\mathrm{Hol}(F)$.

In other words, for a compact leaf F with finite holonomy group, there exists a saturated neighborhood V of F in M and a diffeomorphism from $E := \tilde{F} \times_{\mathrm{Hol}(F)} \mathbb{R}^n$ under which $\mathcal{F}_{|V}$ corresponds to the bundle foliation on E. Here $\tilde{F}$ is a covering space of F associated with $\mathrm{Hol}(F)$.

The following method construction foliations is related to local Reeb stability Theorem. Suppose that a group G acts freely and properly discontinuously on a connected manifold $\tilde{F}$ such that $\tilde{F}/G = F$. Suppose also that G acts on a manifold Q. Now form the quotient space $E := \tilde{F} \times_G Q$, obtained from the product space $\tilde{E} := \tilde{F} \times Q$ by identifying (gy, z) with (y, gz) for any $y \in \tilde{F}$, $g \in G$ and $z \in Q$. Thus E is the orbit space of $\tilde{E}$ w.r.t. a properly discontinuous action of G. It is also Hausdorff, so it is a manifold. The projection $\mathrm{pr}_1 : \tilde{E} \to \tilde{F}$ induces a submersion $\mathrm{pr} : E \to F$, so we have the commutative diagram $(\tilde{F} \to F) \circ \mathrm{pr}_1 = \mathrm{pr} \circ (\tilde{E} \to E)$. The map pr has the structure of a fibre bundle over F with vertical fiber Q. (Fibre bundles which can be obtained in this way are exactly those with discrete structure group). We claim that E admits also horizontal leaves, so that pr maps each leaf to F as a covering projection. Indeed, the foliation $\tilde{\mathcal{F}}$ on $\tilde{E}$, which is given by the submersion $\mathrm{pr}_2 : \tilde{E} \to Q$, is invariant under the action of G, and hence we obtain the quotient foliation $\mathcal{F} = \tilde{\mathcal{F}}/G$ on E. If $z \in Q$ and $G_z \subset G$ is the isotropy group at z of the action by G on Q, then the leaf of E obtained from the leaf $\tilde{F} \times \{z\}$ is naturally diffeomorphic to $\tilde{F}/G_z$, and pr restricted to this leaf is the covering $\tilde{F}/G_z \to F$.

4.2. D-conformal adapted variations of metrics

Denote by $\mathcal{M}$ the space of smooth Riemannian metrics of finite volume on M such that $D_{\mathcal{F}}$ is orthogonal to D. Elements of $\mathcal{M}$ are called *adapted metrics* to the $(D, D_{\mathcal{F}})$-structure on M.

Since the difference of two connections is a tensor, $\partial_t \nabla^t$ is a $(1,2)$-tensor on (M, g_t). Differentiating the formula (1) for the Levi-Civita connection with respect to t yields, see [21],

$$2 \, g_t((\partial_t \nabla^t)(X, Y), Z) = (\nabla^t_X S)(Y, Z) + (\nabla^t_Y S)(X, Z) - (\nabla^t_Z S)(X, Y) \quad (7)$$

for all t-independent vector fields $X, Y, Z \in \Gamma(TM)$. Here $S = \partial_t g$.

Lemma 4. *D-conformal adapted variations of metrics preserve the property "$\mathcal{F}$ is totally geodesic".*

Proof. Let g_t $(t \geq 0)$ be a family of metrics on $(M, \mathcal{F})$ such that $\partial_t g_t = S(g)$, where the tensor $S(g)$ is D-truncated. Using (2), we find for $X \in D$ and $\xi, \eta \in D_{\mathcal{F}}$,

$$
\begin{aligned}
2\,g_t(\partial_t b_{\mathcal{F}}(\xi, \eta), X) &= g_t(\partial_t(\nabla^t_\xi \eta) + \partial_t(\nabla^t_\eta \xi),\ X) \\
&= (\nabla^t_\xi S)(X, \eta) + (\nabla^t_\eta S)(X, \xi) - (\nabla^t_X S)(\xi, \eta) \\
&= -S(\nabla^t_\xi \eta, X) - S(\nabla^t_\eta \xi, X) = -2\,S(b_{\mathcal{F}}(\xi, \eta), X).
\end{aligned}
$$

Assuming $S(g) = s(g)\,\hat{g}$, we have $\partial_t b_{\mathcal{F}} = -s\,b_{\mathcal{F}}$. By the theory of ODEs, if $b_{\mathcal{F}} = 0$ at $t = 0$ then $b_{\mathcal{F}} = 0$ for all $t > 0$.   □

The proof of the next lemma is based on (2) with $S = s\,\hat{g}$.

Lemma 5 (see [21] and [22]). *Let g_t $(t \geq 0)$ be D-conformal metrics on a foliation $(M, \mathcal{F})$. Then*

$$
\partial_t A_N = -\frac{1}{2}\,N(s)\,\widehat{\mathrm{id}}, \qquad \partial_t T^\sharp_N = -s\,T^\sharp_N, \tag{8}
$$

$$
\partial_t H = -\frac{n}{2}\nabla^{\mathcal{F}} s, \qquad \partial_t(\mathrm{div}_{\mathcal{F}}\,H) = -\frac{n}{2}\,\Delta_{\mathcal{F}}\,s. \tag{9}
$$

By Lemma 5, (2) preserves conformal foliations.

Corollary 3. *Let g_t $(t \geq 0)$ be a D-conformal family of metrics on a foliation $(M, \mathcal{F})$. If $\mathcal{F}$ is a conformal foliation w.r.t. g_0 then $\mathcal{F}$ has the same property for all g_t.*

Proof. Let $\mathcal{F}$ be g_0-conformal, hence a self-adjoint operator A_N $(N \in D_{\mathcal{F}})$ is conformal, i.e., proportional to the identity. Applying to $(8)_1$ the theorem on existence/uniqueness of a solution of ODEs, we conclude that A_N is conformal for any $t \geq 0$, hence $\mathcal{F}$ is g_t-conformal.   □

By Proposition 6, the measure of non-umbilicity of D (see Introduction), is preserved by (3).

Proposition 6 (Conservation law). *Let g_t $(t \geq 0)$ be a D-conformal family of metrics on a foliated manifold $(M, \mathcal{F})$. Then the function β_D doesn't depend on t.*

Proof. Using Lemma 5, we calculate

$$
\begin{aligned}
\partial_t \|b\|^2 &= \partial_t \sum_{\alpha=1}^{p} \mathrm{Tr}\,(A^2_{\varepsilon_\alpha}) \\
&= 2\sum_\alpha \mathrm{Tr}\,(A_{\varepsilon_\alpha}\,\partial_t A_{\varepsilon_\alpha}) = -\sum_\alpha \varepsilon_\alpha(s)\mathrm{Tr}A_{\varepsilon_\alpha} = -g(\nabla s, H), \\
\partial_t g(H, H) &= s\,\hat{g}(H, H) + 2g(\partial_t H, H) = -n\,g(\nabla s, H).
\end{aligned}
$$

Hence, $n \, \partial_t \beta_D = \partial_t \, \|b\|_g^2 - \frac{1}{n} \, \partial_t g(H, H) = 0.$ $\qquad\qquad\qquad\qquad\qquad$ $\square$

If one has a solution u_0 to a given non-linear PDE, it is possible to linearize the equation by considering a smooth family $u = u(t)$ of solutions with a variation $v = \partial_t u_{|\,t=0}$. Differentiation of the PDE by t, yields a linear PDE in terms of v. The next lemma concerns the linearization of (3).

Lemma 6. *Let g_t $(t \geq 0)$ be a family of D-conformal totally geodesic metrics on a foliated manifold $(M, \mathcal{F})$, i.e., $\partial_t g = s_t \, \hat{g}$. Then*

$$\partial_t(Sc_{\mathrm{mix}}(g) - \|T\|_g^2) = -\frac{n}{2}\,\Delta_{\mathcal{F}}\,s + g(\nabla s, H).$$

Proof. Differentiating (3) by t, and using Proposition 6 and $\hat{g}(H, H) = 0$, we obtain $\partial_t(Sc_{\mathrm{mix}} - \|T\|^2) = \partial_t(\mathrm{div}_{\mathcal{F}}\,H) - \frac{2}{n}\,g(\partial_t H, H)$. Using Lemma 5, we rewrite this equation in required form. $\qquad\qquad\qquad\qquad\qquad$ $\square$

4.3. *Proofs of Propositions 1–2, Theorem 1 and Corollaries 1–2*

Proof of Proposition 1. Let $g_t = g_0 + h_t$ where $t > 0$. By Lemma 6, the linearization of (3) about g_0 w.r.t. variations $h_t = s_t \hat{g}_0$, where $s_t \in C^1(M)$, is the linear PDE

$$\partial_t s = n\,\Delta_{\mathcal{F}}\,s - 2\,g_0(\nabla s,\,H(g_0)) - 2\,(Sc_{\mathrm{mix}}(g_0) - \|T\|_{g_0}^2 - \Phi)\,s,$$

and $s_{|\,t=0}$ is bounded. The result follows from the theory of linear parabolic PDEs (see [4]) and the finite holonomy assumption (i.e., the local Reeb stability Theorem in Section 4.1). $\qquad\qquad\qquad\qquad\qquad$ $\square$

Proof of Proposition 2. By Proposition 1, (3) admits a unit local solution g_t $(0 \leq t < \varepsilon)$.

(i) By $(9)_1$ with $s = -2(Sc_{\mathrm{mix}} - \|T\|^2 - \Phi)$ and (3), we obtain (2) for H.

(ii) As in the proof of Proposition 4 for $H = \nabla^{\mathcal{F}} \psi$, we reduce (2) to the PDE

$$\partial_t \psi + g(\nabla^{\mathcal{F}} \psi, \nabla^{\mathcal{F}} \psi) - n\,\Delta_{\mathcal{F}}\,\psi = -n^2\,\beta_D.$$

Then applying $\psi = -n \log u$ for a positive function u, we calculate

$$\partial_t \psi = -\frac{n}{u}\,\partial_t u, \quad \nabla^{\mathcal{F}} \psi = -\frac{n}{u}\nabla^{\mathcal{F}} u, \quad \Delta_{\mathcal{F}}\,\psi = -\frac{n}{u}\,\Delta_{\mathcal{F}}\,u + \frac{n}{u^2}\,g(\nabla^{\mathcal{F}} u, \nabla^{\mathcal{F}} u),$$

and obtain

$$\partial_t \psi + g(\nabla^{\mathcal{F}}\psi, \nabla^{\mathcal{F}}\psi) - n\,\Delta_{\mathcal{F}}\,\psi + n^2\,\beta_D = -\frac{n}{u}\left(\partial_t u - n\,\Delta_{\mathcal{F}}\,u - n\,\beta_D\,u\right).$$

Thus, the function $u > 0$, introduced by $H = -n\nabla^{\mathcal{F}}(\log u)$, solves the heat type equation (3).

(iii) By the proof of Proposition 4(b), the solution of (2), $H = -n\nabla^{\mathcal{F}}(\log u)$, approaches as $t \to \infty$ in C^∞ with the exponential rate $\lambda_1 - \lambda_0$ to $\bar{H} = -n\nabla^{\mathcal{F}}(\log e_0)$. Here $e_0 > 0$ is the unit-norm eigenfunction related to the least eigenvalue, λ_0, of the operator (1) on compact leaves. $\quad\square$

Proof of Theorem 1. (i) By Proposition 2(i), the PDE (3) with $u(\cdot, 0) = u_0$ admits a unique solution $u(x, t)$ $(t \geq 0)$ on any leaf. By the maximum principle (see Lemma 1(i) in Section 3.1) we conclude that $u > 0$ for all $t > 0$. Hence there exists a unique global solution g_t $(t \geq 0)$ to (3).

(ii) Let $\Phi = n\,\lambda_0$. By Propositions 2(ii) and 4, the limit vector field $\bar{H}$ is leaf-wise conservative, it depends on the D-conformal class $[g_0]$ only, and is the unique leaf-wise stationary solution of (2):

$$\operatorname{div}_{\mathcal{F}} \bar{H} = \frac{1}{n}\,g(\bar{H}, \bar{H}) + n\,(\beta_D + \lambda_0). \tag{10}$$

Denote by $s_t := -2\,(\operatorname{Sc}_{\mathrm{mix}}(g_t) - \|T\|^2_{g_t} - n\,\lambda_0)$. By (3), we have $s_t = \operatorname{div}_{\mathcal{F}} H - \frac{1}{n}\,g(H, H) - n(\beta_D + \lambda_0)$, where $\operatorname{div}_{\mathcal{F}} H$ and $g(H, H)$ converge as $t \to \infty$ in C^∞ with the exponential rate $\lambda_1 - \lambda_0$. The unique solution of (3) has the form $\hat{g}_t = \hat{g}_{t_0} \exp\left(\int_{t_0}^{t} s_\xi\,\mathrm{d}\xi\right)$ $(t \geq t_0 > 0)$. By Proposition 2 and (10), $\lim_{t\to\infty} s_t = 0$ with the exponential rate $\lambda_1 - \lambda_0$ of convergeness in C^∞-topology. Hence, $|s_t| \leq \tilde{c}\,e^{(\lambda_0 - \lambda_1)t}$ for some real $\tilde{c} > 0$ depending on $[g_0]$. Therefore,

$$\int_t^\infty |s_\xi|\,\mathrm{d}\xi \leq \tilde{c} \int_t^\infty e^{(\lambda_0 - \lambda_1)\xi}\,\mathrm{d}\xi = \frac{\tilde{c}}{\lambda_1 - \lambda_0}\,e^{(\lambda_0 - \lambda_1)\,t}.$$

Denoting $\hat{g}_\infty := \hat{g}_{t_0} \exp(\int_{t_0}^\infty s_\xi\,\mathrm{d}\xi)$ (for some "small" $t_0 > 0$ given due to Proposition 1), we find

$$|(\hat{g}_\infty - \hat{g}_t)(X, X)| \leq |\hat{g}_{t_0}(X, X)| \cdot \Big| \int_0^1 \frac{d}{d\tau} \exp\big((1 - \tau) \int_{t_0}^t s_\xi\,\mathrm{d}\xi$$

$$+ \tau \int_{t_0}^\infty s_\xi\,\mathrm{d}\xi\big)\,\mathrm{d}\tau\Big|$$

$$\leq |\hat{g}_{t_0}(X, X)| \exp\Big(\int_{t_0}^\infty |s_\xi|\,\mathrm{d}\xi\Big) \int_t^\infty |s_\xi|\,\mathrm{d}\xi$$

$$\leq \mu\,e^{(\lambda_0 - \lambda_1)\,t} |\hat{g}_{t_0}(X, X)|, \quad (X \in D)$$

for $t > t_0$ and some constant $\mu > 0$ depending on $[g_0]$. In particular, the metrics g_t are uniformly equivalent, i.e., there exists a constant $c > 0$ such that $c^{-1}\|X\|^2_{g_0} \le \|X\|^2_{g_t} \le c \|X\|^2_{g_0}$ for all points $(x, t) \in M \times [0, \infty)$ and vectors $X \in D_x$. By the above, $\|T\|_{g_t} \ge c^{-2}\|T\|_{g_0}$. Since $\lambda_0(x) \ge -n \max_{F(x)} \beta_D$ for $x \in M$, see Section 2, we obtain

$$\mathrm{Sc}_{\mathrm{mix}}(\bar{g}) = \|T\|^2_{\bar{g}} + n \lambda_0 \ge c^{-4}\|T\|^2_{g_0} - n^2 \max_{F(x)} \beta_D > 0.$$

Similarly, for any milti-index $\alpha = (\alpha_1, \dots, \alpha_p)$ we have $|d^\alpha s_t| \le \tilde{c}_\alpha e^{(\lambda_0 - \lambda_1)t}$ for some real $\tilde{c}_\alpha > 0$ depending on $[g_0]$, hence $\int_t^\infty |d^\alpha s_\xi| \, d\xi \le \frac{c_\alpha}{\lambda_1 - \lambda_0} e^{(\lambda_0 - \lambda_1)t}$. By (3) and the Leibnitz formula, we obtain $\partial_t(d^\alpha \hat{g}) = s_t \, d^\alpha \hat{g} + \sum_{\beta < \alpha} \beta! \, d^{\alpha - \beta} s_t \cdot d^\beta \hat{g}$, where $\beta = (\beta_1, \dots, \beta_p)$ and $\beta! := \frac{|\beta|!}{\beta_1! \cdot \dots \cdot \beta_p!}$ is the multinomial coefficient. Considering the last equality as linear ODE for $d^\alpha \hat{g}$, we get

$$d^\alpha \hat{g}_t = e^{\int_{t_0}^t s_\xi \, d\xi} d^\alpha \hat{g}_{t_0} + \int_{t_0}^t \left(e^{\int_\tau^t s_\xi \, d\xi} \sum_{\beta < \alpha} \beta! \, d^{\alpha - \beta} s_\tau \cdot d^\beta \hat{g}_\tau \right) d\tau.$$

By induction w.r.t. the order of differentiation on leaves, $|\alpha| = \sum_{i=1}^p \alpha_i$, we obtain as above that $\hat{g}_\infty$ is C^∞-smooth and $d^\alpha \hat{g}_t$ converges to $d^\alpha \hat{g}_\infty$ uniformly on F with the exponential rate $\lambda_1 - \lambda_0$. Thus a unique solution g_t of (3) approaches as $t \to \infty$ in C^∞ with the exponential rate $\lambda_1 - \lambda_0$ to a smooth on any leaf metric $\bar{g}$.

(iii) The smoothness of g_t and $\bar{g}$ on M follows from the finite holonomy assumption and results of Section 3.3. Indeed, let F be a leaf. Note that the leaves of a totally geodesic foliation are locally isometric one to another, see for example [19]. By the local Reeb stability Theorem (see Section 4.1), there is a normal neighborhood pr : $V \to F$ of F (with a smooth normal section Q – an open n-dimensional disk) such that $(V, \mathcal{F}_{|V}, \mathrm{pr})$ is a foliated bundle diffeomorphic to $\widetilde{F} \times_{\mathrm{Hol}(F)} Q$. There is a regular covering $\widetilde{\mathrm{pr}} : \widetilde{F} \to F$ with covering group $\mathrm{Hol}(F)$, since this group is finite, $\widetilde{F}$ is a compact manifold. The normal neighborhood V, as a fiber bundle over F, can be pulled back via $\widetilde{\mathrm{pr}}$ to a bundle $\widetilde{V}$ over $\widetilde{F}$ with the same fiber. The standard pull-back construction yields a canonical covering $\widetilde{\mathrm{pr}} : \widetilde{V} \to V$ (a submersion), enabling us to lift the foliation $\mathcal{F}_{|V}$ to a foliation $\widetilde{\mathcal{F}}$ of $\widetilde{V}$, transverse to the fibers and having $\widetilde{F}$ as a leaf. Let $\tilde{x} \in \widetilde{F}$ with $\widetilde{\mathrm{pr}}(\tilde{x}) = x$. Since the covering group is exactly the holonomy group of F, $\widetilde{\mathrm{pr}}_* : \pi_1(\widetilde{F}, \tilde{x}) \to \pi_1(F, x)$ injects $\pi_1(\widetilde{F}, \tilde{x})$ onto the subgroup of $\pi_1(F, x)$, and the leaf $\widetilde{F}$ of $\widetilde{\mathcal{F}}$ has trivial holonomy. By [6, Theorem 2.4.1], $\widetilde{F}$ has a neighborhood in $\widetilde{V}$ that is a foliated disk bundle with all leaves diffeomorphic to $\widetilde{F}$. Let $\tilde{g}_t$ and $\tilde{\bar{g}}$ be the lifts of the metrics g_t and $\bar{g}$ from V onto the product $\widetilde{V}$. The corresponding foliation on $\tilde{V}$ is totally geodesic and the leaves are isometric one to another. Hence the leaf-wise laplacian on $\tilde{V}$ doesn't depend on $q \in Q$, while the lift of potential function, $\tilde{\beta}_D$, smoothly depends on q on $\tilde{V}$. By Proposition 5 (in Section 3.3), $\tilde{e}_0$ and solution to (3), $\tilde{u}_t$ $(t > 0)$, are smooth functions on $\tilde{V}$ (where Q can be replaced by smaller normal section), and they are lifts of leaf-wise smooth functions e_0 and u_t $(t > 0)$ on V or on a smaller neighborhood W of F. By Proposition 4, the vector fields $\tilde{H}_t$ and $\tilde{\bar{H}}$ are smooth on $\tilde{V}$, and they are lifts of smooth vector fields H_t and $\bar{H}$ on a neighborhood of F. By the above we conclude that the metrics g_t and $\bar{g}$ are smooth on M.   □

Proof of Corollary 1. By Corollary 3 (in Section 4.2), $\mathcal{F}$ is conformal w.r.t. the limit metric $\bar{g}$. By Proposition 2, the mean curvature vector of D is a unique global solution to the Burgers equation $\partial_t H + \nabla^{\mathcal{F}} g(H, H) = n \nabla^{\mathcal{F}} (\mathrm{div}_{\mathcal{F}} H)$, see (5). The last is reduced to the heat equation $\partial_t u = n \Delta_{\mathcal{F}} u$, where $H = -n \nabla^{\mathcal{F}} (\log u)$. In this case, we have $\lambda_0 = 0$ and $e_0 = \mathrm{const}$ on the leaves. Hence, $\bar{H} = 0$ (i.e., w.r.t. $\bar{g}$) that yields the vanishing of $\bar{b}$, and we conclude that $\mathcal{F}$ is Riemannian.

By Lemma 3, the mixed scalar curvature of $\bar{g}$ is positive: $\mathrm{Sc}_{\mathrm{mix}}(\bar{g}) = \|T\|_{\bar{g}}^2 > 0$. By (5) and $\bar{b} = 0$, we have $\nabla^{\mathcal{F}} \|T\|_{\bar{g}} = 0$, hence $\mathrm{Sc}_{\mathrm{mix}}(\bar{g})$ is constant on the leaves.   □

Proof of Corollary 2. By Corollary 3, the flow of metrics (3) preserves the twisted product structure. We prove (i) $\Rightarrow$ (ii), the implication (ii) $\Rightarrow$ (i) can be shown similarly. By Proposition 2 with $\beta_D = 0$, the mean curvature vector H of the fibers $M_1 \times \{y\}$ satisfies the Burgers equation (5). By (3) with $\beta_D = 0$ and $H = -n \nabla^{\mathcal{F}} (\log f)$, one may reduce (5) to the heat equation $\partial_t f = n \Delta_{\mathcal{F}} f$ for the warping function $f > 0$ (along the fibers $\{x\} \times M_2$). Thus $\lambda_0 = 0$ and $\lambda_1 > 0$. Consider the fibre bundle $\pi : M \to M_1$ with totally geodesic fibers $F_x = \{x\} \times M_2$. As in the step 1 of the proof of Theorem 1 for the potential function $\psi_0 = -n \log f$ (at $t = 0$) (see also Section 3.2), we conclude that $\bar{H} = 0$ for the limit metric $\bar{g}$. Since the canonical foliation $M_1 \times \{y\}$ is $\bar{g}$-totally umbilical, by the above we have $\bar{b} = 0$, i.e., $M_1 \times \{y\}$ is $\bar{g}$-totally geodesic. By de-Rham decomposition theorem, M is the metric product with respect to $\bar{g} = \bar{f}^2 g_1 \times g_2$.   □

4.4. *One-dimensional case*

We discuss applications of main results to a Riemannian manifold with a unit vector field N. In this case, $\mathrm{Sc}_{\mathrm{mix}}(g)$ coincides with the Ricci curvature

$\mathrm{Ric}_g(N, N)$ of (M, g) in the N-direction.

Case $n = 1$. Let a unit vector field N be orthogonal to a compact totally geodesic foliation $\mathcal{F}$. Hence, the operator (1) coincides with the leaf-wise Laplacian, $\lambda_0 = 0$ and $e_0 = \mathrm{const}$. Certainly, we have $T = 0$ and $\beta_D = 0$, and (3) reads as

$$\mathrm{Ric}_g(N, N) = \mathrm{div}_{\mathcal{F}} H - g(H, H). \tag{11}$$

Assume that the curvature vector H of N-curves is leaf-wise conservative. Let the metric evolves by

$$\partial_t g = -2 \, \mathrm{Ric}_g(N, N) \, \hat{g},$$

with $\Phi = 0$ (see (3)). Then H obeys the homogeneous Burgers equation, (see (2)),

$$\partial_t H + \nabla^{\mathcal{F}} g(H, H) = \nabla^{\mathcal{F}}(\mathrm{div}_{\mathcal{F}} H), \tag{12}$$

where $H = -\nabla^{\mathcal{F}}(\log u)$ and the function $u > 0$ obeys the heat equation on the leaves, (see (3)),

$$\partial_t u = \Delta_{\mathcal{F}} u. \tag{13}$$

For the global solution of (13) we have $\lim_{t \to \infty} u(t, x) = \bar{u}(x) = \int_{F(x)} u_0(x) \, dx \, / \, \mathrm{vol}\, F(x)$. Hence $\bar{H} = \lim_{t \to \infty} H(t, \cdot) = 0$ and $\mathrm{Ric}_{\bar{g}}(N, N) = \lim_{t \to \infty} \mathrm{Ric}_{g_t}(N, N) = 0$.

Case $p = 1$. Let N be a unit vector field tangent to a foliation $\mathcal{F}$ by closed geodesics.

Denote by $h = \mathrm{Tr}_g b$ the mean curvature of the distribution D. Certainly, (3) reads as

$$\mathrm{Ric}_g(N, N) = \|T\|_g^2 + N(h) - \frac{1}{n} h^2 - n \, \beta_D. \tag{14}$$

Suppose that $h = -N(\log u_0)$ for a smooth function $u_0 > 0$ on M. Let the metric evolves by

$$\partial_t g = -2 \, (\mathrm{Ric}_g(N, N) - \|T\|_g^2 - \Phi) \, \hat{g}, \tag{15}$$

see (3). Then h obeys the inhomogeneous Burgers equation on the leaves (circles), see (2),

$$\partial_t h + N(h^2) = n \, N(N(h)) - n^2 \, N(\beta_D), \tag{16}$$

where $h(\cdot, t) = -N(\log u(\cdot, t))$ and the function $u > 0$ obeys the heat Schrödinger equation, see (3),

$$\partial_t u = N(N(u)) + n \, \beta_D \, u \tag{17}$$

with initial condition $u(\cdot, 0) = u_0$. By Theorem 1, the flow of metrics (15) has a unique global solution, and the choice $\Phi = n \lambda_0$ yields the exponential convergence of a solution g_t as $t \to \infty$ with the rate $\lambda_1 - \lambda_0$. Moreover, there exists real $c > 0$ such that $\text{Ric}_{\bar{g}}(N, N) > 0$ when $g_0 \in [g]_c$ and

$$\|T(x)\|_{g_0}^2 > n^2 c^4 \max_{F(x)} \beta_D \quad (x \in M). \tag{18}$$

If the leaves have finite holonomy group then g_t $(t > 0)$ and $\bar{g}$ are smooth on M.

4.5. Applications to surfaces

The results of the section can be easily generalized for foliated hypersurfaces.

A *partially geodesic semisurface* is a surface (M^2, g) with two orthogonal families of curves, one of which consists of geodesics, see [2]. The metric in biregular foliated coordinates (x, ϑ) is $g = dx^2 + G \, d\vartheta^2$, where G is a positive function. A surface (M^2, g) with a geodesic unit vector field N is a partially geodesic semisurface, and ∂_x is the N-derivative.

Let $k, K \in C^2(M)$ be the curvature of ϑ-curves and the gaussian curvature of M^2, respectively. The co-nullity and Jacobi operators (see Section 4.1) have the form $C(X) = k \cdot X$ and $R_N(X) = K \cdot X$ for $X \perp N$, respectively. Equation (3) with $\Phi = 0$ reduces to (4). By Lemmas 5–6, we obtain PDEs

$$\partial_t K = K_{,xx} - 2\, k\, K_{,x}, \qquad \partial_t k = K_{,x}. \tag{19}$$

For $n = 1$, (2) reads as the Riccati equation $k_{,x} = k^2 + K$ along N-curves (the leaves). Substituting $K = k_{,x} - k^2$ into $(19)_2$, we obtain the Burgers equation for k (along N-curves)

$$\partial_t k = k_{,xx} - (k^2)_{,x}, \tag{20}$$

which also follows from (2) with $\beta_D = 0$. Finally, we recover the metric as $\hat{g}_t = \hat{g}_0 \exp\left(-2 \int_0^t K \, dt\right)$.

The following class of partially geodesic semisurfaces extends the class of surfaces of revolution.

Definition 5. (see [2]) A surface $M^2 \subset \mathbb{R}^3$ is said to be *parallel curved* (PC) if there exists a plane P (called a *base plane* of M^2) such that at each point of M^2, at least one principal direction is parallel to P. A PC surface is called *canonical* if there exist smooth functions f_i, ρ, h of one variable satisfying $(f_1')^2 + (f_2')^2 = 1$ on I, $(\rho')^2 + (h')^2 = 1$ on $[0, l]$, and M^2 (after a rigid motion such that a base plane P is XOY) is the image of the map $r(x, \vartheta) = [f_1(\vartheta) + \rho(x)f_2'(\vartheta), \; f_2(\vartheta) - \rho(x)f_1'(\vartheta), \; h(x)]$.

The first fundamental form of a canonical PC surface is a special case of twisted products:

$$g = dx^2 + \left(1 + k_0(\vartheta)\rho(x)\right)^2 d\vartheta^2, \tag{21}$$

where $k_0(\vartheta)$ is the curvature of a plane curve defined by (f_1, f_2). A canonical PC surface is represented as a disjoint union of x-curves (plane curves, geodesics on M^2) which are congruent in $\mathbb{R}^3$ with one another and tangent to the principal directions of M^2. A line of curvature on a PC surface without umbilical points which is not contained in any base plane is a geodesic.

A surface of revolution is PC: any plane orthogonal to an axis of rotation is a base plane and profile curves are congruent geodesics. The following proposition generalizes Example 3(c).

Proposition 7. *Let $M_t^2 \subset \mathbb{R}^3$ $(t \geq 0)$ be a smooth family of canonical PC surfaces*

$$r_t(x, \vartheta) = [f_1(\vartheta)+\rho(x,t)f_2'(\vartheta), \; f_2(\vartheta)-\rho(x,t)f_1'(\vartheta), \; h(x,t)], \quad \vartheta \in I, x \in [0,l]$$

without umbilics and with $K \neq 0$. Then the following properties are equivalent:

(i) *the induced metrics $g_t = dx^2 + \left(1 + k_0(\vartheta)\rho(x,t)\right)^2 d\vartheta^2$ solve $\partial_t g = -2\,K(g)\,\hat{g}$, see (4);*

(ii) *the function ρ satisfies the heat equation $\rho_{,t} = k_0(\vartheta)\rho_{,xx}$ (when $k_0(\vartheta) > 0$);*

(iii) *the geodesic curvature k of ϑ-curves obeys the Burgers equation $k_{,t} = k_0(\vartheta)(k_{,xx} - (k^2)_{,x})$.*

In any of cases (i) – (iii), the metrics g_t converge as $t \to \infty$ in C^∞ with the exponential rate λ_1 to the flat metric $\bar{g} = dx^2 + \left(1 + k_0(\vartheta)\bar{\rho}\right)^2 d\vartheta^2$ of the canonical PC surface $\bar{r}(x, \vartheta) = [f_1(\vartheta) + \bar{\rho}f_2'(\vartheta), \; f_2(\vartheta) - \bar{\rho}f_1'(\vartheta), \; \bar{h}]$, where $\bar{\rho} = \rho_1 + x\rho_2, \; \bar{h} = h_1 + xh_2, \; \rho_i, h_i \in \mathbb{R}$ and $\rho_2 = \rho_1 + \int_0^l \rho(x,0)\,dx$, $h_2^2 = 1 - \rho_2^2$.

Proof. This is similar to Corollary 2. One may show that $K = -Q_{,xx}/Q$, and $k = -Q_{,x}/Q$, where $Q = 1 + k_0(\vartheta)\rho(x,t)$. Assuming (4), from GF equation $\partial_t(Q^2) = -2KQ^2$ we deduce PDE $\rho_{,t} = k_0(\vartheta)\rho_{,xx}$. Finally, using Riccati equation $k_{,x} = k^2 + K$, we obtain the Burgers equation for k. The convergence $g_t \to \bar{g}$ and the formula for $\bar{r}$ follow from the standard heat equation theory. $\square$

Example 4. The induced metric on a surface of revolution in $\mathbb{R}^3$ is a special class of warped products, see Definition 1. We are looking for a one-parameter family M_t^2 of such surfaces, which are foliated by profile curves, and the induced metric g_t obeys (4). The profile of M_0^2 parameterized as in Example 3(c) is XZ-plane curve $\gamma_0 = r(\cdot, 0)$ (a leaf), and ϑ-curves are circles in $\mathbb{R}^3$. One may assume that $\rho > 0$ and $|\rho_{,x}| < 1$ at $t = 0$. Let x be the natural parameter of $\gamma_t = r(\cdot, t)$. Thus $N = r_{,x}$ is the unit normal to ϑ-curves on M_t^2. The geodesic curvature, k, of ϑ-curves obeys the Burgers equation (20), while the radius, ρ, of ϑ-curves (as Euclidean circles) satisfies the heat equation, both functions are related by the Cole-Hopf transformation $k = -(\log \rho)_{,x}$. The induced metric $g_t = dx^2 + \rho^2 d\vartheta^2$ on M_t has the rotational symmetric form, its gaussian curvature is $K = -\rho_{,xx}/\rho$. The flow of metrics $\partial_t g = -2K \hat{g}$ reduces to $\rho_{,t} = -K\rho$. Since $(\rho_{,x})_{,t} = (\rho_{,x})_{,xx}$, by the maximum principle, we have the inequality $|\rho_{,x}| < 1$ for all $t \geq 0$. When such a solution $\rho(x, t)$ ($t \geq 0$) is known, we find $h = \int \sqrt{1 - (\rho_{,x})^2}\, dx$.

Suppose that the boundary conditions are $\rho(0, t) = \rho_1$, $\rho(l, t) = \rho_2$ and $h(0, t) = h_1$, where $\rho_2 \geq \rho_1 > 0$ and $t \geq 0$. By the heat equation theory, the solution ρ approaches as $t \to \infty$ to a linear function $\bar{\rho} = x\rho_1 + (l - x)\rho_2 > 0$. Also, h approaches as $t \to \infty$ to a linear function $\bar{h} = xh_1 + (l-x)h_2$, where h_2 may be determined from the equality $(\rho_2 - \rho_1)^2 + (h_2 - h_1)^2 = l^2$. The curves γ_t are isometric one to another for all t (with the same arc-length parameter x). The limit curve $\lim_{t \to \infty} \gamma_t = \bar{\gamma} = [\bar{\rho}, \bar{h}]$ is a line segment of length l. Thus, M_t approach as $t \to \infty$ to the flat surface of revolution $\bar{M}$ – the patch of a cone or a cylinder generated by $\bar{\gamma}$.

5. Appendix: Operators in Banach spaces

The section contains results on operators in Banach spaces, which are used in Section 3.

5.1. *An isolated eigenvalue of a linear operator with parameter*

Let $\mathcal{O}$ be an open neighborhood of origin (e.g., a ball of a small radius) in a real Banach space Q.

Proposition 8. *Let $\{\mathcal{H}(q)\}_{q \in \mathcal{O}}$ be a family self-adjoint linear operators acting in a complex Hilbert space E and satisfying the following conditions:*

(a) all $\mathcal{H}(q)$ have a common domain of definition $\mathcal{D} \subseteq E$;
(b) $\mathcal{P}(q) := \mathcal{H}(q) - \mathcal{H}(0)$ is bounded for any $q \in \mathcal{O}$, and $\mathcal{P}(\cdot) \in C^j(\mathcal{O}, \mathcal{B}(E))$
for some $j \in \mathbb{N}$.

Suppose that $\mathcal{H}(0)$ has a simple eigenvalue λ_0 which is an isolated point of the spectrum of $\mathcal{H}(0)$, and $e_0 \in S_1$ is an eigenvector of $\mathcal{H}(0)$, corresponding to λ_0. Then there exist open neighborhoods $U \subseteq \mathbb{R}$ of λ_0 and $W \subseteq \mathcal{O}$ of origin (which does not depend on j) and functions

$$e(\cdot) \in C^j(\mathcal{W}, E), \qquad \lambda(\cdot) \in C^j(\mathcal{W}, \mathbb{R}), \tag{1}$$

such that $\lambda(0) = \lambda_0$, $e(0) = e_0$, and for any $q \in \mathcal{W}$ the real $\lambda(q)$ is a unique in U point of the spectrum of $\mathcal{H}(q)$, it is a simple eigenvalue of $\mathcal{H}(q)$, and $e(q) \in S_1$ is an eigenvector related to $\lambda(q)$.

Proof. Let Γ be a circle of small radius in the complex plane $\mathbb{C}$ not intersecting with $\sigma(\mathcal{H}(0))$ and surrounding only $\lambda_0 \in \sigma(\mathcal{H}(0))$. Hence there exists an open neighborhood $\mathcal{W}_1 \subseteq \mathcal{O}$ of origin such that the claims (i) and (ii) of Lemma 8 are valid. The above implies that the Riesz projection

$$P(q) = -\frac{1}{2\pi i} \oint_\Gamma (\mathcal{H}(q) - \mu\,\mathrm{id})^{-1}\,\mathrm{d}\mu \qquad (q \in \mathcal{W}_1)$$

onto the invariant subspace of $\mathcal{H}(q)$ corresponding to the part of its spectrum lying inside of Γ ([15, Introduction, Sect. 4]) has the property: the mapping $q \to P(q)$ belongs to $C^j(\mathcal{W}_1, \mathcal{B}(E))$. In particular, there exists an open neighborhood $\mathcal{W} \subseteq \mathcal{W}_1$ of origin such that $\|P(q) - P(0)\|_{\mathcal{B}(E)} \leq \frac{1}{2}$ for any $q \in \mathcal{W}$, hence $\dim(\mathrm{Im}\,P(q)) = \dim(\mathrm{Im}\,P(0))$, see [1, Chapt. III, Sect. 34]. Since λ_0 is a simple eigenvalue of $\mathcal{H}(0)$, $P(q)$ is an orthogonal projection onto one-dimensional subspace of E for any $q \in \mathcal{W}$, that is each operator $\mathcal{H}(q)$ has a unique point $\lambda(q)$ of the spectrum inside of the circle Γ, it is a simple eigenvalue of $\mathcal{H}(q)$, it is real because $\mathcal{H}(q)$ is self-adjoint, $\mathrm{Im}\,P(q)$ is the eigenspace of $\mathcal{H}(q)$ corresponding to $\lambda(q)$, and $\lambda(0) = \lambda_0$. Since the mapping $q \to P(q)e_0$ belongs to $C^j(\mathcal{W}, E)$ and

$$\|P(q)e_0\|_E \geq \|P(0)e_0\|_E - \|(P(q) - P(0))e_0\|_E \geq 1/2$$

for any $q \in \mathcal{W}$, then the vector-function $e(q) = \frac{P(q)\,e_0}{\|P(q)\,e_0\|_E}$ obeys $(1)_1$. Evidently, $e(0) = e_0$ and $e(q) \in S_1$ for any $q \in \mathcal{W}$. The inclusion $(1)_2$ follows from the obvious equality $(\lambda(q) - \gamma)((\mathcal{H}(q) - \gamma\,\mathrm{id})^{-1}e(q), e(q))_E = 1$, where $\mathrm{Im}\,\gamma \neq 0$, the inclusion $(1)_1$ and the fact that by Lemma 8 the mapping $q \to (\mathcal{H}(q) - \gamma\,\mathrm{id})^{-1}$ belongs to $C^j(\mathcal{W}, \mathcal{B}(E))$. □

In the proof of Proposition 8 we have used the following lemmas:

Lemma 7. *Let* $\{L(q)\}_{q \in \mathcal{O}}$ *be a family of linear bounded operators acting in a real or complex Banach E. Denote by $V^j := C^j(\mathcal{O}, \mathcal{B}(E))$ and suppose that*

(a) $L(\cdot) \in V^j$ for some $j \in \mathbb{N}$;   (b) $r := \|L(\cdot)\|_{V^0} < 1$.

Then for any $q \in \mathcal{O}$ the operator $\mathrm{id} + L(q)$ is continuously invertible, the operator function $M(q) := (\mathrm{id} + L(q))^{-1}$ belongs to V^j, and for some $c_j(r) > 0$ we have

$$\|M(\cdot)\|_{V^j} \leq c_j(r) \left(1 + \|L(\cdot)\|_{V^j}\right)^j. \tag{2}$$

Proof. As is known, under condition (b) the operator $\mathrm{id} + L(q)$ is continuously invertible for any $q \in \mathcal{O}$, and the von Neumann expansion (which converges in the $\mathcal{B}(E)$-norm) is valid:

$$M(q) = \sum\nolimits_{n=0}^{\infty} (-1)^n L(q)^n. \tag{3}$$

Using condition (a), let us calculate the differentials of the terms of this series:

$$d_v^m L(q)^n \tag{4}$$

$$= \sum_{\substack{1 \leq k_1 < \\ \dots < k_s \leq n}} \sum_{\substack{(S_1,\dots,S_s) \\ \in \Pi_{s,m}}} L_1(q) \dots d_{S_1} L_{k_1}(q) \dots d_{S_s} L_{k_s}(q) \dots L_n(q)_{\,|L_i(q) \equiv L(q)} \; {\scriptstyle (i=1,\dots n)}$$

for $v = (v_1, \dots, v_m) \in Q$, where $\Pi_{s,m}$ is the set of ordered partitions $(S_1, \dots, S_s)$ of $(v_1, \dots, v_m)$ formed by s subsequences, and if $S_i = (v_{j_1}, \dots, v_{j_{r(i)}})$ then we denote $d_{S_i} := d_{v_{j_1}, \dots, v_{j_{r(i)}}}^{r(i)}$. In this case we have $\sum_{i=1}^{s} \sum_{a=1}^{r(i)} v_{j_a} = m$. Using (a)–(b), we obtain from (4) the estimate:

$$\|d^m L(q)^n\|_{\mathcal{B}^j(E)} \leq (1 + \|L(\cdot)\|_{V^j})^m \, 2^{m^2} \begin{cases} n^m r^{n-m} & \text{for } n > m, \\ m^m & \text{for } 0 \leq n \leq m \end{cases} \tag{5}$$

for any $q \in \mathcal{O}$. Thus the series in (3) converges in the V^j-norm. Hence $M(\cdot) \in V^j$ and (2) is valid with a coarse estimate $c_j(r) = 2^{j^2}((j+1)j^j + \sum_{n=1}^{\infty}(n+j)^j r^n)$. $\quad\square$

Lemma 8. *Let* $\{\mathcal{H}(q)\}_{q \in \mathcal{O}}$ *be a family self-adjoint linear operators acting in a complex Hilbert space E such that conditions (a)–(b) of Proposition 8 are satisfied. Suppose that the inequality*

$$\inf_{\mu \in \Gamma} \mathrm{dist}(\mu, \sigma(\mathcal{H}(0))) > 0 \tag{6}$$

holds for a subset $\Gamma \subset \mathbb{C}$. Then there exists an open neighborhood $\mathcal{W} \subseteq \mathcal{O}$ of origin (which does not depend on the integer j from condition (b) of Proposition 8) such that:

(i) $\sigma(\mathcal{H}(q)) \cap \Gamma = \emptyset$ *for any* $q \in \mathcal{W}$;

(ii) the mapping $\mu \to (\mathcal{H}(\cdot) - \mu\,\mathrm{id})^{-1}$ *(where* $\mu \in \Gamma$*) belongs to* $C(\Gamma, C^j(\mathcal{W}, \mathcal{B}(E)))$.

Proof. Denote by $V^j := C^j(\mathcal{W}, \mathcal{B}(E))$. Since the operator $\mathcal{H}(0)$ is self-adjoint, we obtain $\|(\mathcal{H}(0) - \mu\,\mathrm{id})^{-1}\|_{\mathcal{B}(E)} \leq 1/\mathrm{dist}(\mu, \sigma(\mathcal{H}(0)))$. Hence, and in view of (6), we have

$$\bar{H} := \sup_{\mu \in \Gamma} \|(\mathcal{H}(0) - \mu\,\mathrm{id})^{-1}\|_{\mathcal{B}(E)} < \infty. \tag{7}$$

The following obvious representation holds:

$$\mathcal{H}(q) - \mu\,\mathrm{id} = (\mathcal{H}(0) - \mu\,\mathrm{id})(\mathrm{id} + L(q, \mu)) \quad (q \in \mathcal{O}), \tag{8}$$

where $L(q, \mu) := (\mathcal{H}(0) - \mu\,\mathrm{id})^{-1}\mathcal{P}(q)$ and $\mathcal{P}(q) = \mathcal{H}(q) - \mathcal{H}(0)$. By condition (b) of Proposition 8, $L(\cdot, \mu) \in V^j$ for any $\mu \in \Gamma$, and there exists an open neighborhood $\mathcal{W} \subseteq \mathcal{O}$ of origin such that

$$\sup_{\mu \in \Gamma} \|L(\cdot, \mu)\|_{V^0} \leq 1/2. \tag{9}$$

By Lemma 7, $\mathrm{id} + L(q, \mu)$ is continuously invertible for any $q \in \mathcal{W}$ and $\mu \in \Gamma$, and we have

$$(\mathrm{id} + L(\cdot, \mu))^{-1} \in V^j \qquad \forall\, \mu \in \Gamma. \tag{10}$$

Then, in view of (8), we conclude that claim (i) is valid and

$$(\mathcal{H}(q) - \mu\,\mathrm{id})^{-1} = (\mathrm{id} + L(q, \mu))^{-1}(\mathcal{H}(0) - \mu\,\mathrm{id})^{-1} \quad (q \in \mathcal{W},\ \mu \in \Gamma). \tag{11}$$

The above and (10) imply $(\mathcal{H}(\cdot) - \mu\,\mathrm{id})^{-1} \in V^j$ for $\mu \in \Gamma$. Observe that (7) yields

$$\sup_{\mu \in \Gamma} \|L(\cdot, \mu)\|_{V^j} \leq \bar{H}\|\mathcal{P}(\cdot)\|_{V^j}. \tag{12}$$

In view of the equality

$$(\mathcal{H}(0) - \mu\,\mathrm{id})^{-1} - (\mathcal{H}(0) - \mu'\,\mathrm{id})^{-1} = (\mu - \mu')(\mathcal{H}(0) - \mu\,\mathrm{id})^{-1}(\mathcal{H}(0) - \mu'\,\mathrm{id})^{-1}$$
$$(\mu, \mu' \in \Gamma)$$

and the property (7), we obtain

$$(\mu \to (\mathcal{H}(0) - \mu\,\mathrm{id})^{-1}) \in C(\Gamma, \mathcal{B}(E)). \tag{13}$$

Then, by (b) of Proposition 8, the mapping $\mu \to L(\cdot, \mu)$ belongs to $C(\Gamma, V^j)$. Hence

$$(\mu \to L(\cdot, \mu)^n) \in C(\Gamma, V^j), \quad n \in \mathbb{N}. \tag{14}$$

On the other hand, in view of (9), the von Neumann expansion is valid:

$$(\mathrm{id} + L(q,\mu))^{-1} = \sum_{n=0}^{\infty} (-1^n)\, L(q,\mu)^n, \qquad (15)$$

which converges in $\mathcal{B}(E)$-norm for any $q \in \mathcal{W}$ and $\mu \in \Gamma$. Note that for any $n \in \mathbb{N}$ and $m \in \{1, 2, \dots, j\}$, the partial differential $\partial_q^m (L(q,\mu))^n$ obeys the inequality (5) with $L(q,\mu)$ instead of $L(q)$ and $r = \frac{1}{2}$. Thus, in view of (12) and (14), the series in (15) converges in $C(\Gamma, V^j)$-norm. Then, using (11) and (13), we obtain the desired claim (ii). $\qquad\qquad\square$

5.2. *More lemmas*

Again, let $\mathcal{O}$ be an open neighborhood of origin in a real Banach space Q.

Lemma 9. *Let $\{\mathcal{H}(q)\}_{q \in \mathcal{O}}$ be a family of linear operators acting in a real or complex Banach space E, and G_1, G_2 be Banach spaces such that $G_2 \hookrightarrow G_1 \hookrightarrow E$. Suppose that $\mathcal{H}(0)$ is continuously invertible and the following conditions are satisfied:*

(a) $\mathcal{H}(0)_{|G_1}^{-1} \in \mathcal{B}(G_1, G_2)$;

(b) the operator $\mathcal{P}(q) := \mathcal{H}(q) - \mathcal{H}(0)$ has the properties

$$\mathcal{P}(\cdot) \in C(\mathcal{O}, \mathcal{B}(E)), \qquad \mathcal{P}(\cdot)_{|G_1} \in C^j(\mathcal{O}, \mathcal{B}(G_1)), \qquad (16)$$

for some integer $j \geq 0$. Then there exists an open neighborhood $\mathcal{W} \subseteq \mathcal{O}$ of origin (which does not depend on j) such that $\mathcal{H}(\cdot)_{|G_1}^{-1} \in C^j(\mathcal{W}, \mathcal{B}(G_1, G_2))$.

Proof. From $(16)_1$ we find that there exists an open neighborhood $\mathcal{W}_1 \subseteq \mathcal{O}$ of origin such that $\sup_{q \in \mathcal{W}_1} \|L(q)\|_{\mathcal{B}(E)} < 1$, where $L(q) := \mathcal{H}(0)^{-1}\mathcal{P}(q)$. By Lemma 7, for any $q \in \mathcal{W}_1$ the operator $\mathrm{id}_E + L(q)$ is continuously invertible in E. As in the proof of Lemma 8, we get $\mathcal{H}(q)^{-1} = (\mathrm{id}_E + L(q))^{-1}\mathcal{H}(0)^{-1}$. Using the von Neumann expansion, we find

$$\mathcal{H}(q)^{-1} = \mathcal{H}(0)^{-1} + \sum_{n=1}^{\infty} (-1)^n L(q)^n \mathcal{H}(0)^{-1}$$
$$= \mathcal{H}(0)^{-1} - \mathcal{H}(0)^{-1}\mathcal{P}(q)(\mathrm{id}_E + L(q))^{-1}\mathcal{H}(0)^{-1}. \quad (17)$$

Conditions (a) and $(16)_2$ imply $L(\cdot)_{|G_1} \in C^j(\mathcal{O}, \mathcal{B}(G_1, G_2))$, hence in view of the embedding $G_2 \hookrightarrow G_1$, we have $L(\cdot)_{|G_1} \in C^j(\mathcal{O}, \mathcal{B}(G_1))$. The last fact and Lemma 7 yield that there exists a neighborhood $\mathcal{W} \subseteq \mathcal{W}_1$ of origin such that the operator $(\mathrm{id}_{G_1} + L(q))_{|G_1}$ is continuously invertible in G_1 for any $q \in \mathcal{W}$, and

$$(\mathrm{id}_{G_1} + L(q))_{|G_1}^{-1} = (\mathrm{id}_E + L(q))_{|G_1}^{-1} \in C^j(\mathcal{W}, \mathcal{B}(G_1)). \qquad (18)$$

Using condition (a) and the embedding $G_2 \hookrightarrow G_1$, we conclude that $\mathcal{H}(0)^{-1}_{\rceil G_1} \in \mathcal{B}(G_1)$. This fact, the representation (17), condition (a) and the inclusions $(16)_2$, (18) imply the desired claim. $\qquad\square$

Lemma 10. *Let* $\varphi \in C^k$ $(k \geq 0)$. *Then the multiplication operator* $(\mathcal{M}_\varphi u) := \varphi \cdot u$ $(u \in H^k)$ *belongs to* $\mathcal{B}(H^k)$ *and*

$$\|\mathcal{M}_\varphi\|_{\mathcal{B}(H^k)} \leq c_k \|\varphi\|_{C^k} \tag{19}$$

for some constant $c_k > 0$ *which doesn't depend on* φ.

Proof. Let $k = 0$. We have for any $u \in H^0 = L_2$:

$$\|\mathcal{M}_\varphi u\|_0 = \Big(\int_F \varphi^2(x)\, u^2(x)\, \mathrm{d}x \Big)^{1/2} \leq \|\varphi\|_C \, \|u\|_0, \tag{20}$$

that is (19) is proved for the case $k = 0$. Now consider the case $k > 0$. Recall that the Sobolev space H^k consists of all functions $u \in L_2$, whose weak derivatives up to the order k belong to L_2 and the norm in H^k is defined as follows: $\|u\|^2_k = \sum_{|\alpha| \leq k} \|d^\alpha u\|^2_0$. Let us take $u \in H^k$. Using the Leibnitz rule for the weak derivative $d^\alpha(\varphi \cdot u) = \sum_{\beta \leq \alpha} \beta!\, d^{\alpha-\beta}\varphi \cdot d^\beta u$, where $\beta!$ are the multinomial coefficients, and carrying out the estimates, similarly to (20) we get the estimate

$$\|\mathcal{M}_\varphi u\|^2_k = \sum_{|\alpha| \leq k} \|d^\alpha(\varphi \cdot u)\|^2_0 \leq \|\varphi\|^2_{C^k} \sum_{|\alpha| \leq k,\, \beta \leq \alpha} (\beta!)^2 \sum_{\beta \leq \alpha} \|d^\beta u\|^2_0,$$

that proves the desired estimate (19), in which $c_k > 0$ doesn't depend on φ. $\qquad\square$

Lemma 11. *Let a real-valued function* $\varphi(x,q)$ $(x \in F, q \in \mathcal{O})$ *satisfies the conditions:*

 (a) $\varphi(\cdot, q)$ *belongs to* C^k *for any* $q \in \mathcal{O}$;

 (b) the mapping $\mathcal{G} : q \to \varphi(\cdot, q) \in C^k$ *belongs to* $C^j(\mathcal{O}, C^k)$ *for some* $j \in \mathbb{N}$.

Then $\mathcal{M}_{\varphi(\cdot,q)} \in \mathcal{B}(H^k)$ $(q \in \mathcal{O})$ *and* $\mathcal{S} : q \to \mathcal{M}_{\varphi(\cdot,q)} \in \mathcal{B}(H^k)$ *belongs to* $C^j(\mathcal{O}, \mathcal{B}(H^k))$.

Proof. By Lemma 10, the mapping $\mathcal{R} : g \to \mathcal{M}_g$ is a linear bounded operator from C^k into $\mathcal{B}(H^k)$. Then the representation $\mathcal{S} = \mathcal{R} \circ \mathcal{G}$ completes the proof. $\qquad\square$

References

[1] N.I. Akhieser and I.M. Glazman, Theory of Linear Operators In Hilbert Space, Dover Publications, INC, New York, 1993.

[2] N. Ando, *A two-dimensional Riemannian manifold with two one-dimensional distributions*, Kyushu J. Math., **59** (2005), 285–299.

[3] M. Ashbaugh, *The Fundamental Gap*, in Low eigenvalues of Laplace and Schrödinger operators, ARCC Workshop, Palo Alto, California, May 22–26, 2006.

[4] T. Aubin, Some nonlinear problems in Riemannian geometry, Springer, 1998.

[5] A. Bejancu and H. Farram, *On totally geodeasic foliations with bundle-like metric*, J. Geom., **85** (2006), 7–14.

[6] A. Candel and L. Conlon, Foliations I, II, AMS, Providence, 2000.

[7] M. Czarnecki and P.G. Walczak, *Extrinsic geometry of foliations*, in Foliations 2005, World Scientific Publ., NJ, 2006, 149–167.

[8] B. Chow and D. Knopf, The Ricci Flow: An Introduction, AMS, 2004.

[9] P. Dombrowski, *Jacobi fields, totally geodesic foliations and geodesic differential forms*, Resultate Math., **1** (2) (1978), 156–194.

[10] N. Dunford and J.T. Schwartz, Linear Operators, Part 1 and 2, Intersc. Publ., New York, 1963.

[11] R. Escobales, *Riemannian submersions with totally geodesic fibers*, J. Diff. Geom., **10** (1975), 253–276.

[12] E. Ghys, *Classification of totally geodesic foliations of codimension one*, Comment. Math. Helv., **58** (4) (1983), 543–572.

[13] W. Kirsch and B. Simon, *Approach to equilibrium for a forced Burgers equation*, J. Evol. Equ., **1** (4) (2001), 411–419.

[14] M.A. Krasnoselskii, Positive solutions of operator equations, Gröningen, Noordhoff, 1964.

[15] S.G. Krein, Linear Differential Equations in Banach Space, Transl. of Math. Monographs, **29**, AMS, Providence, 1972.

[16] P.D. Lax and A.N. Milgram, *Parabolic equations, Contributions to the theory of partial differential equations*, Annals of Math. Studies, Princeton University Press, Princeton, N. J., **33** (1954), 167–190.

[17] R. Ponge and H. Reckziegel, *Twisted products in pseudo-Riemannian geometry*, Geom. Dedicata, **48** (1993), 15–25.

[18] A. Ranjan, *Structural equations and an integral formula for foliated manifolds*, Geom. Dedicata, **20** (1986), 85–91.

[19] V. Rovenski, Foliations on Riemannian Manifolds and Submanifolds, Birkhäuser, 1998.

[20] V. Rovenski, *Extrinsic geometric flows on codimension-one foliations*, J. of Geom. Analysis, (2012), DOI 10.1007/s12220-012-9297-1.

[21] V. Rovenski and P. Walczak, Topics in Extrinsic Geometry of Codimension-One Foliations, Springer-Verlag, 2011.

[22] V. Rovenski and R. Wolak, *Deforming metrics of foliations*, Centr. Eur. J. Math., **11** (6) (2013), 1039–1055.

[23] P. Sachdev, Nonlinear Diffusive Waves, Cambridge University Press, Cambridge, 1987.

[24] I. Vaisman, *Conformal foliations*, Kodai Math. J., **2** (1979), 26–37.

[25] P. Walczak, *An integral formula for a Riemannian manifold with two orthogonal complementary distributions*, Colloq. Math., **58** (1990), 243–252.

[26] P. Walczak, *Conformally defined geometry on foliated Riemannian manifolds*, in Foliations 2005, World Sci. Publ., Hackensack, NJ, 2006, 431–439.

Received November 20, 2012.

FOLIATIONS 2012
ed. by Paweł WALCZAK *et al.*
World Scientific, Singapore, 2013
pp. 205–213

Tautness and the Godbillon-Vey class of foliations

Paweł G. Walczak[*]

Katedra Geometrii, Wydział Matematyki i Informatyki
Uniwersytet Łódzki, Łódź, Poland
e-mail: pawelwal@math.uni.lodz.pl

1. Introduction

The classical *Godbillon-Vey class* was defined first for foliations of codimension 1 [6]. If $\mathcal{F}$ is such a foliation on a compact manifold M and $\mathcal{F}$ is defined by the equation $\omega = 0$ for some 1-form ω, then its Godbillon-Vey class $\mathrm{gv}(\mathcal{F})$ lies in $H^3(M, \mathbb{R})$ and is represented by the form $\eta \wedge d\eta$, where η is such a 1-form on M that $d\omega = \eta \wedge \omega$. In codimension $q > 1$, a transversely oriented foliation $\mathcal{F}$ can be defined by the equation $\omega = 0$, where ω is a q-form. Again, $d\omega = \eta \wedge \omega$ for a 1-form η, the form $\eta \wedge (d\eta)^q$ is closed and represents the Godbillon-Vey class of $\mathcal{F}$: $\mathrm{gv}(\mathcal{F}) \in H^{2q+1}(M, \mathbb{R})$.

The history of studies of the Godbillon-Vey cohomology class is long and rich of interesting results, see, for example, [7]. Among the other results, let us mention the Duminy [4] (see [9] for a new proof and [3] for an exposition of the Duminy's work) theorem saying that if a codimension-one foliation $\mathcal{F}$ has the non-vanishing Godbillon-Vey class, then $\mathcal{F}$ contains a resilient leaf; this fact combined with the results of [5] (see also [18]) about entropy of foliations implies that any codimension-one foliation with non-zero Godbillon-Vey class has non-zero geometric entropy. Moreover,

---

[*]The author was supported by the Polish NSC grant N 6065/B/H03/2011/40

Thurston [17] gave a "helical wobble" description of the Godbillon-Vey invariant while Reinhart and Wood [10] provided a formula describing it in terms of Riemannian geometry. In codimension $q > 1$ the situation is less clear. Still in 2005, Hurder [8] decided to include in the list of open problems the following one (called by him "classical"):

Problem 1. Let $\mathcal{F}$ be a codimension $q > 1$ foliation with $\mathrm{gv}(\mathcal{F}) \neq 0$. What can be said about the geometry and dynamics of $\mathcal{F}$?

Our goal here is to provide a partial answer to this question.

2. A formula

Assume that $\mathcal{F}$ is a transversely oriented codimension q foliation of a compact oriented manifold M of dimension $n = 2q + 1$. Equip M with the Riemannian structure $g = \langle \cdot, \cdot \rangle$ and denote by ∇ the Levi-Civita connection on (M, g). Assume that the orthogonal complement $\mathcal{F}^\perp$ of $\mathcal{F}$ in TM is integrable and denote by $H^\perp$ its mean curvature vector:

$$H^\perp = \sum_{i=1}^{q} \left(\nabla_{E_i} E_i \right)^\top,$$

where $(E_1, \ldots, E_q)$ is a local, positive oriented, orthonormal frame of $\mathcal{F}^\perp$ and $v = v^\top + v^\perp$ denotes the decomposition of a vector $v \in TM$ into the components, respectively, tangent and orthogonal to $\mathcal{F}$. Denote also by ω the volume form of $\mathcal{F}^\perp$:

$$\omega(X_1, \ldots X_q) = \det[\langle X_i, E_j \rangle; i, j \leq q]$$

for any vector fields $X_1, \ldots, X_q$ on M.

Elementary calculation shows that

$$d\omega(X, E_1, \ldots, E_q) = -\langle X, H^\perp \rangle \tag{1}$$

for any X on M. Formula (1) may be called the *Rummler formula* since it appeared and was used already in 1979 in the paper [13]. Formula (1) shows that one may take

$$\eta = \langle H^\perp, \cdot \rangle$$

as the 1-form defining the Godbillon-Vey invariant.

With this η, one has $d\eta(X, Y) = \langle \nabla_X H^\perp, Y \rangle - \langle X, \nabla_Y H^\perp \rangle$ for all X and Y.

Assume now that we work within a set where $\eta \neq 0$. Take as before a positive oriented orthonormal frame $E_1, \ldots, E_q$ of vectors orthogonal to

$\mathcal{F}$ and complete it with the fields $E_{q+1}, \ldots, E_{2q}, E_{2q+1}$ in such a way that $E_{2q+1} = H^{\perp}/\|H^{\perp}\|$ and $E_1, \ldots, E_{2q+1}$ form a positive oriented orthonormal frame on M. Since $\eta(E_i) = 0$ for all $i \leq 2q$ and $d\eta(E_i, E_j) = 0$ for all $i, j \leq q$,

$$(\eta \wedge (d\eta)^q)(E_1, \ldots, E_{2q+1})$$
$$= \eta(E_{2q+1}) \cdot \sum_{\sigma \in S_q} \operatorname{sgn} \sigma \cdot d\eta(E_1, E_{q+\sigma_1}) \cdot \ldots \cdot \eta(E_q, E_{q+\sigma_q}),$$

where S_q denotes the group of all permutations of the set $\{1, \ldots, q\}$. In other words,

$$(\eta \wedge (d\eta)^q)(E_1, \ldots, E_{2q+1}) = \|H^{\perp}\| \cdot \det[\langle \nabla_{E_i} H^{\perp}, E_{q+j} \rangle - \langle E_i, \nabla_{E_{q+j}} H^{\perp} \rangle]$$

or

$$(\eta \wedge (d\eta)^q)(E_1, \ldots, E_{2q+1}) = \|H^{\perp}\|^2 \cdot \det[\langle \nabla_{E_i} N, E_{q+j} \rangle - \langle E_i, \nabla_{E_{q+j}} N \rangle],$$

where $N = E_{2q+1}$ is the *"principal normal"* of the foliation $\mathcal{F}^{\perp}$.

Next, let us denote by $B_{\mathcal{F}}$ the second fundamental form of (the leaves) of the foliation $\mathcal{F}$:

$$B_{\mathcal{F}}(X, Y) = (\nabla_X Y)^{\perp}$$

for arbitrary vector fields X and Y tangent to $\mathcal{F}$. Then, $\langle E_i, \nabla_{E_{q+j}} N \rangle = -\langle B_{\mathcal{F}}(E_{q+j}, N), E_i \rangle$ and we get the following.

Theorem 1. *The Godbillon-Vey class of a foliation $\mathcal{F}$ of a Riemannian manifold (M, g) ($\dim M = 2q + 1$, $\dim \mathcal{F} = q + 1$) with the orthogonal foliation $\mathcal{F}^{\perp}$ can be represented by the form $\eta \wedge (d\eta)^q$ given by*

$$(\eta \wedge (d\eta)^q)(E_1, \ldots, E_{2q+1})$$
$$= \|H^{\perp}\|^2 \cdot \det[\langle \nabla_{E_i} N, E_{q+j} \rangle + \langle B_{\mathcal{F}}(E_{q+j}, N), E_i \rangle], \qquad (2)$$

where $E_1, \ldots, E_{2q+1}$ is a local orthonormal frame such that $E_i \perp \mathcal{F}$ for $i \leq q$ and N is a unit vector field tangent to $\mathcal{F}$ parallel to the mean curvature vector $H^{\perp}$ of $\mathcal{F}^{\perp}$. $\square$

Corollary 1. *If $\mathcal{F}$ on (M, g) has integrable normal distribution defining a minimal ($H^{\perp} = 0$) foliation $\mathcal{F}^{\perp}$, then $\operatorname{gv}(\mathcal{F}) = 0$.* $\square$

Remark 1. Note that formula (2) in Theorem 1 for $q = 1$ coincides with that of [10]. Indeed, if $q = 1$ and $T = E_1$ is the unit normal of $\mathcal{F}$, then $H^{\perp} = \nabla_T T$ and $\|H^{\perp}\|$ is equal to the curvature of the trajectories of T, the leaves of $\mathcal{F}^{\perp}$, while the term $\langle \nabla_T N, E_2 \rangle$ can be interpreted as the torsion of these trajectories.

Remark 2. One can also define the *second fundamental form B* of an arbitrary, probably non-integrable, distribution D:

$$B(X,Y) = \frac{1}{2}(\nabla_X Y + \nabla_Y X)^\perp, \quad X,Y \in D,$$

where $v^\perp$ is the orthogonal to D component of a vector $v \in TM$. If so, the *mean curvature vector H* of D can be also defined:

$$H = \text{tr}(B) = \sum_i B(E_i, E_i),$$

where (E_i) is a local orthonormal frame of D. Again, if $\mathcal{F}$ is a foliation and $D = \mathcal{F}^\perp$ is the orthogonal distribution, the form η defining (together with its differential) the Godbillon-Vey class of $\mathcal{F}$ can be chosen as before: $\eta(X) = \langle V, H^\perp \rangle$, $X \in TM$, $H^\perp$ being the mean curvature vector of D. Therefore, the integrability of $\mathcal{F}^\perp$ in Corollary 1 can be deleted and one can formulate the following.

Corollary 2. *If $\mathcal{F}$ on (M,g) has minimal ($H^\perp = 0$) normal distribution, then* $\text{gv}(\mathcal{F}) = 0$. □

3. Tautness

Following [16], we will say that a foliation $\mathcal{F}$ on a compact manifold M is *geometrically taut* whenever there exists a Riemannian structure g on M for which all the leaves become minimal (that is, of vanishing mean curvature) submanifolds of (M,g). Using the tools from [15], Sullivan was able to prove that geometric tautness is equivalent to a topological condition – called *topological tautness* in [16] – expressed in terms of currents as follows.

Recall now that k-*currents* are continuous (with respect to the weak $*$-topology) functionals on the Frechet space D^k of k-forms, that is the space D_k of k-currents is the dual of D^k: $D_k = (D^k)^*$.

Given an oriented foliation $\mathcal{F}$ on M, denote by $C_\mathcal{F}$ the convex closed cone in the space D_k of k-currents ($k = \dim \mathcal{F}$) on M generated by all the elements of the form $v_1 \wedge \ldots \wedge v_k$, where $(v_1, \ldots, v_k)$ is a positive oriented frame of the space $T_x\mathcal{F}$ tangent to $\mathcal{F}$ at x, $x \in M$. Denote also by $B_\mathcal{F}$ the closed linear subspace of D_k generated by the boundaries of all the $(k+1)$-currents of the form $w \wedge v_1 \wedge \ldots \wedge v_k$, where $v_i \in T_x\mathcal{F}$, $w \in T_xM$ and $x \in M$. The foliation $\mathcal{F}$ is said to be *topologically taut* whenever $C_\mathcal{F}$ and $B_\mathcal{F}$ intersect trivially, that is $C_\mathcal{F} \cap B_\mathcal{F} = \{0\}$.

Formula (1) shows directly that geometrical tautness implies topological tautness. Indeed, if the leaves of $\mathcal{F}$ have zero mean curvature with respect to a Riemannian structure g, then the volume form of the leaves vanishes on $B_{\mathcal{F}}$ and is strictly positive on $C_{\mathcal{F}} \smallsetminus \{0\}$. The proof of the converse can be found in [16]. In this proof, (1) the classical Hahn-Banach Theorem, (2) the Schwartz Theorem (see [14]) saying that any continuous functional on D_k represents a k-form and (3) the existence of a compact base of the cone $C_{\mathcal{F}}$ are used. With this tools in hands, one can find a (pure) k-form ω on M which is positive on $C_{\mathcal{F}}$ and vanishes on $B_{\mathcal{F}}$, and a Riemannian structure g for which ω becomes the volume form of the leaves; again from (1) it follows that the mean curvature of the leaves with respect to g is identically zero.

The form ω in the Sullivan's proof is defined by arbitrary extension to D_k of the functional vanishing identically on $B_{\mathcal{F}}$ and being constant equal 1 on a base of the cone $C_{\mathcal{F}}$. The orthogonal complement $\mathcal{F}^{\perp}$ of $\mathcal{F}$ with respect to the Riemannian structure g in this proof coincides with the kernel of ω, so cannot be controlled at all. This forces us to adopt the following definitions.

A pair $\mathcal{P} = (\mathcal{F}, \mathcal{F}p)$ of complementary, pairwise transverse, oriented foliations on a manifold M is *geometrically taut* whenever there exists a Riemannian structure g on M for which $\mathcal{F}p = \mathcal{F}^{\perp}$ and all the leaves of $\mathcal{F}$ are minimal submanifolds of (M, g). The pair $\mathcal{P}$ is *topologically taut* whenever the cone $C_{\mathcal{F}}$ intersects trivially the smallest closed linear subspace $P_{\mathcal{P}}$ of D_k ($k = \dim \mathcal{F}$) containing $B_{\mathcal{F}}$ and all the currents of the form $w \wedge v_1 \wedge \ldots \wedge v_{k-1}$, where $w \in T_x \mathcal{F}p$, $v_i \in T_x M$ and $x \in M$.

Theorem 2. *A pair $\mathcal{P} = (\mathcal{F}, \mathcal{F}p)$ is geometrically taut iff it is topologically taut.*

Proof. If $\mathcal{P}$ is geometrically taut and g is a Riemannian structure making all the leaves of $\mathcal{F}$ minimal and $\mathcal{F}p$ orthogonal to $\mathcal{F}$, then the volume form ω of $\mathcal{F}$ on (M, g) is strictly positive on $C_{\mathcal{F}} \smallsetminus \{0\}$ and equal identically to zero on $P_{\mathcal{F}}$. Therefore, $C_{\mathcal{F}} \cap P_{\mathcal{P}} = \{0\}$ and $\mathcal{P}$ is topologically taut.

Assume now that the pair $\mathcal{P}$ is topologically taut. Since, as we mentioned before, the cone $C_{\mathcal{F}}$ has a compact base B, the Hahn-Banach Theorem implies the existence of a continuous linear functional $\lambda : D_k \to \mathbb{R}$ such that $\lambda = 0$ on $P_{\mathcal{P}}$ and $\lambda = 1$ on B. The Schwarz Theorem says that λ represents a k-form ω: $z(\omega) = \lambda(z)$ for any $z \in D_k = (D^k)^*$. Since ω is positive on $C_{\mathcal{F}}$, there exists a Riemannian structure g on M for which ω is the volume form of $\mathcal{F}$. Since ω vanishes on $B_{\mathcal{F}}$, the leaves of $\mathcal{F}$ are minimal submanifolds of (M, g). Since $T\mathcal{F}p$ is contained in the kernel of ω, $T\mathcal{F}p \subset T\mathcal{F}^{\perp}$. Comparing dimensions one gets $T\mathcal{F}p = T\mathcal{F}^{\perp}$ and $\mathcal{F}p = \mathcal{F}^{\perp}$. Therefore, $\mathcal{P}$ is geometrically taut. $\square$

Theorem 2 and Corollary 1 yield the following.

Corollary 3. *If $\mathcal{F}$ is a $(q+1)$-dimensional foliation on a compact manifold M of dimension $2q + 1$ and $gv(\mathcal{F}) \neq 0$, then there exists no foliation $\mathcal{F}p$ transverse to $\mathcal{F}$ such that the pair $\mathcal{P}^{\pitchfork} = (\mathcal{F}p, \mathcal{F})$ is topologically taut.*   □

Remark 3. One can expect that the integrability of $\mathcal{F}p$ in Corollary 3 is not needed as it was not needed in Corollary 1. This is rather not true: in the proof of equivalence of topological and geometrical tautnesses, one has apply the Sullivan's [16] *purification* of differential forms. This operation can be defined for arbitrary distributions, but if the distribution under consideration is not integrable, purification does not enjoy properties needed to prove such equivalence.

4. A generalization

Recall that (see, for example, [2], Chapters 6 and 7) the totality of all exotic characteristic classes of a codimension-q foliation $\mathcal{F}$ on a manifold M is obtained from a homomorphism

$$\lambda_{\mathcal{F}} : H^*(\mathrm{WO}_q) \to H^*(M, \mathbb{R}),$$

where WO_q is the tensor product of the graded algebra $\mathbb{R}_q(c_1, \ldots, c_q)$ of the truncated polynomials and the graded exterior algebra $\Lambda(y_1, \ldots, y_r)$, r being the largest odd number less that q; here, $\deg(c_i) = 2i$ and $\deg(y_i) = 2i - 1$. The algebra WO_q admits a unique linear operator d of degree 1 such that $d(c_i) = 0$, $d(y_i) = c_i$, $d(a \cdot b) = d(a) \cdot b + (-1)^{\deg a} a \cdot d(b)$ and $d^2 = 0$. $H^*(WO_q)$ above denotes the cohomology algebra of the cochain complex (WO_q, d).

The *generalised Godbillon-Vey classes* are the ones of the form $\lambda_{\mathcal{F}}([y_1 \cdot a])$, $a \in \mathrm{WO}_q$. Among them, one can find the standard Godbillon-Vey class: $gv(\mathcal{F}) = (-1)^{q+1}\lambda_{\mathcal{F}}([y_1(c_1)^q])$. If, as before, $\mathcal{F}$ is defined by the equation $\omega = 0$ for a q-form ω and $d\omega = \eta \wedge \omega$, then the *Godbillon class* $g(\mathcal{F})$ of $\mathcal{F}$ is defined as the linear map

$$g(\mathcal{F}) : H^k(M, \mathcal{F}) \to H^{k+1}(M, \mathbb{R}), \quad g(\mathcal{F})[\alpha]_{\mathcal{F}} = [\eta \wedge \alpha];$$

here $[\alpha]_{\mathcal{F}}$ denotes the cohomology class of a k-form $\alpha \in A^*(M, \mathcal{F}) = A^*(M) \wedge \{\omega\}$ (with respect to the standard exterior differential d). It is rather easy to show (see, for example Corollary 7.1.4 in [2]) that the condition $g(\mathcal{F}) = 0$ implies vanishing of all the generalized Godbillon-Vey classes of $\mathcal{F}$. Combining this fact with our Theorems 1 and 2 one can improve our Corollary 3 and get the following.

Corollary 4. *If $\mathcal{F}$ is a $(q+1)$-dimensional foliation on a compact manifold M of dimension $2q+1$ and some of generalized Godbillon-Vey classes of $\mathcal{F}$ do not vanish, then there exists no foliation $\mathcal{F}p$ transverse to $\mathcal{F}$ such that the pair $\mathcal{P}^{th} = (\mathcal{F}p, \mathcal{F})$ is topologically taut.* □

5. Invariants

In [11] (see also [12]), we introduced geometric invariants for pairs (more generally, n-tuples, $n = 1, 2, \ldots$) of square matrices (or, rather, endomorphisms). Given such matrices $A_1, \ldots, A_n$ one can consider the polynomial $P(A_a, \ldots, P_n)$ of variables $t_1, \ldots, t_n$ given by

$$P(A_1, \ldots, A_n)(t_1, \ldots t_n) = \det(I + t_1 A_1 + \ldots + t_n A_n).$$

The coefficients $\sigma_\lambda = \sigma_\lambda(A_1, \ldots, A_n)$, $\lambda = (\lambda_1, \ldots, \lambda_n)$ and λ_i's being non-negative integers, at $t_1^{\lambda_1} \cdot \ldots \cdot t_n^{\lambda_n}$ are invariants of our system of matrices. It is known that they can be expressed in terms of traces of matrices of the system and their products. In [11], we applied this idea to the Weingarten operator A of the leaves and the curvature operator $R_N = R(\cdot, N)N$ on a foliated Riemannian manifold $(M, \mathcal{F}, g)$, $\operatorname{codim}\mathcal{F} = 1$, N being a unit normal; integrating some expressions in terms of σ_λ's we obtained formulae (on, for simplicity, foliated locally symmetric spaces) generalizing those of Brito, Langevin and Rosenberg [1] for foliated space forms.

Our formula (2) here, allows to express the form $\eta \wedge d\eta)^q$ representing the Godbillon-Vey class in terms of invariants σ_λ of two liner transformations: $A = \nabla^{\mathcal{F}} N$, N being – as in Section 2 – the unit vector field parallel to the mean curvature vector $H^\perp$, and $B = B_{\mathcal{F}}(\cdot, N)$; hereafter, $\nabla^{\mathcal{F}}$ denotes the connection in the bundle $T\mathcal{F}$ generated by the Levi-Civita connection on (M, g). These maps transform one of the spaces $\hat{T}_x\mathcal{F} = \{v \in T_x\mathcal{F}; v \perp N(x)\}$, $T_x\mathcal{F}^\perp$, into another and can be represented in positive oriented orthonormal frames by square matrices of size $q \times q$. With this notation, formula (2) reads as

$$\eta \wedge (d\eta)^q = \det(A + B) \cdot \Omega, \tag{1}$$

Ω being the volume form on (M, g). Using our invariants $\sigma_{k,l}$, $k, l \geq 0$, for the matrices A and B, one can transform (1) into the following form:

$$\eta \wedge (d\eta)^q = \sum_{k+l=q} \sigma_{k,l}(A, B) \cdot \Omega. \tag{2}$$

Indeed, one has

$$\det(I + t(A + B)) = 1 + t \cdot \operatorname{tr}(A + B) + \ldots + t^q \cdot \det(A + B)$$

and

$$\det(I + tA + sB) = \sum_{k,l} t^k s^l \cdot \sigma_{k,l}(A, B);$$

putting $s = t$ and comparing the coefficients at t^q we get

$$\det(A + B) = \sum_{k+l=q} \sigma_{k,l}(A, B).$$

Our result can be expressed in the form of

Theorem 3. *The Godbillon-Vey class of a foliation $\mathcal{F}$ of a Riemannian manifold (M, g) (dim $M = 2q + 1$, dim $\mathcal{F} = q + 1$) with the orthogonal foliation $\mathcal{F}^\perp$ can be represented by the form $\eta \wedge (d\eta)^q$ given by (1), where $A = \nabla^{\mathcal{F}} N : \hat{T}\mathcal{F} \to T\mathcal{F}^\perp$ and $B = B_{\mathcal{F}}(\cdot, N) : T\mathcal{F}^\perp \to \hat{T}\mathcal{F} \to T\mathcal{F}^\perp$ are linear transformations depending on the extrinsic geometry of the pair $\mathcal{P} = (\mathcal{F}, \mathcal{F}^\perp)$.* □

Formula (2) allows to search for geometrical conditions which imply non-vanishing of the Godbillon-Vey class of $\mathcal{F}$. For example, one can obtain immediately the following.

Corollary 5. *In the situation considered in Theorem 3, if the mean curvature vector $H^\perp$ is $\nabla^{\mathcal{F}}$-parallel and non-zero at some points, and the second fundamental form $B_{\mathcal{F}}(\cdot, N)$ is either positive or negative definite, then $\mathrm{gv}(\mathcal{F}) \neq 0$.* □

References

[1] F. Brito, R. Langevin and H. Rosenberg, *Intégrales de courboure sur des variétés feuilletées*, J. Diff. Geom., **16** (1981), 19–50.

[2] A. Candel and L. Conlon, Foliations II, Amer. Math. Soc., Providence 2003.

[3] J. Cantwell and L. Conlon, *Endsets of exceptional leaves; a theorem of G. Duminy*, in Foliations, Geometry and Dynamics, World. Sci. Publ. 2002, 225–261.

[4] G. Duminy, *L'invariant de Godbillon-Vey d'un feuilletage se localise dans les feuilles resort*, unpublished preprint, Univ. Lille I, 1982.

[5] E. Ghys, R. Langevin and P. Walczak, *Entropie géométrique des feuilletages*, Acta. Math., **160** (1988), 105–142.

[6] C. Godbillon and J. Vey, *Un invariant des feuilletages de codimension 1*, C. R. Acad. Sci. Paris, **273** (1971), 92–95.

[7] S. Hurder, *Dynamics and the Godbillon-Vey class: a history and survey*, in Foliations, Geometry and Dynamics, World. Sci. Publ. 2002, 29–60.

[8] S. Hurder, *Problem set*, in Foliations 2005, World Sci. Publ. 2006, 441–475.

[9] S. Hurder and R. Langevin, *Dynamics and the Godbillon-Vey class of C^1-foliations*, preprint, 2003.

[10] B. Reinhart and J. Wood, *A metric formula for the Godbillon-Vey invariant for foliations*, Proc. Amer. Math. Soc., **38** (1973), 427–430.

[11] V. Rovenski and P. Walczak, *Integral formulae on foliated symmetric spaces*, Math. Ann., 2011, DOI 10.1007/s00208-011-0637-4

[12] V. Rovenski, P. Walczak, *Topics in Extrinsic Geometry of Codimension-One Foliations*, Springer Briefs in Mathematics, Springer Verlag 2011.

[13] H. Rummler, *Quelques notions simples en géométrie riemannienne et leur applications aux feuilletages compacts*, Comm. Math. Helv., **54** (1979), 224–239.

[14] L. Schwartz, *Théorie des distributions*, Hermann, Paris 1966.

[15] D. Sullivan, *Cycles for the dynamical study of foliated manifolds and complex manifolds*, Invent. Math., **36** (1976), 225–255.

[16] D. Sullivan, *A homological characterization of foliations consisting of minimal surfaces*, Comment. math. Helv., **54** (1979), 218–223.

[17] W. Thurston, *Noncobordant foliations of S^3*, Bull. Amer. Math. Soc., **78** (1972), 511–514.

[18] P. Walczak, Dynamics of foliations, groups and pseudogroups, Monografie Matematyczne **64**, Birkhäuser, Basel 2004.

Received September 9, 2012.

FOLIATIONS 2012
ed. by Paweł WALCZAK *et al.*
World Scientific, Singapore, 2013
pp. 215–233

Local and global stability of leaves of conformal foliations*

NINA I. ZHUKOVA

Department of Mechanics and Mathematics
Nizhny Novgorod State University, Russia
e-mail: n.i.zhukova@rambler.ru

1. Basic concepts and results

One of the main goals of this work is to apply the previous results of the author [29, 30], and to prove new theorems on local and global leaf stability of conformal foliations of codimension $q > 2$. We also remind our results about local and global stability of compact leaves of foliations with quasi analytical holonomy pseudogroup admitting an Ehresmann connection and corresponding results of other authors.

Local stability of leaves and foliations

The notion of stability of leaves of foliations was introduced by Ehresmann and Reeb, the founders of the theory of foliations.

Remind that a subset of foliated manifold is called *saturated* if it may be represented as a union of some leaves of the foliation.

Definition 1. A leaf L of a foliation (M, F) is said to be *proper* if it is an

---

*This work was supported by the Federal Target Program 'Scientific and Scientific-Pedagogical Personnel', Project No 14.B37.21.036, and the Russian Federation Ministry of Education and Science, Project N 1.1907.2011.

embedded submanifold of the foliated manifold M. A foliation is proper, if every its leaf is proper. A leaf L is called closed if L is a closed subset of M.

Definition 2. A leaf L of a foliation (M, F) of codimension q is said to be locally stable in sense of Ehresmann and Reeb, if there exists a family of its saturated neighbourhoods W_β, $\beta \in \mathcal{B}$, with the following properties:

(1) there exists a locally trivial fibration $f_\beta : W_\beta \to L$, $\beta \in \mathcal{B}$, with a q-dimensional disk D^q as the typical fiber, whose fibers are transversal to the leaves of the foliation $(W_\beta, \mathcal{F}_{W_\beta})$;

(2) for some $\delta \in \mathcal{B}$ the traces of these neighbourhoods form a base of the topology of a fiber of the fibration $f_\delta : W_\delta \to L$ over $x \in L$ at the point x.

A foliation is refer to be *locally stable* if each its leaf is locally stable.

According to the well-known theorem of Reeb [23, 24], any compact leaf of a foliation with finite holonomy group is locally stable.

The leaf stability of Riemannian foliations

Blumenthal and Hebda [4] introduced a notion of Ehresmann connection for a smooth foliation (M, F) as a smooth q-dimensional distribution $\mathfrak{M}$ on M transverse to (M, F) with the vertical-horizontal property (the precise definition see in Section 2). We showed ([31], Proposition 2), that a complete Cartan foliation admits an Ehresmann connection. It is known examples of Riemannian foliations with an Ehresmann connection whose are not (transversally) complete (Example 1). Thus, the existence of an Ehresmann connection for a Cartan foliation (with fix transverse Cartan geometry) does not imply the completeness of this foliation in general.

The Proposition 1 describes the structure of a saturated neighbourhood of a proper leaf L, and $\Gamma(L, x)$, $x \in L$, is the germ holonomy group usually used in the foliation theory [24].

Proposition 1. *Let L be a proper leaf of a Riemannian foliation (M, F) with an Ehresmann connection $\mathfrak{M}$. Then there exist a bundle like metric g on M relatively which $\mathfrak{M}$ is orthogonal to (M, F) and a family of saturated tubular neighbourhood W_β of the radius $\beta \in (0, r]$, $r > 0$, with the orthogonal projection $f_\beta =: W_\beta \to L$, where $f_\beta = f_r|_{W_\beta}$, satisfying the following conditions:*

(1) the neighbourhood W_β is a smooth fibre space with the projection f_β : $W_\beta \to L$, and its structure group is the germ holonomy group $\Gamma(L, a)$,

$a \in L$, of L. The the typical fibre is $D_\beta(a) = exp_a(D(0, \beta))$, where $D(0, \beta) \subset \mathfrak{M}_a$, the q-dimensional disk of radius β, on which $\Gamma(L, a)$ naturally acts by isometries;

(2) the distribution $\mathfrak{M}_\beta = \mathfrak{M}|_{W_\beta}$ is an integrable Ehresmann connection for the submersion $f_\beta : W_\beta \to L$;

(3) the germ holonomy group of an arbitrary leaf $L(z) \subset W_\beta$, $z \in f_\beta^{-1}(a)$, is isomorphic to the stationary subgroup Γ_z of the group $\Gamma(L, a)$ at point z;

(4) the restriction $f_\beta|_{L(z)} : L(z) \to L$ is the covering map, and its set of sheets is bijective to the orbit $\Gamma(L, a) \cdot z$ of the point z under the action of $\Gamma(L, a)$.

The following assertion was proved with the use of Proposition 1.

Theorem 1. *Let (M, F) be a Riemannian foliation of an arbitrary codimension $q \geq 1$ admitting an Ehresmann connection. Then the following three conditions for a leaf L are equivalent:*

(i) L *is locally stable leaf;*
(ii) L *is a proper leaf;*
(iii) L *is a closed leaf.*

For transversally complete Riemannian foliations Theorem 1 and assertions equivalent to Proposition 1 were proved by the author in [32]. Under an additional assumption about the existence of a complementary topological foliation, the local stability of a proper leaf of Riemannian foliation has been proved by Ehresmann [10]. For parallel foliations on a complete Riemannin manifold the equivalence of conditions (i)-(iii) of Theorem 1 was proved in [15], where the proof is considerably simpler, due to specificity of the case. In [1] it was proved that a proper leaf of a Riemannin foliation on manifold with a complete bundle like metric is covered by all near leaves.

Theorem 2. *Let (M, F) be a Riemannian foliation of codimension $q \geq 1$ admitting an Ehresmann connection. If there exists a closed leaf L of (M, F) with a finite (germ) holonomy group $\Gamma(L, x)$, $x \in L$, then:*

(1) *any its leaf L_α is closed subset of M with a finite holonomy group $\Gamma(L_\alpha, x_\alpha)$, $x_\alpha \in L_\alpha$, and L_α is a locally stable leaf;*
(2) *the leaf space M/F is a smooth q-dimensional orbifold.*

Theorem 2 may be proved by analogy with the author's proof of similar Theorem 2 in [32]. Here we give a new proof of this statement.

Corollary 1. *Let (M, F) be a Riemannian foliation of codimension $q \geq 1$ with an Ehresmann connection. If any its leaf is a closed subset of M, then (M, F) is locally stable and the leaf space M/F is a smooth q-dimensional orbifold.*

In the case, when (M, F) is a Riemannian foliation on a Riemannian manifold with complete bundle like metric the statement of Corollary 1 was proved by Reinhart [21].

Theorem 3. *Let (M, F) be a Riemannian foliation of codimension $q \geq 1$ admitting an Ehresmann connection. If there exists a closed leaf L of (M, F) with a finite fundamental group $\pi_1(L, x)$, then any its leaf L_α is closed with a finite fundamental group $\pi_1(L_\alpha, x_\alpha)$, $x_\alpha \in L_\alpha$, and (M, F) is a locally stable Riemannian foliation.*

Based on statements of this section, Theorem 1 in [30] and the paper [28] we ask the following.

Question: *For a Riemannian foliation (M, F) with an Ehresmann connection $\mathfrak{M}$ there is a bundle like metric g such that $\mathfrak{M}$ is a orthogonal distribution to (M, F). Does there exist a transversally complete bundle like metric $\widetilde{g}$, which is $\mathfrak{M}$-conformal to g?*

Criterions of the local stability of leaves of conformal foliations

Using Theorems 1 and 2 and results of our previous paper [30] we prove the following two criterions of the local leaf stability for conformal foliations.

Theorem 4. *Let (M, F) be a conformal foliation of codimension $q > 2$ admitting an Ehresmann connection. Then a leaf L of (M, F) is locally stable if and only if L is a proper leaf with inessential holonomy group (in sense of Section 2).*

Theorem 5. *Let (M, F) be a proper non-Riemannian conformal foliation of codimension $q > 2$ admitting an Ehresmann connection. Then the following three conditions for a leaf L of (M, F) are equivalent:*

(i) L is locally stable;
(ii) L is an unclosed leaf;
(iii) L has a finite holonomy group $\Gamma(L, x)$.

The problem of local stability of compact foliations

A foliation is called compact, if every its leaf is compact. Epstein [11] proved that any leaf of a compact foliation (M, F) has a finite holonomy group iff the leaf space M/F is Hausdorff. Reeb showed that a codimension one compact foliation has a Hausdorff leaf space. Millett [18] put out the conjecture that all holonomy groups of a compact foliation on a compact manifold are finite. As it was said, according to the famous Reeb's theorem a compact leaf with a finite holonomy group is locally stable. Therefore the Millett's conjecture is called *a problem of local stability*. Now it is known that for $q = 2$ the Millett's conjecture is valid unlike the case $q = 3$. If the foliated manifold M is not compact, the analog of the Millett's conjecture is not true for compact foliations (M, F) of codimension 2. Different criterions of local stability of a compact foliations were proved [9, 18, 33, 34]. Among them there is Rummer's characterization of a compact locally stable foliation by the existence of a Riemannian metric with respect to which every leaf is a minimal submanifold. Epstein stated that (M, F) is a compact locally stable foliation iff there exists a Riemannian bundle like metric g on M such that the volume function of leaves is locally bounded.

The leaf local stability takes an important place in works on partially hyperbolic diffeomorphisms with compact central foliations [5, 13].

Definition 3. Pseudogroup of local diffeomorphisms $\mathcal{H}$ of a manifold N is quasi analytical, if for any open subset U in N and an element $h \in \mathcal{H}$, the condition $h|_U = id_U$ implies $h = id_{D(h)}$, where $D(h)$ is the connected domain of definition of h containing U.

The results of our works ([33], Theorem 5 and [34], Theorem 8.1) imply the following criterion of the local leaf stability of compact foliations (M, F) without assumption of compactness of M.

Theorem 6. *All holonomy groups of a compact foliation (M, F) are finite if and only if it satisfies the following two conditions:*

(1) there exists an Ehresmann connection for (M, F);
(2) the holonomy pseudogroup $\mathcal{H}(M, F)$ of this foliation is quasi analytical.

The effectivity of this criterion is confirmed by the following corollary.

Corollary 2. *Compact complete Cartan foliations and compact complete G-foliations of a finite type are locally stable. Leaf spaces of those foliations are smooth orbifolds.*

In the case when M is compact, Corollary 2 implies Theorem 1 of Wolak [26] for complete compact G-foliations of a finite type.

Lawson put out the following problem ([17], Problem 14):

Characterize the foliations of compact manifolds in which every leaf is compact.

Theorem 7 decides this problem for conformal foliations (M, F) of codimension $q > 2$ without assumption of compactness of the foliated manifold M.

Theorem 7. *Any compact conformal foliation of codimension $q > 2$ is a locally stable Riemannian foliation, the leaf space of which is a smooth q-dimensional orbifold.*

The analogous theorem for a compact transversely holomorphic foliation of codimention 2 was proved by Walczak [25].

Remark 1. If all leaves of a conformal foliation (M, F) of codimension $q > 2$ are closed subsets of M, then (M, F) is a Riemannian foliation, which is not local stable in general (see Example 2).

Global stability of a compact leaf of foliations with quasi analytical pseudogroup

For a leaf L of a foliation (M, F) with an Ehresmann connection $\mathfrak{M}$, Blumenthal and Hebda introduced a holonomy group $H_{\mathfrak{M}}(L, x)$ (its definition is given in Section 2).

Our results from [33, 34] implies the following statement.

Proposition 2. *Let (M, F) be a foliation with an Ehresmann connection $\mathfrak{M}$ and L be any its leaf. The natural group epimorphism $\chi : H_{\mathfrak{M}}(L, x) \to \Gamma(L, x)$ of holonomy groups is isomorphism if and only if the holonimy pseudogroup $\mathcal{H} = \mathcal{H}(M, F)$ of this foliation is quasi analytical.*

Application of Theorem 1 of Blumenthal – Hebda [4] and Proposition 2 allowed us to obtain the following assertion about the global stability of some compact leaves.

Theorem 8. *Let (M, F) be a smooth foliation with an Ehresmann connection and quasi analytical holonomy pseudogroup. Then the existence a compact leaf L with a finite germ holonomy group $\Gamma(L, x)$ (or finite fundamental group $\pi_1(L, x)$) guarantees compactness of every leaf L_α of this foliation and finiteness of the holonomy group $\Gamma(L_\alpha, x_\alpha)$, $x_\alpha \in L_\alpha$, (or finite fundamental group $\pi_1(L_\alpha, x_\alpha)$) and the local stability of (M, F).*

In particular, Theorem 8 implies the global stability of a compact leaf with finite germ holonomy group of complete Cartan foliations. For complete G-foliations of finite type the analogous result belongs to Wolak [26].

Global leaf stability of conformal foliations

Let (M, F) be a foliation. A *saturated set* is a union of leaves.

Theorem 9. *Let (M, F) be a conformal foliation of codimension $q > 2$ admitting an Ehresmann connection. If there exists a closed leaf L of (M, F) with a finite holonomy group $\Gamma(L, x)$ (or a finite fundamental group $\pi_1(L, x)$)), then any its leaf L_α is closed with a finite holonomy group $\Gamma(L_\alpha, x_\alpha)$, $x_\alpha \in L_\alpha$, (respectively, a finite fundamental group $\pi_1(L_\alpha, x_\alpha)$) and (M, F) is a locally stable Riemannian foliation, the leaf space of which is a smooth q-dimensional orbifold.*

Corollary 3 ([2]). *Let (M, F) be a complete conformal foliation of codimension $q > 2$. If there exists a compact leaf with a finite holonomy group, then any its leaf is compact with a finite holonomy group.*

Corollary 3 belonging to Blumenthal [2] was a unique known result about the leaf stability of a conformal foliation.

Corollary 4. *Let (M, F) be a conformal foliation of codimension $q > 2$ admitting an Ehresmann connection. If any its leaf is a closed subset of M, then (M, F) is a locally stable Riemannian foliation.*

Remark 2. We constructed an example of a complete transversally affine foliation (M, F) of an arbitrary codimension $q \geq 2$ with an Ehresmann connection such that (M, F) satisfies conditions of both Theorem 9 and Corollary 4, but it is not locally stable (Example 3). Thus, statements on stability of noncompact leaves (Theorems 9 and Corollary 4) can not be generalized to all complete Cartan foliations unlike statements on compact leaves (Corollary 2 and Theorem 8)

The following assertion about global stability of a compact leaf with a finite germ holonomy group was proved by us without assumption of completeness or the existence of an Ehresmann connection of the foliation (M, F).

Theorem 10. *If a conformal foliation (M, F) of codimension $q > 2$ on a compact manifold M has a compact leaf L with a finite holonomy group*

$\Gamma(L)$, *then any its leaf is compact with a finite holonomy group and* (M, F) *is a locally stable compact Riemannian foliation.*

The analogous theorem for holomorphic foliations of codimension k on compact complex Kaehler manifolds was proved by Pereira [20].

Remark 3. Theorems 3, 8–10 are some analogous of the well-known Reeb global stability theorem [23], according to which a smooth codimension one foliation (M, F) of a closed manifold M containing a compact leaf with a finite fundamental group has only compact leaves with finite fundamental groups.

2. Cartan foliations. Holonomy and completeness

Ehresmann connection for foliations

Remind the notion of an Ehresmann connection belongs to Blumenthal and Hebda [4]. At that we use a term *a vertical-horizontal homotopy* introduced earlier by Hermann [14].

Let $(M, \mathcal{F})$ be a foliation of arbitrary codimension $q \geq 1$. A distribution $\mathfrak{M}$ on a manifold M is called *transversal* to a foliation $\mathcal{F}$ if for any $x \in M$ the equality $T_x M = T_x \mathcal{F} \oplus \mathfrak{M}_x$ holds, where $\oplus$ stands for a direct sum of vector spaces. Vectors from $\mathfrak{M}_x$, $x \in M$, are called horizontal. A piecewise smooth curve σ is horizontal (or $\mathfrak{M}$-horizontal) if each of its smooth segments is an integral curve of the distribution $\mathfrak{M}$. A distribution TF tangent to leaves of the foliation (M, F) is called vertical. One says that a curve h is vertical if h is contained in the leaf of the foliation (M, F).

A *vertical-horizontal homotopy* (v.h.h. for shot) is a piecewise smooth map $H : I_1 \times I_2 \to M$, where $I_1 = I_2 = [0, 1]$, such that for any $(s, t) \in I_1 \times I_2$ the curve $H|_{I_1 \times \{t\}}$ is horizontal and the curve $H|_{\{s\} \times I_2}$ is vertical. A pair of curves $(H|_{I_1 \times \{0\}}, H|_{\{0\} \times I_2})$ is called *a base of the v.h.h.* H. Two paths (σ, h) with common origin $\sigma(0) = h(0)$, where σ is a horizontal path and h is vertical one, are called an *admissible pair of paths*.

A distribution $\mathfrak{M}$ transversal to a foliation (M, F) is called an *Ehresmann connection for* (M, F) if for any admissible pair of paths (σ, h) there exists a v.h.h. with a base (σ, h).

Let $\mathfrak{M}$ be an Ehresmann connection for a foliation (M, F). Then for any admissible pair of paths (σ, h) there exists a unique v.h.h. H with base (σ, h). We say that $\widetilde{\sigma} := H|_{I_1 \times \{1\}}$ is the result of the *transfer of the path* σ *along* h *with respect to the Ehresmann connection* $\mathfrak{M}$. It is denoted by $\sigma \xrightarrow{h} \widetilde{\sigma}$. Take any point $x \in M$. Denote by Ω_x the set of horizontal curves with the origin at x. An action of the fundamental group $\pi_1(L, x)$ of the leaf $L = L(x)$ on the set Ω_x is defined by the following a way:

$$\Phi_x : \pi_1(L, x) \times \Omega_x \to \Omega_x : ([h], \sigma) \mapsto \widetilde{\sigma},$$

where $[h] \in \pi_1(L, x)$ and $\widetilde{\sigma}$ is the result the transfer of σ along h relatively $\mathfrak{M}$. The quotient group $H_{\mathfrak{M}}(L, x) = \pi_1(L, x)/Ker(\Phi_x)$ of the kernel $Ker(\Phi_x)$ of the action Φ_x in $\pi_1(L, x)$ is a *group of $\mathfrak{M}$-holonomy of a leaf L* [4].

Cartan foliations

Notions belonging to Cartan geometry can be found in [16] and [6]. The definition of Cartan geometry $\xi = (P(N, H), \omega)$ of type (G, H) is equivalent to specifying the following objects:

(1) a Lie group G and its closed Lie subgroup H with Lie algebras $\mathfrak{g}$ and $\mathfrak{h}$, respectively;
(2) a principal H-bundle $\pi : P \to M$;
(3) a $\mathfrak{g}$-values 1-form ω on P called *a Cartan connection* having the following properties:

    (i) $\omega(A^*) = A$ for any $A \in \mathfrak{p}$, where A^* is the fundamental vector field corresponding to A;

    (ii) $R_a^*\omega = Ad_H(a^{-1})\omega$, $\forall a \in H$, where Ad_H is the adjoint representation of the Lie subgroup H in the Lie algebra $\mathfrak{g}$ of G;

    (iii) for any $u \in P$ the map $\omega_u : T_u(P) \to \mathfrak{g}$ is bijection.

Further we assume that Cartan geometry ξ of a type (G, H) is effective, i.e., the left action of the group G on G/H is effective. At that the Blumenthal's definition of a Cartan foliation [3] and our one [31] are equivalent.

Let N be q-dimensional manifold and M be a smooth n-dimensional manifold, $0 < q < n$. Unlike M the connectedness of the topological space N is not assumed. An N-cocycle is the set $\{U_i, f_i, \{k_{ij}\}\}_{i,j \in J}$ such that:

(1) The family $\{U_i, i \in J\}$ forms an open cover of M.
(2) The mappings $f_i : U_i \to N$ are submersions into N with connected fibers.
(3) If $U_i \cap U_j \neq \emptyset$, $i, j \in J$, then a diffeomorphism $k_{ij} : f_j(U_i \cap U_j) \to f_i(U_i \cap U_j)$ is well-defined and satisfies the equality $f_i = k_{ij} \circ f_j$.

Definition 4. Let a foliation (M, F) be given by an N-cocycle $\{U_i, f_i, \{k_{ij}\}\}_{i,j \in J}$. If the manifold N admits an effective Cartan geometry such that every local diffeomorphism k_{ij} is an isomorphism of the

Cartan geometries induced on open subsets $f_i(U_i \cap U_j)$ and $f_j(U_i \cap U_j)$, then we refer to (M, F) as a *Cartan foliation defined by the* (N, ξ)-*cocycle* $\{U_i, f_i, \{k_{ij}\}\}_{i,j \in J}$.

At the beginning we represent in the following statement about different interpretations of the holonomy groups of Cartan foliations, which was established in the previous work of the author ([30], Proposition 5).

Proposition 3. *Let (M, F) be an arbitrary Cartan foliation defined by (N, ξ)-cocycle $\{U_i, f_i, \{k_{ij}\}\}_{i,j \in J}$ and $\pi : \mathcal{R} \to M$ be the projection of the foliated H-bundle over (M, F) with lifted foliation $(\mathcal{R}, \mathcal{F})$. For each leaf $L = L(x)$ of (M, F) consider the leaf $\mathcal{L} = \mathcal{L}(u)$, where $u \in \mathcal{R}$, $\pi(u) = x \in U_i$, of the lifted foliation $(\mathcal{R}, \mathcal{F})$ and $v = f_i(x)$. Then the germ holonomy group $\Gamma(L, x)$ of L is isomorphic to the following groups:*

- *the subgroup $H(\mathcal{L}) := \{a \in H \mid R_a(\mathcal{L}) = \mathcal{L}\}$ of H;*
- *the group of covering transformations of the regular covering $\pi|_{\mathcal{L}} : \mathcal{L} \to L$.*
- *the group of germs at point v of local isomorphisms from the holonomy isotropy subpseudogroup $\mathcal{H}_v(N, \xi)$.*

If in conditions of the Proposition 3 we consider an other point $u' \in \pi^{-1}(x)$ and the leaf $\mathcal{L}' = \mathcal{L}'(u')$, then the group $H(\mathcal{L}')$ must be conjugated to $H(\mathcal{L})$ in H. Therefore, the following definition makes sense.

Definition 5. Refer to the holonomy group of a leaf L of a Cartan foliation as *relatively compact* or *inessential* if the corresponding subgroup $H(\mathcal{L})$ of the Lie group H is relatively compact. Otherwise the holonomy group of a leaf is called *essential*.

Completeness

Let $\mathfrak{M}$ be a smooth q-dimensional distribution on M transverse to a Cartan foliation (M, F) of codimension q and $\widetilde{\mathfrak{M}}$ be a smooth distribution on $\mathcal{R}$ transverse to the lifted foliation $(\mathcal{R}, \mathcal{F})$ such that $\pi_{*u}(\widetilde{\mathfrak{M}}_u) = \mathfrak{M}_{\pi(u)}$, $u \in \mathcal{R}$. Denote by $\mathfrak{X}(\mathcal{R})$ the set of smooth vector fields on $\mathcal{R}$ and by $\mathfrak{X}_{\widetilde{\mathfrak{M}}}(\mathcal{R})$ the subset of smooth vector fields tangent to $\widetilde{\mathfrak{M}}$. A Cartan foliation (M, F) is called *complete* (or $\mathfrak{M}$-*complete*) if any $\omega_{\mathcal{R}}$-constant vector field $X \in \mathfrak{X}_{\widetilde{\mathfrak{M}}}(\mathcal{R})$ is complete, where $\omega_{\mathcal{R}}$ is $\mathfrak{g}$-valued base 1-form on $\mathcal{R}$ induced by Cartan connection ω [2, 31]. As was proved by us in [31] (Proposition 2), if (M, F) is a $\mathfrak{M}$-complete Cartan foliation, then $\mathfrak{M}$ is an Ehresmann connection for (M, F). It is naturally true for conformal and Riemannian foliations. An $\mathfrak{M}$-complete Riemannian foliation with a bundle like metric, where $\mathfrak{M}$ is the complementary orthogonal distribution to this foliation, is called a transversally complete one.

3. Proofs of statements for Riemannian foliations

Proof of Proposition 1. Let (M, F) be a Riemannian foliation with an Ehresmann connection $\mathfrak{M}$ defined by N-cocycle $\{U_i, f_i, \{k_{ij}\}\}_{i,j \in J}$ and $\mathfrak{M}$ be a transverse distribution to (M, F). Then, as known (see, for instance [30], Proposition 1) there exists such Riemannian metric g on M and g_N on N, that:

(i) the distribution $\mathfrak{M}$ is orthogonal to the foliation (M, F), and every submersion $f_i : U_i \to V_i = f_i(U_i)$ is Riemannian;

(ii) any geodesic γ on the Riemannian manifold (M, g), which is tangent to $\mathfrak{M}$ at one point, is tangent to $\mathfrak{M}$ at every point;

(iii) for every admissible pair of paths of the form (σ, h), where σ is $\mathfrak{M}$-horizontal geodesic, the result $\tilde{\sigma}$ of the transfer $\sigma \xrightarrow{h} \tilde{\sigma}$ is also $\mathfrak{M}$-horizontal geodesic of the same length as σ, i.e. $l(\tilde{\sigma}) = l(\sigma)$.

The metric g is a bundle like one in terminology of Reinhart [22]. Denote by d the distance function of the Riemannian manifold (M, g).

Suppose now that a leaf L is proper. Let S be a connected open relatively compact subset in the leaf L. Then ([19], p. 73) there are such $\varepsilon > 0$ and an open contractible neighbourhood V_ε satisfying the following properties:

(i) For any $y \in V_\varepsilon$ there exists a unique $x =: f(y) \in S$ and a unique vector $X \in \mathfrak{M}_x$ such that $y = exp_x(X)$ and $||X||_x = d(y, S)$.

(ii) The orthogonal projection $f : V_\varepsilon \to S$ thus defined is trivial fibration whose typical fiber is the open disc $D(0, \varepsilon)$ in $\mathfrak{M}_a$, where a is a fixed point in S.

One say that V_ε is *a tubular neighbourhood of S* with radius ε. As a leaf L is proper, according to ([24], Theorem 4.11) without loss generality we assume that $L \cap V_\varepsilon = S$.

There exists a normal convex neighbourhood $B(a, 2r) \subset V_\varepsilon$ at a with radius $2r$. The set of convex neighbourhoods $B(a, \beta)$, $\beta \in (0, r]$ forms a base of the topology of M at point a. Put $W_\beta := \cup L_\alpha$, where $L_\alpha \cap B(a, \beta) \neq \emptyset$. Then every W_β is an open saturated neighbourhood of L ([24], Theorem 4.10).

We shall use notation $D_\beta(a) = exp_a(D(0, \beta))$, where $D(0, \beta) \subset \mathfrak{M}_a$. Remark that $D_\beta(a)$ is the diffeomorphism image of $D(0, \beta)$. Show that the family of neighbourhoods $\{W_\beta \mid \beta \in (0, r]\}$ satisfies to Definition 2.

Define a map $f_\beta : W_\beta \to L$ by the following a way. Take any point $z \in W_\beta$. According to the definition of W_β there is a point $y \in L(z) \cap D_\beta(a)$. There is a unique horizontal geodesic γ jointing $a = \gamma(0)$ with $y = \gamma(1)$, and $\gamma(s) \in D_\beta(a)$, $\forall s \in [0, 1]$. Connect y with z by a piecewise smooth path h in the leaf $L(z)$. Then (γ^{-1}, h) is an admissible pair of paths and there exists the transfer $\gamma^{-1} \xrightarrow{h} \widetilde{\gamma}^{-1}$. Put $f_\beta(z) = \widetilde{\gamma}(0) \in L$. Show that this definition takes meaning.

Let h' be an other piecewise smooth path in the leaf $L(z)$ connecting $h'(0) = y$ with $h'(1) = z$ and $\gamma^{-1} \xrightarrow{h'} \gamma'^{-1}$. Then $f_\beta(z) = \gamma'(0) \in L$. Consider the transfer $\gamma'^{-1} \xrightarrow{h^{-1}} \widehat{\gamma}^{-1}$, then $\widehat{\gamma}^{-1}(1) = y$. According the above property (iii) $l(\widehat{\gamma}) = l(\gamma)$. Therefore $d(\gamma(0), \widehat{\gamma}(0)) \leq d(a, y) + d(y, \widehat{\gamma}(0)) = 2l(\gamma) < 2\tau$ and $\widehat{\gamma}(0) \in V_\varepsilon \cap L(z) = S$. So it is necessary that $\gamma(0) = \widehat{\gamma}(0) = y$ and $\widehat{\gamma} = \gamma$. It implies the equality $\widetilde{\gamma} = \gamma'$. Thus f_β does not depend of the choice of the path connecting y with z.

Consider the case, when there is an other point $y' \in L(z) \cap D_\beta(a)$. Let k be a path connecting y' with z in L and σ be a horizontal geodesic in $D_\beta(a)$ joints a with y'. By analogy with the above arguments we see that the result of the transfer of γ^{-1} along $h \cdot k^{-1}$ is equal to σ. Therefore the result of the transfers of γ^{-1} along h and σ^{-1} along k are coincided, i.e. $\widetilde{\gamma} = \widetilde{\sigma}$ and $f_\beta(z) = \widetilde{\gamma}(0) = \widetilde{\sigma}(0)$. Thus, the map $f_\beta : W_\beta \to L$ is really defined. It is not difficult to check that $f_\beta : W_\beta \to L$ is a surjective submersion. Note that the foliation $F_\beta = F|_{W_\beta}$ is an integrable Ehresmann connection for the submersion f_β. Therefore $f_\beta : W_\beta \to L$ is the projection of the locally trivial fibration whose typical fiber is the open q-dimensional disc $D_\varepsilon(a)$ on which the holonomy group $\Gamma(L, a)$ naturally acts by isometries. Thus, the statements (1) and (2) are valid.

Remark that $(W_\beta, F|_{W_\beta})$ is a suspended foliation with $f_\beta : W_\beta \to L$ as transverse fibration. This foliation is defined by suspension of the natural group homomorphism $\pi_1(L, a) \to \Gamma(L, a)$, where $\Gamma(L, a)$ is considered as a subgroup of the diffeomorphism group of $D_\beta(a)$. Therefore, thanks to the quasi analyticity of $\Gamma(L, a)$ the properties (3) and (4) also take place. $\square$

Proof of Theorem 1. It is follows from Definition 2 that a locally stable leaf L is proper. In conformity with Proposition 1 the family of saturated neighbourhoods W_β, $\beta \in (0, \varepsilon]$, of a proper leaf L satisfies Definition 2, i.e.,

(i) is equivalent to (ii).

Let L be a proper leaf of a Riemannian foliation (M, F) admitting an Eresmann connection. We proved ([30], Theorem 1) that the holonomy pseudogroup of Riemannian foliation admitting an Eresmann connection is complete and the closure of any its leaf is a minimal set. As a nontrivial minimal set contains only improper leaves it is necessary that L be a closed leaf. As it is well known (see for instance [19], p. 22), every closed leaf is proper. Thus, (ii) is equivalent to (iii).   □

Proof of Theorem 2. Let $\mathfrak{M}$ be an Ehresmann connection for a Riemannian foliation (M, F). Consider the foliated H-bundle $\pi : \mathcal{R} \to M$, where H is $O(q)$ (or $H = SO(q)$, if (M, F) is a transversally orientable Riemannian foliation), with the lifted foliation $(\mathcal{R}, \mathcal{F})$ over the given conformal foliation (M, F). Observe that the induced distribution $\widetilde{\mathfrak{M}} = \{\widetilde{\mathfrak{M}}_u \mid u \in \mathcal{R}\}$, where $\widetilde{\mathfrak{M}}_u = \{X \in T_u\mathcal{R} \mid \pi_{*u}(X) \in T_xM, x = \pi(u)\}$ is an Ehresmann connection for $(\mathcal{R}, \mathcal{F})$.

Let L be a closed leaf with a finite holonomy group $\Gamma(L, x)$. Take an arbitrary point $u \in \pi^{-1}(x)$. Denote by $\mathcal{L}$ the leaf of $(\mathcal{R}, \mathcal{F})$ passing through u. In accordance with Proposition 2, the finiteness of $\Gamma(L, x)$ implies that the restriction $\pi|_{\mathcal{L}} : \mathcal{L} \to L$ is a finite sheet covering. As the closed leaf L is proper, $\mathcal{L}$ is a proper leaf of the lifted foliation. Thus the e-foliation $(\mathcal{R}, \mathcal{F})$ with an Ehresmann connection has a proper leaf. Thanks this, by analogy with the proof of Proposition 4.4 from the Conlon's work [8] it is not difficult to show that leaves of $(\mathcal{R}, \mathcal{F})$ are coincided with fibres of a locally trivial bundle $\pi_b : \mathcal{R} \to W$. Therefore every leaf of the lifted foliation is proper and closed.

Take any point $y \in M$. Let $y \in L_\alpha$ and $\mathcal{L}_\alpha$ is a leaf of $(\mathcal{R}, \mathcal{F})$ over L_α, i.e. $\mathcal{L}_\alpha \subset \pi^{-1}(L_\alpha)$. Therefore the intersection $\pi^{-1}(y) \cap \mathcal{L}_\alpha$ is discrete and closed subset of $\pi^{-1}(y)$. The fiber $\pi^{-1}(y)$ is compact, because it is diffeomorphic to the compact Lie group H. It implies the finitness of the set $\pi^{-1}(y) \cap \mathcal{L}_\alpha$. As the group of the covering transformations of the regular covering $\pi_{\mathcal{L}_\alpha} : \mathcal{L}_\alpha \to L_\alpha$ is bijective to the set $\pi^{-1}(y) \cap \mathcal{L}_\alpha$, it is finite. In accordance with Proposition 3 the holonomy group $\Gamma(L_\alpha, y)$ of L_α is also finite. The formula

$$R^W : W \times H \to W : (w, a) \mapsto \pi_b(R_a(u)), \forall(w, a) \in W \times H, \forall u \in \pi_b^{-1}(w),$$

defines a smooth action of the group H on the base manifold W (see for example, [31], Proposition 4). Observation that the stationary group H_v, $v \in W$, of the action R^W is isomorphic to the group of covering transformations of the covering $\pi|_{\mathcal{L}_\alpha} : \mathcal{L}_\alpha \to L_\alpha$, where $\mathcal{L}_\alpha = \pi_b^{-1}(v)$, implies the finiteness of all stationary groups. Therefore the orbit space W/H of this action is a smooth q-dimensional orbifold.

Denote by $f : M \to M/F : L \mapsto [L]$ and $f^W : W \to W/H$ the corresponding projections. The map

$$\kappa : M/F \to W/H : [L] \mapsto \pi_b(\pi^{-1}(f^{-1}[L])), \ \forall [L] \in M/F,$$

is well defined and is a bijection. Remind that both f and f^W are open mappings. The submersions π and π_b are also open maps. Thank this, $\kappa : M/F \to W/H$ is a homeomorphism of the topological spaces. Thus we have a natural identification of M/F with N/H through κ, hence M/F is a q-dimensional orbifold. $\square$

Proof of Corollary 1. Let (M, F) be a Riemannian foliation all leaves of which are closed subsets of M. By Theorem of Epstein-Millett-Tishler [12] there exists a saturated dense G_δ subset of M formed by leaves without holonomy. Therefore there exists a closed leaf L of (M, F) with the trivial holonomy group. The application of Theorem 2 finishes the proof. $\square$

Proof of Theorem 3. Suppose that there exists a closed leaf L with a finite fundamental group $\pi_1(L, x)$. As the holonomy group $\Gamma(L, x)$ is a group homomorphism image of $\pi_1(L, x)$, the group $\Gamma(L, x)$ is finite. According to Theorem 2 every leaf L_α of (M, F) is closed and has a finite holonomy group $\Gamma(L_\alpha, x_\alpha)$, where $x_\alpha \in L_\alpha$, and (M, F) is a locally stable foliation.

We shall use notations introduced above. Let $\pi : \mathcal{R} \to M$ be the projection of the foliated H-bundle over M, where H is equal to $O(q)$ or $SO(q)$. Using Proposition 3 we see that $\pi|_{\mathcal{L}} : \mathcal{L} \to L$ is a finite sheet covering map onto L, so the fundamental group $\pi_1(\mathcal{L}, u)$, $\pi(u) = x$, is also finite. Therefore the universal covering map $f : \mathcal{L}^0 \to \mathcal{L}$ is a finite sheet covering.

Let L_α be any leaf of (M, F), $x_\alpha \in L_\alpha$ and $u_\alpha \in \pi^{-1}(x_\alpha)$. Denote by $\mathcal{L}_\alpha$ the leaf of lifted foliation $(\mathcal{R}, \mathcal{F})$ passing through u_α. In accordance with Proposition 3 the group of covering transformations of the regular covering map $\pi|_{\mathcal{L}_\alpha} : \mathcal{L}_\alpha \to L_\alpha$ is isomorphic to the group $\Gamma(L_\alpha, x_\alpha)$, hence it is a finite sheet covering map.

In the proof of Theorem 2 we have showed that the existence of an Ehresmann connection for (M, F) implies the existence of an Ehresmann connection for the lifted foliation $(\mathcal{R}, \mathcal{F})$. Therefore the leaves $\mathcal{L}$ and $\mathcal{L}_\alpha$ are diffeomorphic. So the universal covering map $f_\alpha^0 : \mathcal{L}_\alpha^0 \to \mathcal{L}_\alpha$ is also a finite sheet covering. Therefore $\pi|_{\mathcal{L}_\alpha} \circ f_\alpha^0 : \mathcal{L}_\alpha^0 \to L_\alpha$ is the finite sheet universal covering map. It implies the finiteness of the fundamental group $\pi_1(L_\alpha, x_\alpha)$. $\square$

4. Proofs of other theorems and their corollaries

Proof of Theorem 4. If (M, F) is a Riemannian foliation, the assertion of Theorem 3 follows from Theorem 1.

Further by an attractor of a foliation (M, F) we understand a nonempty closed saturated subset $\mathcal{M}$ of M admitting an open saturated neighbourhood $\mathcal{U}$ such that the closure in M of any leaf from $\mathcal{U} \setminus \mathcal{M}$ contains $\mathcal{M}$. The neighbourhood $\mathcal{U}$ is called a basin of $\mathcal{M}$ and denoted by $\mathcal{A}ttr(\mathcal{M})$. If, moreover, $\mathcal{A}ttr(\mathcal{M}) = M$, then the attractor $\mathcal{M}$ is called *global*.

Assume now that (M, F) is non-Riemannian conformal foliation of codimension $q > 2$ with an Ehresmann connection $\mathfrak{M}$. By our Theorem 4 from [30], (M, F) is complete. Therefore, in accordance with Theorem 5 proved by us in [29] (M, F) has the following properties:

- there exists a global attractor $\mathcal{M}$ of this foliation, that is either one closed leaf or the union of two closed leaves, or the nontrivial minimal set;
- the induced foliation (M_0, F_{M_0}), where $M_0 = M \setminus \mathcal{M}$, is Riemannian.

Note that the restriction $\mathfrak{M}_0$ of the distribution $\mathfrak{M}$ onto the open saturated subset M_0 is an Ehresmann connection for the Riemannian foliation (M_0, F_{M_0}).

Assume that L is a local stable leaf of (M, F). Agreeably to Definition 2 any local stable leaf is proper and is not an attractor. If $\mathcal{M}$ is nontrivial minimal set, then any leaf from $\mathcal{M}$ is improper. Therefore, the leaf L belongs to the Riemannian foliation (M_0, F_{M_0}) admitting an Ehresmann connection $\mathfrak{M}_0$. So in accordance with our Theorem 3 from [30] the holonomy group of the leaf L is inessential.

Converse, suppose that L is a proper leaf with inessential holonomy group of non-Riemannian conformal foliation (M, F) of codimension $q > 2$. Then by Theorem 5 from [30] it is necessary $L \subset M_0$. Thus, L is a proper leaf of the Riemannian foliation (M_0, F_{M_0}) with an Ehresmann connection. According to Theorem 1 L is a local stable leaf. $\quad\square$

Proof of Theorem 5. Consider a proper non-Riemannian conformal foliation (M, F) of codimension $q > 2$. Then there exists a global attractor $\mathcal{M}$ which is either a closed leaf with an essential holonomy group or the union of two closed leaves with essential holonomy groups ([30], Theorem 6).

Put $M_0 = M \setminus \mathcal{M}$. As it was observed in the proof of Theorem 3, in this case (M_0, F_{M_0}) is a Riemannian foliation with an Ehresmann connection on the open saturated subset M_0 of M. Therefore (M_0, F_{M_0}) is proper foliation.

Application of Theorem 1 and Corollary 1 allowed us to state that every leaf L of (M_0, F_{M_0}) is locally stable and closed in M_0 with a finite holonomy group. Hence the closure $Cl(L)$ of L in M is equal to $Cl(L) = L \cup \mathcal{M}$ and L is locally stable unclosed leaf of (M, F) with a finite holonomy group. Thus, $L \subset M_0$ iff L is an unclosed leaf or, equivalent, L has a finite holonomy group.

The remark that L is a local stable leaf of (M, F) iff $L \subset M_0$ leads to the finish of the proof.   □

Proof of Theorem 7. As (M, F) is a compact conformal foliation (M, F) of codimension $q > 2$, it has not an attractor. According to Theorem 2 proved by us in [29] in this case (M, F) is a Riemannian foliation. It is well known (see for instance, [19], Proposition 3.7) a codimension q compact Riemannian foliation is locally stable, and its leaf space M/F admits a structure of smooth q-dimensional orbifold.   □

Proof of Theorem 9. According to our Theorem 4 proved in [30] for non-Riemannian conformal foliation (M, F) of codimension $q > 2$ the existence of an Ehresmann connection is equivalent to its completeness. Therefore Theorem 5 of [30] implies that if a conformal foliation (M, F) admitting an Ehresmann connection has a closed leaf with a finite holonomy group $\Gamma(L, x)$ or a finite fundamental group $\pi_1(L, x)$, then (M, F) is a Riemannian foliation. So the statements of Theorem 9 follow from Theorem 2 and Theorem 3 respectively.   □

Proof of Corollary 4. If all leaves of conformal foliation (M, F) of codimension $q > 2$ are closed, then (M, F) has not attractors. Thus (M, F) is a Riemannian foliation satisfying Corollary 1.   □

Proof of Theorem 10. Let (M, F) be a conformal foliation of codimension $q > 2$ on a compact manifold M. Emphasize that a finite holonomy group is inessential. Therefore in accordance with Theorem 4 proved by the author in [29] the existence of a compact leaf with a finite holonomy group implies that (M, F) is a Riemannian foliation. As M is compact, a bundle like metric g on M relatively the foliation (M, F) is complete. Then the q-dimensional distribution orthogonal to TF is an Ehresmann connection for this foliation. Therefore the assertion of Theorem 10 follows from Theorem 9 (and also from Theorem 8).   □

5. Examples

Example 1 ([27]). Let $E^3 = E^1 \times E^2$ be an 3-dimensional Euclidian space. Put $M = E^3 \setminus (E^1 \times \{(0,0)\})$, where $(0,0) \in E^2$. Then the foliation (M, F), where $F = \{E^1 \times \{(x,y)\} \mid (x,y) \in E^2 \setminus (0,0)\}$, is not transversally complete Riemannian foliation admitting an Ehresmann connection.

Example 2. Consider a product of circles $S^1 \times S^1$. Let $p : S^1 \times S^1 \to S^1 : (x,y) \mapsto x$ be the canonical projection onto the first multiplier. Put $M = (S^1 \times S^1) \setminus \{(a,b)\}$, where $(a,b) \in S^1 \times S^1$ and $p_M = p|_M$. Let (M,F) be the foliation formed by fibres of the submersion $p_M : M \to S^1$.

The foliation (M, F) is Riemannian, and every its leaf is a closed subset of M with the trivial holonomy group. This foliation has a compact leaf diffeomorphic to the circle S^1 with finite holonomy group. As there exists a noncompact leaf $L_0 = p_M^{-1}(a)$, the foliation (M, F) is not locally stable. Remark that (M, F) does not admit an Ehresmann connection. This example shows that the existence of an Ehresmann connection is the essential condition in Theorems 1–3, 6 and Corollary 1.

Example 3. Denote by f_A the linear transformation of the plane R^2 having the matrix $A = \begin{pmatrix} 1 & 1 \\ 0 & 1 \end{pmatrix}$ in the canonical basis. Define an action of the group the integral numbers $\mathbb{Z}$ on the product $R^1 \times R^2 \times R^m \cong R^{3+m}$, where m is an arbitrary nonnegative integer number, by the following formula

$$\Phi : \mathbb{Z} \times R^1 \times R^2 \times R^m \to R^1 \times R^2 \times R^m : (n, t, z, w) \mapsto (t \frown n, (f_A)^n(z), w)$$

for all $n \in \mathbb{Z}$ and $(t, z, w) \in R^1 \times R^2 \times R^m$. As the action Φ of $\mathbb{Z}$ is proper and free, the $(m + 3)$-manifold of orbits $M = R^1 \times_{\mathbb{Z}} (R^2 \times R^m)$ is defined. Let $p : R^1 \times R^2 \times R^m \to M$ be the projection on the orbit space. We get a foliation (M, F) of codimension m+2 covered via p by the trivial foliation $F_{tr} = \{R^1 \times \{(z,w)\} \mid (z,w) \in R^2 \times R^m\}$. Note that the other trivial foliation $F'_{tr} = \{\{t\} \times R^2 \times R^m \mid t \in R^1\}$ of the product $R^1 \times R^2 \times R^m$ is projected by p onto a simple foliation (M, F') transversal to (M, F). Therefore the distribution $\mathfrak{M} = TF'$ tangent to (M, F') is an integrable Ehresmann connection for (M, F). Moreover, it is not difficult to see that (M, F) is $\mathfrak{M}$-complete transversally affine foliation. Observe that all leaves of (M, F) are closed subsets in M.

Let $r : R^1 \times R^2 \times R^m \to R^2 \times R^m$ be the projection onto the multiplier. For any point $v \in M$ there exists a point $(z, w) \in r(p^{-1}(v)) \in R^2 \times R^m$. Emphasize that a leaf $L = L(v)$ is compact and diffeomorphic to the circle S^1 iff $z = (x, 0) \in R^2$ in the the coordinates defined by the canonical basis of R^2. Note that any neighbourhood of a point $(z, w) = ((x, 0), w)$ in $R^2 \times R^m$ contains a subset of the form $\{((x + ny, x), w) \mid n \in \mathbb{Z}\}$ for some $y > 0$. So does not exist a neighbourhood of a point $((x, 0), w)$ invariant relatively the action Φ of the group $\mathbb{Z}$ and belonging to an ε-neighbourhood of $((x, 0), w)$ in the usual topology in $R^2 \times R^m$. It means that any compact leaf $L = L(v)$ is not stable unlike noncompact leaves diffeomorphic to R^1.

Thus, the constructed foliation (M, F) is not locally stable.

References

[1] N.D. Bachmetova, V.A. Igoshin and G.A. Onosova, *O sobstvennom sloe sloenija Reihardta*, Materiali VIII nauchnoi conf. molodich uchenich mech.-mat. fak. and NII mech. Dep. VINITI N 1846-84. Dep., (1984), 110–125 (in Russian).

[2] R.A. Blumenthal, *Stability theorems for conformal foliations*, Proc. AMS, **91** (3) (1984), 485–491.

[3] R. Blumenthal, *Cartan Submersions and Cartan foliations*, Ill. Math. J., **31** (1987), 327–343.

[4] R.A. Blumenthal and J. J. Hebda, *Complementary distributions which preserve the leaf geometry and applications to totally geodesic foliations*, Quart. J. Math. Oxford, **35** (1984), 383–392.

[5] D. Bohnet, *Partially hyperbolic diffeomorphisms with compact central foliation with finite holonomy*, Diss. Univ. Hamburg, (2011).

[6] A. Cap and J. Slovak, *Parabolic Geometries I: Background and General Theory*, AMS Publishing House Math. Surveys Monogr., **154** (2009).

[7] P.D. Carrasco, *Compact dynamical foliations*, Ph.D. thesis, University of Toronto, (2010).

[8] L. Conlon, *Transversally parallelizable foliations of codimension two*, Trans. Amer. Math. Soc., **194** (1971), 79–102.

[9] R. Edwards, K. Millett and D. Sullivan, *Foliations with all leaves compact*, Topology, **16** (1977), 13–32.

[10] C. Ehresmann, *Structures feuilletees*, Proc. of the Fifth Canadian Math. Congress, (1961), 109–172.

[11] D.B. Epstein, *Foliations with all leaves compact*, Ann. Inst. Fourier, **26** (1) (1976), 265–282.

[12] D. Epstein, K. Millett and D. Tishler, *Leaves without holonomy*, J. London Math. Soc., **16** (2) (1977), 548–552.

[13] A. Gogolev, *Partially hyperbolic diffeomorphisms with compact central foliations*, JMD, **4** (2011), 747–769.

[14] R. Hermann, *On the differential geometry of foliations*, Ann. of Math., **72** (2) (1960), 445457.

[15] V.A. Igoshin and Y.L. Shapiro, *Stability of leaves of foliations with a compatible Riemannian metric*, Mat. Zametki, **27** (5) (1980), 767–778.

[16] S. Kobayashi, Transformation groups in differential geometry, Berlin–Heidelberg–New York: Springer Verlag, 1972.

[17] H.B. Lawson, *Foliations,* Bull. Amer. Math. Soc., **80** (3) (1974), 369-418.

[18] K.C. Millett, *Compact foliations,* Lect. Notes Math., **484** (1975), 277–287.

[19] P. Molino, Riemannian Foliations, Progress in Math. 73, Birkhauser Boston, 1988.

[20] J.V. Pereira, *Global stability for holomorphic foliations on Kaehler manifolds,* Qual. Theory Dyn. Syst., **2** (2001), 381–384.

[21] B. Reinhart, *Closed metric foliations,* Michigan Math. J., **8** (1) (1961), 7–9.

[22] B. Reinhart, *Foliated manifolds with bundle-like metrics,* Ann. of Math., **69** (1959), 119–132.

[23] G. Reeb, *Sur certaines proprietes topologiques des varietes feuilletees,* Actualites Sci., **1183**, Hermann, Paris, 1952.

[24] I. Tamura, Topology of foliations. Iwanami Shoten, Japan, 1979.

[25] P. Walczak, *Local stability of holomorphic and transversely holomorphic foliations* Bill. Soc. Sci. Let. de Lodz., **36** (28) (1986), 1–7.

[26] R.A. Wolak, *Leaves of foliations with a transverse geometric structure of finite type,* Publicacions Mat., **33** (1989), 153–162.

[27] R.A. Wolak, *Graphs, Ehresmann connections and vanishing cycles,* Differential Geometry and applications. Proc. Conf., 1995, Brno, Czech Republic, Masaric Uni., (1996), 345–352.

[28] Y.H. Lan, *Existence of complete metric of Riemannian foliations,* Math. J. Toyama Univ., **15** (1992), 35–38.

[29] N.I. Zhukova, *Attractors and an analog of the Lichnerowicz conjecture for conformal foliations,* Siberian Math. J., **52** (3) (2011), 436–450.

[30] N.I. Zhukova, *Global attractors of complete conformal foliations,* Sbornik: Mathematics, **203** (3) (2012), 380-405.

[31] N.I. Zhukova, *Minimal Sets of Cartan Foliations,* Proceedings of the Steklov Institute of Math., **256** (2007), 105–135.

[32] N. Zukova, *On the stability of leaves of Riemannian foliations,* Ann. Global Anal. Geom., **5** (3) (1987), 261-271.

[33] N.I. Zhukova, *The graph of a foliation with Ehresmann connection and stability of leaves,* Iz. VUZ. Mat., **2** (1994), 78–81.

[34] N.I. Zhukova, *Properties of graphs of Ehresmann foliations,* Vestnik Nizhegorodskogo Univ., Ser. Mat., **1** (2004), 77–91 (in Russian).

Received January 16, 2013.

FOLIATIONS 2012
ed. by Paweł WALCZAK *et al.*
World Scientific, Singapore, 2013
pp. 235–255

Problem set

STEVEN HURDER

Department of Mathematics, University of Illinois at Chicago
Chicago, IL 60607-7045, USA
e-mail: hurder@uic.edu

1. Introduction

This is a collection of problems from the theory of foliations and related areas, proposed by the participants in the conference "Foliations 2012" held in Lodz, Poland during June 25–30, 2005.

"Foliation problem sets" have a long tradition: Stanford 1976, compiled by Mark Mostow and Paul Schweitzer [71]; Rio de Janeiro 1976, compiled by Paul Schweitzer [81]; Rio de Janeiro 1992, compiled by Rémi Langevin [47]; Santiago do Compostela 1994, compiled by Xosé Masa and Enriqué Macias-Virgós [56]. There was no general problem set published after the meeting Warsaw 2000, although the survey [39] formulated open problems in the area of foliation dynamics and secondary classes, and the unpublished problem set [40] was prepared for the Conference Geometry and Foliations" held in Kyoto, Japan in September 2003.

2. Jesús Álvarez López

Furstenberg type structure theorem for distal foliated spaces

A celebrated work of H. Furstenberg [23] shows that an ergodic distal action on a Borel probability space is isomorphic to an inverse limit of a tower of equicontinuous factor actions. The Furstenberg Structure Theory was extended to minimal Borel actions on compact spaces in the works of Lindenstrauss [55] and Akin, Auslander and Glasner [1]. The notion of a distal pseudogroup action was defined by Hurder in [43], [15, section 4] and [44, section 3], and by A. Biś and P. Walczak in [3]. The notion of a compactly generated pseudogroup was defined by A. Haefliger in [33], as a generalization of the pseudogroups obtains from nice sections to foliations of compact manifolds.

PROBLEM 2.1. Is there a (Borel) version of the Furstenberg Structure Theory for compactly generated pseudogroups whose action is distal?

Perhaps the main point of this problem is the concept of completeness for pseudogroups, as proven by É. Salem [32, 80] and [69, Appendix D] for isometric actions. The reason is that, for distal actions, a key ingredient of Furstenberg's proof is played by the Ellis group (see [1]), which is defined by taking pointwise limits of the transformations. However, a sequence of local transformations of the pseudogroup may have smaller and smaller domains, obtaining a point domain at the limit. A pseudogroup version of completeness would guarantee that the transformations of the pseudogroup can be locally extended to large enough open sets, so that an analogous version of the "Ellis pseudogroup" would exist, and then Furstenberg's arguments can be easily adapted. But distality does not seem to imply completeness, and therefore some additional argument is needed. Perhaps a weaker notion of completeness, which holds for all compactly generated distal pseudogroups, could be enough to extend Furstenberg's arguments.

3. Andrzej Biś

Multifractal analysis of dynamics of foliated manifolds

In the study of chaos and complexity of foliated manifolds, one often encounters invariant sets with very complicated geometry which reflect where the dynamics concentrate on. The multifractal analysis approach arises both in dimension theory and dynamical systems [72], and essentially concerns decomposition of a set of fractal nature into subsets with certain properties. This approach can be applied to foliation theory.

Geometric entropy was defined for a C^1-foliation $\mathcal{F}$ of a compact manifold M by Ghys-Langevin-Walczak in [26], and it has since become one of the principal numerical invariants for the dynamics of foliations (see Walczak [93] or Hurder [43, 44]). Let $\mathcal{T} \subset M$ be a complete transversal to $\mathcal{F}$ with $\mathcal{G}_{\mathcal{F}}$ the induced compactly generated pseudogroup acting on $\mathcal{T}$, with a fixed generating set $\mathcal{G}_{\mathcal{F}}^{(1)}$. Given a subset $X \subset \mathcal{T}$, we say that $S = \{x_1, \ldots, x_\ell\} \subset X$ is (k, ϵ)-separated for $\mathcal{G}_{\mathcal{F}}$ and X if

$$\forall \, x_i \neq x_j \, , \, \exists \, g \in \mathcal{G}|X \text{ such that } \|g\| \leq k \,\&\, d_{\mathcal{T}}(g(x_i), g(x_j)) \geq \epsilon.$$

Here, $\|g\| \leq k$ means that g can be written as a composition of at most k elements of $\mathcal{G}_{\mathcal{F}}^{(1)}$. Then set

$$h(\mathcal{G}_{\mathcal{F}}, \mathcal{G}_{\mathcal{F}}^{(1)}, X, k, \epsilon) = \max \#\{S \mid S \subset X \text{ is } (k, \epsilon) - \text{separated}\}.$$

When $X = \mathcal{T}$, set $h(\mathcal{G}_{\mathcal{F}}, \mathcal{G}_{\mathcal{F}}^{(1)}, k, \epsilon) = h(\mathcal{G}_{\mathcal{F}}, \mathcal{G}_{\mathcal{F}}^{(1)}, \mathcal{T}, k, \epsilon)$.

Definition [Ghys, Langevin, Walczak [26]]: Let $\mathcal{G}_{\mathcal{F}}$ be a C^r-pseudogroup for $r \geq 1$, with generating set $\mathcal{G}_{\mathcal{F}}^{(1)}$. The *geometric entropy* of $\mathcal{G}_{\mathcal{F}}$ on $X \subset \mathcal{T}$ is

$$h(\mathcal{G}, \mathcal{G}_{\mathcal{F}}^{(1)}, X) = \lim_{\epsilon \to 0} \left\{ \limsup_{k \to \infty} \frac{\ln\{h(\mathcal{G}_{\mathcal{F}}, X, k, \epsilon)\}}{k} \right\}.$$

The *geometric entropy* of $\mathcal{F}$ is defined to be $h(\mathcal{F}) \equiv h(\mathcal{G}_{\mathcal{F}}, \mathcal{G}_{\mathcal{F}}^{(1)}, \mathcal{T}) < \infty$.

Brin and Katok introduced in [8, 28] a notion of local measure-theoretic entropy for maps. The concept of local entropy, as adapted to geometric entropy, was introduced in [43, Definition 13.3]. The set X in the above definition is not assumed to be saturated, so take $X = B(x, \delta) \subset \mathcal{T}$, the open δ-ball about $x \in \mathcal{T}$, to obtain a measure of the amount of "expansion" by the pseudogroup in an open neighborhood of x. Perform the same double limit process as used above for the sets $B(x, \delta)$, but then also let the radius of the balls tend to zero, to obtain:

Definition: The *local geometric entropy* of $\mathcal{G}_{\mathcal{F}}$ at x is

$$h_{loc}(\mathcal{G}_{\mathcal{F}}, \mathcal{G}_{\mathcal{F}}^{(1)}, x) = \lim_{\delta \to 0} \left\{ \lim_{\epsilon \to 0} \left\{ \limsup_{n \to \infty} \frac{\ln\{h(\mathcal{G}_{\mathcal{F}}, \mathcal{G}_{\mathcal{F}}^{(1)}, B(x, \delta), k, \epsilon)\}}{k} \right\} \right\}.$$

The local entropy determines the geometric entropy [43, Proposition 13.4]:

$$h_{loc}(\mathcal{G}_{\mathcal{F}}, \mathcal{G}_{\mathcal{F}}^{(1)}, x) \leq h(\mathcal{G}_{\mathcal{F}}, \mathcal{G}_{\mathcal{F}}^{(1)}, \mathcal{T}), \quad h(\mathcal{G}_{\mathcal{F}}, \mathcal{G}_{\mathcal{F}}^{(1)}, \mathcal{T}) = \sup_{x \in \mathcal{T}} h_{loc}(\mathcal{G}_{\mathcal{F}}, \mathcal{G}_{\mathcal{F}}^{(1)}, x).$$

In contrast with the extensively studied properties of local entropy for a single transformation (see [28, 83]), there are many open questions about the property of local geometric entropy (see [4]).

Definition: Assume that $h(\mathcal{G}_{\mathcal{F}}, \mathcal{G}_{\mathcal{F}}^{(1)}, \mathcal{T}) = A > 0$. For $\alpha \in [0, A]$, we say that a point $x \in X$ is a *point of α-entropy* if $h_{loc}(\mathcal{G}_{\mathcal{F}}, \mathcal{G}_{\mathcal{F}}^{(1)}, x) = \alpha$. Define

$$K_\alpha = \{x \in \mathcal{T} \mid h_{loc}(\mathcal{G}_{\mathcal{F}}, x) = \alpha\}.$$

Then $K_\alpha \neq \emptyset$ for at least some values of α.

PROBLEM 3.1. What is the topological structure of the sets K_α?

PROBLEM 3.2. When K_α is not empty? That is, find properties of the foliation $\mathcal{F}$ which imply there exists $x \in \mathcal{T}$ such that $h_{loc}(\mathcal{G}_\mathcal{F}, \mathcal{G}_\mathcal{F}^{(1)}, x) = \alpha$.

PROBLEM 3.3. Characterize the function $\alpha \mapsto$ Hausdorff Dimension(K_α).

Define an n-ball of radius r, centered at $x \in \mathcal{T}$, to be the set

$$B_n^\mathcal{G}(x, r) := \{y \in \mathcal{T} \mid d(h(x), h(y)) < r$$
$$\text{for all } h \in \mathcal{G}_\mathcal{F} \text{ such that } \|h\| \leq n \text{ and } x, y \in Dom(h)\}.$$

Definition [4] : For any $x \in X$ and a Borel probability measure μ on $\mathcal{T}$, the quantity

$$h_\mu^\mathcal{G}(x) = \lim_{\epsilon \to 0} \limsup_{n \to \infty} \left\{ -\frac{1}{n} \log \mu(B_n^\mathcal{G}(x, \epsilon)) \right\}$$

is called a *local upper μ-measure entropy* at the point x, with respect to $(\mathcal{G}_\mathcal{F}, \mathcal{G}_\mathcal{F}^{(1)})$, and

$$h_{\mu, \mathcal{G}}(x) = \lim_{\epsilon \to 0} \liminf_{n \to \infty} \left\{ -\frac{1}{n} \log \mu(B_n^\mathcal{G}(x, \epsilon)) \right\}$$

is called a *local lower μ-measure entropy* at the point x, with respect to $(\mathcal{G}_\mathcal{F}, \mathcal{G}_\mathcal{F}^{(1)})$.

For a Borel probability measure μ on $\mathcal{T}$, define

$$L_\alpha := \{x \in \mathcal{T} \mid h_\mu^\mathcal{G}(x) = \alpha\}, \qquad M_\alpha := \{x \in \mathcal{T} \mid h_{\mu, \mathcal{G}}(x) = \alpha\},$$

which yields the decompositions (see [4]), $\mathcal{T} = \bigcup_\alpha M_\alpha = \bigcup_\alpha L_\alpha$.

PROBLEM 3.4. What are properties of the families of sets $\{L_\alpha\}$ and $\{M_\alpha\}$?

PROBLEM 3.5. What is the relation between the sets $\{K_\alpha\}$, $\{L_\alpha\}$ and $\{M_\alpha\}$?

4. Hélène Eynard-Bontemps

Surface group representations

Let $\mathcal{D}_+^r$ denote the set of orientation-preserving, C^r diffeomorphisms of the unit interval. The following result is shown in [20].

Theorem: *For $r \geq 2$, any two representations of $\mathbb{Z}^2$ in $\mathcal{D}_+^r$ can be connected by a continuous path of representations of $\mathbb{Z}^2$ in $\mathcal{D}_+^1$. In other words, every two pairs of commuting C^r diffeomorphisms on the interval can be connected by a continuous path of commuting C^1 diffeomorphisms. The result extends to $\mathbb{Z}^k$ representations (i.e., k-tuples of commuting diffeomorphisms), but unfortunately, the diffeomorphisms in the connecting path are not shown to be more smooth than C^1.*

This result was extended in [7] to show:

Theorem: *The space of C^∞ orientation-preserving actions of $\mathbb{Z}^n$ on $[0,1]$ is connected. Similarly, the group of non-free actions of $\mathbb{Z}^2$ on the circle is connected.*

Now let $\sum_g$ denote a compact surface with genus $g \geq 2$, and let $\Gamma_g = \pi_1(\sum_g, x_0)$ denote its fundamental group for some basepoint x_0.

The space of representations of Γ_g into $\text{Diff}_+^r(\mathbb{S}^1)$ corresponds to the C^r-actions of Γ_g on the circle. The properties of these representation spaces have been extremely well-studied, and the topology of the representation variety is connected to the values of the euler class associated to the circle bundle over $\sum_g$ obtained [30, 61, 84]. Much less is known in the case of actions on the interval $[0,1]$. Let

$$\mathcal{R}(\Gamma_g, \text{Diff}_+^\infty([0,1])\} \ = \ \{\rho \colon \Gamma_g \to \text{Diff}_+^\infty([0,1])\}$$

be the representation variety with the C^∞-topology.

PROBLEM 4.1. What are the path-components of the space of representations $\mathcal{R}(\Gamma_g, \text{Diff}_+^\infty([0,1])\}$?

PROBLEM 4.2. Find invariants which distinguish the path components of $\mathcal{R}(\Gamma_g, \text{Diff}_+^\infty([0,1])\}$.

5. Steven Hurder

Exceptional minimal sets

Let $\mathcal{F}$ be a C^r-foliation of a closed manifold M of codimension-q, for $r \geq 1$. Let $\mathfrak{M} \subset M$ be a minimal set for $\mathcal{F}$. That is, $\mathfrak{M}$ is a closed subset which is a union of leaves, and every leaf in $\mathfrak{M}$ is dense. We say that $\mathfrak{M}$ is *exceptional* if its intersection with every transversal to $\mathcal{F}$ is totally disconnected.

The construction of exceptional minimal sets for codimension-1 foliations by Sacksteder [78] and Rosenberg and Roussarie [74], along with Sacksteder's theorem that the Denjoy minimal set does not exist for C^2-foliations [79], can be viewed as the beginnings of the modern study of foliation dynamics. The works [10, 11, 45, 46, 58] all consider the structure of exceptional minimal sets in codimension-1. There remain open questions in this case, none-the-less, as discussed in the problem sets [39, 40].

However, the situation for higher codimension is essentially completely open.

PROBLEM 5.1. Classify the exceptional minimal sets for codimension-q, for $q \geq 2$ and $r \geq 2$.

In codimension-1, there is a dichotomy for the derivative of the holonomy maps on a minimal set, in that they either have slow (subexponential) growth, or exponential growth [78, 38, 44]. For higher codimensions, the transverse dynamics can be more complicated, as evidenced by the constructions of solenoidal minimal sets for flows of 3-manifolds, and hyperbolic minimal sets which have the transverse structure of a horseshoe. The work of Clark and Hurder [16] constructed solenoidal minimal sets for foliations with leaf dimensions at least 2, and this work has many references to prior works. A basic problem remains to find new constructions of such sets.

On the other hand, the work on the structure of laminations arising from tiling spaces has led to new topological classification results for exceptional minimal sets [14] and part of the question is to what extent does the fact that $\mathfrak{M}$ arises from a C^2-dynamical system restrict its topological type?

6. Victor Kleptsyn

Anosov questions

The following is the simplest case of a well-known existence problem:

PROBLEM 6.1. Let M^4 be a simply connected closed 4-manifold. Show that there does not exist an Anosov diffeomorphism of M.

The second problem was asked by Anosov in the 1960's. A *polynomial foliation* on $\mathbb{C}^2$ is defined by complex differential equations:

$$\dot{X} = P_n(X, Y)$$
$$\dot{Y} = Q_n(X, Y)$$

where P_n, Q_n are homogeneous polynomials on degree n.

PROBLEM 6.2. For the generic such polynomial foliation, is the generic leaf simply connected?

7. Rémi Langevin

Conformal geometry of foliations

The Riemannian geometry of foliations studies properties of the ambient manifold M and the leaves of a foliation $\mathcal{F}$ on it, such as intrinsic and extrinsic invariants of the curvature for the leaves. In contrast, the conformal geometry of a foliation [51, 48] studies properties derived from properties of the geodesics, such as their intersections, and properties of pencils (families of geodesic rays based at a point) and evolutes of the leaves. Both of these approaches to the geometry of foliations can then be related to other properties of the foliation, such as the total curvature of leaves [6, 49, 50, 76], minimality for the leaves [77] and the mean curvature flow [92], the dynamics of the leafwise geodesic flow [90] and the foliation entropy defined in [26, 93].

In all of the works of this nature, it is helpful to have an ambient Riemannian metric on the manifold M such that the leaves of the foliation $\mathcal{F}$ have the "most simple" geometry. This suggest the very general (though vague) problem:

PROBLEM 7.1. Let $\mathcal{F}$ be a C^∞ foliation on a closed manifold M. What is the simplest (or nicest) Riemannian metric on M which results in the simplest extrinsic geometry for the leaves?

The study of Dupin and Canal foliations in [2, 51, 52] provide motivating examples for this question.

The following is a very old problem, that remains as attractive and inaccessible as always.

PROBLEM 7.2. Let $\mathcal{F}$ be a codimension-one foliation of a closed oriented 3-manifold M. Give an interpretation of the Godbillon-Vey invariant $GV(\mathcal{F}) \in \mathbb{R}$ for $\mathcal{F}$ in terms of the conformal geometry of $\mathcal{F}$.

Thurston gave an intuitive interpretation of $GV(\mathcal{F})$ in terms of the total "helical wobble" of the leaves [85, 73]. Gelfand offered an interpretation via the simplexes swept out by planes parallel to the leaves. Only the result of Reinhart and Wood [73] is precise, and not just an intuitive statement. But it is not a conformal invariant, as it is derived from the extrinsic properties of the immersions of the leaves. The problem is to find a precise calculation of the number $GV(\mathcal{F})$ in terms of conformal invariants, if possible.

8. Yoshifumi Matsuda

Rotation numbers

Let $\text{Diff}^{\omega}_+(\mathbb{S}^1)$ denote the group of orientation-preserving, real analytic diffeomorphisms of the circle. For each $\varphi \in \text{Diff}^{\omega}_+(\mathbb{S}^1)$, the Poincaré rotation number $\rho(\varphi) \in \mathbb{R}/\mathbb{Z}$. If φ has finite order, or more generally has a periodic orbit, then $\rho(\varphi) \in \mathbb{Q}/\mathbb{Z}$.

Let $\Gamma \subset \text{Diff}^{\omega}_+(\mathbb{S}^1)$ be a finitely-generated subgroup. Let $\text{rot}(\Gamma) \subset \mathbb{R}/\mathbb{Z}$ denote the set of Poincaré rotation numbers for the elements of Γ.

This work [57] studied the case when $\text{rot}(\Gamma)$ is a finite set, and showed that if such a group is nondiscrete with respect to the C^1-topology then it has a finite orbit. This implies that if such a group has no finite orbit then each of its subgroups contains either a cyclic subgroup of finite index or a nonabelian free subgroup.

PROBLEM 8.1. Suppose that $\text{rot}(\Gamma) \subset \mathbb{Q}/\mathbb{Z}$ then must $\text{rot}(\Gamma)$ be a finite set?

This is known to be true if $\Gamma \subset \mathbf{PSL}(2, \mathbb{R})$ considered as acting on $\mathbb{S}^1$, which follows by Selberg's Lemma. But the results of Ghys-Sergiescu in [25] show that this is false, if we consider the embedding they obtain of the Thompson group $T \subset \text{Diff}^{\infty}_+(\mathbb{S}^1)$, so that the analytic assumption is necessary.

PROBLEM 8.2. Does Γ always contain a finite-index subgroup which is torsion-free?

This is known to be true for subgroups $\Gamma \subset \mathbf{GL}(n, \mathbb{K})$ where $\mathbb{K}$ is a field with characteristic 0 by Selberg's Lemma. It is false for the Thompson group $T \subset \text{Diff}^{\infty}_+(\mathbb{S}^1)$.

These two problems ask whether $\text{Diff}^{\omega}_+(\mathbb{S}^1)$ is more similar to a finite dimensional Lie group, or whether it contains subgroups with properties of the Thompson group T as embedded in $\text{Diff}^{\infty}_+(\mathbb{S}^1)$.

9. Gaël Meigniez

Lie foliations modeled on G

A smooth foliation $\mathcal{F}$ on a closed manifold M is said to be a Lie foliation modeled on a connected Lie group G [17, 21] if there exists a covering $\pi: \widetilde{M} \to M$ and a submersion $\Pi: \widetilde{M} \to G$ such that the lifted foliation $\widetilde{\mathcal{F}}$ under the covering map has leaves equal to the fibers of the map Π. A Lie foliation is necessarily a transversally complete Riemannian foliation, so the Molino structure theory for Riemannian foliations can be applied to obtain many results, as in [12, 27, 68, 69].

PROBLEM 9.1. Classify the Lie foliations whose leaves have dimension two, and are covered by the hyperbolic plane $\mathbb{H}^2$.

For example, one family of such foliations is obtained from taking a finitely-generated group Γ and a discrete, cocompact smooth action ρ on the product space $\mathbb{H}^2 \times G$, where the action is assumed to preserve the product structure, then $M \cong \mathbb{H}^2 \times G/\rho$ has a Lie G-foliation $\mathcal{F}_\rho$ induced from the product foliation on the covering $\mathbb{H}^2 \times G$.

10. Eva Miranda

Symplectic foliations

Consider a regular Poisson manifold M with a smooth codimension-one foliation $\mathcal{F}$ admitting a defining 1-form α. Using the existence of this 1-form, V. Guillemin and E. Miranda and A.R. Eva introduced in [29] two invariants in the leafwise cohomology groups: the Reeb class in leafwise cohomology $[c_\mathcal{F}] \in H^1(M, \mathcal{F})$ (cf. [36]), and a class $[\sigma_\mathcal{F}] \in H^2(M, \mathcal{F})$ derived from a global 2-form on M which restricts to leafwise symplectic forms.

Theorem: *Let $\mathcal{F}$ be a regular Poisson manifold M with a smooth codimension-one foliation $\mathcal{F}$. Suppose that the invariants $[c_\mathcal{F}] = 0$ and $[\sigma_\mathcal{F}] = 0$, and $\mathcal{F}$ has a compact leaf, then M is a symplectic mapping torus.*

PROBLEM 10.1. Can we characterize the regular Poisson manifolds M with smooth codimension-one foliation $\mathcal{F}$ which have no compact leaf? Is there a type of rigidity result for such geometric structures?

For example, an application of Tischler's Theorem [54, 86] and Ghys' work in [27] may prove useful.

11. Yoshuhiko Mitsumatsu

11.1. *Leafwise symplectic foliations on* $\mathbb{S}^5$

In a celebrated work [53], H. Blaine Lawson, Jr. constructed smooth codimension-one foliations on the 5-sphere $\mathbb{S}^5$. We quote from [64]:

It was achieved by a beautiful combination of the complex and differential topologies and was a breakthrough in an early stage of the history of foliations. The foliation is composed of two components. One is a tubular neighbourhood of a 3-dimensional nil-manifold and the other one is, away from the boundary, foliated by Fermat-type cubic complex surfaces. As the common boundary leaf, there appears one of Kodaira-Thurston's 4-dimensional nil-manifolds. As each Fermat cubic leaf is spiraling to this boundary leaf, its end is diffeomorphic to a cyclic covering of Kodaira-Thurston's nil-manifold.

Meerssemann and Verjovsky in [66, 67] considered the existence of leafwise complex and symplectic structures on Lawsons foliations as well as on slightly modified ones. Then in [64] it was proved:

Theorem: *Lawson's foliation on the 5-sphere* $\mathbb{S}^5$ *admits a leafwise symplectic structure.*

For $\frac{1}{p} + \frac{1}{q} + \frac{1}{r} < 1$, the polynomial

$$f(Z_0, Z_1, Z_2) = Z_0^p + Z_1^q + Z_2^r + Z_0 Z_1 Z_2 = 0$$

defines a cusp singularity of the hypersurface at the origin in $\mathbb{C}^3$.

PROBLEM 11.1. For the Milnor fibration of this singularity, does a construction similar to that in [64] work to produce a leafwise symplectic foliation on $\mathbb{S}^5$? Or at least, does the Milnor fibre admit an end-periodic symplectic structure?

Note that the link is a *Solv* 3-manifold, hence it fibres over $\mathbb{S}^1$ and even fits into Mori's framework, so that an isotopic family of associated contact structure converges to the associated spinnable foliation.

If the answer to Problem 11.1 is "yes", can we further generalize it to higher dimensional case [65]?

PROBLEM 11.2. Let F be the Fermat cubic surface. Does $F \times F$ admit an end-periodic symplectic structure? Does $F \times \mathbb{C}$ admit an end-periodic symplectic structure? In general, one can ask, if W and V are open symplectic manifolds which are periodic on their ends, does $W \times V$ admit such a structure also?

2-calibration and tautness

A codimension-one foliated manifold $(M, \mathcal{F})$ with a closed 2-form ω which restricts to a symplectic form on each leaf is said to be *2-calibrated*. If M

is a closed manifold, then $\mathcal{F}$ is necessarily taut.

PROBLEM 11.3. Does there exist a 2-calibrated codimension-1 foliation on $\mathbb{S}^5$?

A tangential symplectic structure on a foliation is the same thing as a *regular Poisson* structure.

PROBLEM 11.4. Find a (classical) physical procedure to obtain a regular Poisson structure which can not be lifted to a global closed 2-form, that is, they are not 2-calibrated.

PROBLEM 11.5. Is any closed transversal to a taut codimension-1 foliation not null-homotopic? Is this true in the 2-calibrated case?

PROBLEM 11.6. Is it possible to destroy the compact leaf of Lawson's foliation without destroying a leafwise symplectic structure? (cf. Gaël Meigniez's h-principle [60].)

PROBLEM 11.7. Does there exists an integral closed symplectic manifold (W, ω) of dimension ≥ 4 such that the associated unit $\mathbb{S}^1$-bundle of the pre-quantization admits a foliated bundle structure?

The 2-dimensional case is well-known to have a positive solution. The construction of higher dimensional examples would provide explicit examples of the following result:

Theorem [Morita [70]] *The universal euler class* $e \in H^2(B\mathit{Diff}_+^\infty(\mathbb{S}^1); \mathbb{R})$ *of flat $\mathbb{S}^1$-bundles has all its powers non-trivial. That is,*

$$\forall n \in \mathbb{N}, \quad e^n \neq 0 \in H^{2n}(B\mathit{Diff}_+^\infty(\mathbb{S}^1); \mathbb{R}).$$

There is no known explicit construction of a cycle (a foliated $\mathbb{S}^1$-bundle) with $e^n \neq 0$ for $n \geq 2$.

PROBLEM 11.8. Find a (topological) obstruction to the existence of leafwise symplectic foliations.

If the ambient closed manifold M has trivial $H^2(M, \mathbb{R})$, then the existence of a compact leaf for $\mathcal{F}$ is an easy obstruction. This is the situation for $M = \mathbb{S}^5$, for example.

Moduli for leafwise complex structures

For Lawson's foliation in [53] (the first of three simple elliptic hypersurface singularities), the boundary of the tube component is *Kodaira's primary surface* and admits a complex structure. Its moduli space is well understood.

Meerssemann and Verjovsky in [66, 67] analysed the moduli space of leafwise complex structures on foliated manifolds, and in particular for Lawson's foliation (see also G. Deschamps2010 [18]). They obtained the following results:

Theorem: *The moduli space for the 3-dimensional Reeb component exactly coincides with that of the boundary torus through the restriction map.*

Theorem: *The tube component of Lawson's foliation does not admit a leafwise complex structure, that is, its moduli space is empty.*

PROBLEM 11.9. Find non-trivial examples of leafwise Kähler foliations. Conversely, find a (topological) obstruction to their existence.

PROBLEM 11.10. Prove that the tube component for a cusp singularity does not admit a leafwise complex structure.

For the cusp singularities, first we should find a complex structure for the boundary. If it does not admit any, then already the problem is solved.

Flat surface bundles

Consider the representation space $\pi_0(Hom(\pi_1(\sum_g), Diff_+^\infty(\mathbb{S}^1)))$. Here are three problems about this space.

PROBLEM 11.11. Assume that e divides $(2 - 2g) = \chi(\sum_g)$, and set $d = (2 - 2g)/e$. Consider the component of the representation variety which contains the d-fold covering of the Anosov foliation arising from the geodesic Anosov flow. Does it consist of topologically conjugate foliations?

PROBLEM 11.12. Let $\mathcal{F}_0$ and $\mathcal{F}_2$ be foliations associated to representations in different connected components, but with the same euler numbers. Does there exists a continuous family $\mathcal{F}_t$ for $0 \le t \le 1$ of taut foliations of the total space between $\mathcal{F}_0$ and $\mathcal{F}_1$?

PROBLEM 11.13. Let $\mathcal{F}_0$ and $\mathcal{F}_2$ be foliations associated to representations in different connected components. Does there exist a foliation $\mathcal{F}$ of codimension 2 on $W^4 = [0,1] \times M$ which is tangent to the boundary $\partial W = \{0,1\} \times M$ and restricts to $\mathcal{F}_i$ on $\{i\} \times M$ for $i = 0, 1$? If so, give a geometric construction of $\mathcal{F}$.

12. Hiraku Nozawa

Secondary invariants

Let $n \geq 3$ and (M, g_0) a compact hyperbolic n-manifold with constant sectional curvature. That is, we assume there is a uniform lattice $\Gamma \subset \mathrm{Isom}(\mathbb{H}^n)$ such that $M \cong \mathbb{H}^n/\Gamma$. Let g be a metric on M which is C^∞-close to a metric of constant negative curvature g_0 on M. Then the stable foliation $\mathcal{F}_g$ of the geodesic flow of g on the unit tangent sphere bundle SM of g_0 are $C^{1,\alpha}$ for $\alpha = 1 - \epsilon$ where $\epsilon < 1/n$ by results of Hasselblatt [34]. Thus, the extended Godbillon-Vey class $\mathrm{GV}(\mathcal{F}_g) \in H^{2n+1}(SM; \mathbb{R})$, defined as in [37], is well-defined.

PROBLEM 12.1. Compute the Godbillon-Vey number $\mathrm{GV}(\mathcal{F}_g)$ of $\mathcal{F}_g$.

In the case of $n = 2$, so-called Mitsumatsu Defect formula that computes $\mathrm{GV}(\mathcal{F}_g)$ was obtained by Mitsumatsu [62] and Hurder-Katok [37]. The answer of the problem will give an example of families of $C^{1,\alpha}$-foliations which are topological conjugate but have different Godbillon-Vey numbers in higher codimension, yet the foliations are topologically conjugate by the stability of Anosov flows.

13. Paul Schweitzer

Space of foliations

For a closed manifold N, and $I = [0,1]$, let $\mathrm{Fol}_c^r(N \times I)$ be the space of codimension-one, C^r-foliations on $N \times I$ with the C^r-topology for $1 \leq r \leq \infty$, and all leaves compact submanifolds, with the boundary being leaves as well. This space is easily seen to be locally contractible for $r = \infty$.

PROBLEM 13.1. Show that $\mathrm{Fol}_c^\infty(N \times I)$ is connected, which is equivalent to it being path connected.

PROBLEM 13.2. Is each connected component of the space $\mathrm{Fol}_c^\infty(N \times I)$ contractible?

Note that a closed manifold M has a compact codimension-one foliation, then some double covering of M it fibers over $\mathbb{S}^1$ with the leaves as fibers.

For the case $N = \mathbb{S}^2$ the space $\mathrm{Fol}^\infty(N \times I)$ is known to be connected, by Cerf's Theorem [13] and contractible by Hatcher's proof of the Smale Conjecture [35]. On the other hand, for some $3 \leq n \leq 5$, for the sphere $\mathbb{S}^n$, $\mathrm{Fol}^\infty(\mathbb{S}^n \times I)$ is not contractible, as a consequence of Burghelea's results on the homotopy types of diffeomorphism groups of compact manifolds.

PROBLEM 13.3. Let $\sum_g$ be the closed oriented surface of genus $g \geq 1$. Show that $\mathrm{Fol}_c^r(\sum_g \times [0,1])$ is contractible, for $r \geq 1$.

Ends of leaves

There is an extensive knowledge of the structure of codimension-one foliations on compact manifolds. However, some problems remain a just as much a puzzle as always. One general set of open questions concern the ends of non-compact leaves.

PROBLEM 13.4. Must an isolated end of a non-compact leaf of a codimension-one foliation of a compact manifold be periodic?

Here is a more specific version of this question:

PROBLEM 13.5. Can a contractible 3-manifold that is not periodic at infinity occur as a such a leaf with isolated ends?

For foliations with leaves of dimension 4, there is an analogous question, which has been asked at foliation meetings for many years. Recall that an exotic $\mathbb{R}^4$ is a smooth manifold X which is homeomorphic to Euclidean $\mathbb{R}^4$, but is not diffeomorphic to $\mathbb{R}^4$. There are in fact continuous families of such exotic beasts (see Gompf [31], Furuta and Ohta [24]). There are also constructions of continuous families of exotic differentiable structures of open 4-manifolds (see [5, 19]). All of these manifolds are not the coverings of a compact manifold, and it seems just as unlikely that they could be leaves of a foliation.

PROBLEM 13.6. Show that an exotic $\mathbb{R}^4$ cannot be quasi-isometric to a leaf of a C^1-foliation of a compact manifold M in codimension-one.

Though the conclusion seems very plausible, there is as yet no solution known. One supposes that the same sort of restriction is true for the other constructions of exotic open 4-manifolds, and also in higher codimension $q > 1$, but again, no results in this direction seem to be known.

14. Takashi Tsuboi

Commutator width

Let M^{2n} be a closed, even dimensional C^∞-manifold. It is known that the connected component $\mathrm{Diff}^\infty(M)_0$ of the identity of $\mathrm{Diff}^\infty(M)$ is a perfect group, so that every element $\varphi \in \mathrm{Diff}^\infty(M)_0$ can be written as a product of commutators. The *commutator length* of φ is the least number of commutators required. The *commutator width* of $\mathrm{Diff}^\infty(M)_0$ is the uniform upper bound for all of the commutator lengths, which may possibly be infinite. For example, in the case $n = 2$ it is not known if the commutator width is finite.

PROBLEM 14.1. Estimate the commutator width of $\mathrm{Diff}^\infty(M)_0$.

It is known that the commutator width is finite for $2n \geq 6$ and may depend on the manifold M, as shown in [88]. It is also known that the commutator width is finite, and does not depend on M when M admits a handle decomposition without handles of the middle index [87].

A continuum is a non-empty, compact connected metric space $\mathfrak{M}$.

PROBLEM 14.2. Characterize the continua $\mathfrak{M}$ such that every homeomorphism of $\mathfrak{M}$ is a commutator. That is, the commutator width of $\mathrm{Homeo}(\mathfrak{M})$ is one.

For example, it is known that this is true if $\mathfrak{M} = \mathbb{S}^n$ (the group of orientation preserving homeomorphisms), or $\mathfrak{M}$ is a Menger space [89].

15. Vladimir Rovenski

Totally geodesic foliations

Let M^{n+p} be a connected manifold, endowed with a p-dimensional foliation $\mathcal{F}$, i.e., a partition of M into p-dimensional submanifolds. A foliation $\mathcal{F}$ on a Riemannian manifold (M, g) is *totally geodesic* if the leaves (of $\mathcal{F}$) are totally geodesic submanifolds. A Riemannian metric g on $(M, \mathcal{F})$ is called *totally geodesic* if $\mathcal{F}$ is totally geodesic with respect to g. We have the g-orthogonal decomposition $TM = D_{\mathcal{F}} \oplus D$, where the distribution $D_{\mathcal{F}}$ (dim $D_{\mathcal{F}} = p$) is tangent to $\mathcal{F}$.

Let $\{e_i, \varepsilon_\alpha\}_{i \leq n, \alpha \leq p}$ be a local orthonormal frame on TM adapted to D and $D_{\mathcal{F}}$. The *mixed scalar curvature* is the following function on M, see [75, 91] etc:

$$\mathrm{Sc}_{\mathrm{mix}} = \sum_{i=1}^{n} \sum_{\alpha=1}^{p} K(e_i, \varepsilon_\alpha),$$

where $K(e_i, \varepsilon_\alpha)$ is the sectional curvature of the plane spanned by the vectors e_i and ε_α.

The integral formula with total $\mathrm{Sc}_{\mathrm{mix}}$, see [91], gives us decomposition criteria for foliations with an integrable orthogonal distribution D under the constraints on the sign of $\mathrm{Sc}_{\mathrm{mix}}$, for example:

(1) If $\mathcal{F}$ and $\mathcal{F}^{\perp}$ are complementary orthogonal totally umbilical and totally geodesic foliations on a closed oriented Riemannian manifold M with $\mathrm{Sc}_{\mathrm{mix}} \geq 0$, then M splits along the foliations.

(2) A compact minimal foliation $\mathcal{F}$ on a Riemannian manifold M with an integrable orthogonal distribution and $\mathrm{Sc}_{\mathrm{mix}} \geq 0$ splits along the foliations.

The basic question that we want to address is the following.

PROBLEM 15.1. Which foliations admit a totally geodesic metric with $\mathrm{Sc}_{\mathrm{mix}} > 0$?

Notice that a change of initial metric along orthogonal (to $\mathcal{F}$) distribution D preserves the property "$\mathcal{F}$ is totally geodesic". Let $\pi : M \to B$ be a fiber bundle with compact fibers. One may deform the metric g along D on a neighborhood of a fiber to obtain a bundle-like totally geodesic metric $\tilde{g}$ (which in general is not D-conformal to g) on the fiber. If there is a section $\xi : B \to M$ then the deformation can be done globally (on M), and π becomes a Riemannian submersion with totally geodesic fibers. In this case, the mixed sectional curvature is non-negative, see O'Neill's formula $K(X, V)|X|^2|V|^2 = |A_X V|^2$ ($X \in D$, $V \in D_{\mathcal{F}}$), moreover, if D is nowhere integrable then $\mathrm{Sc}_{\mathrm{mix}} > 0$ ($\mathrm{Sc}_{\mathrm{mix}} \equiv 0$ when D is integrable).

Due to above, we ask the following, which is a special case of Problem 15.2.

PROBLEM 15.2. Given a Riemannian manifold (M, g) with a totally geodesic foliation $\mathcal{F}$, does there exist a D-conformal to g metric $\tilde{g}$ on M such that $\mathrm{Sc}_{\mathrm{mix}} > 0$?

Problem 15.2 for Riemannian metrics on a fiber bundle was studied in [75], where it was shown:

Example: For any $n \geq 2$ and $p \geq 1$, there exists a fiber bundle with a closed $(n + p)$-dimensional total space and a compact p-dimensional fiber, having a totally geodesic metric of positive mixed scalar curvature.

To show this, consider the Hopf fibration $\tilde{\pi}\colon \mathbb{S}^3 \to \mathbb{S}^2$ of a unit sphere $\mathbb{S}^3$ by great circles (closed geodesics). Let $\tilde{F}$ and $\tilde{B}$ be closed Riemannian manifolds with dimensions, respectively, $(p - 1)$ and $(n - 2)$. Let $M = \tilde{F} \times \mathbb{S}^3 \times \tilde{B}$ be the metric product, and $B = \mathbb{S}^2 \times \tilde{B}$. Then $\pi : M \to B$ is a fibration with a totally geodesic fiber $F = \tilde{F} \times \mathbb{S}^1$. Certainly, $\mathrm{Sc}_{\mathrm{mix}} = 2 > 0$.

References

[1] E. Akin, J. Auslander and E. Glasner, The topological dynamics of Ellis actions, Mem. Amer. Math. Soc., **195**, 2008.

[2] A. Bartoszek and P. Walczak, *Foliations by surfaces of a peculiar class*, Ann. Polon. Math., **94** (2008), 89–95.

[3] A. Biś and P. Walczak, *Entropy of distal groups, pseudogroups, foliations and laminations*, Ann. Polon. Math., **100** (2011), 45–54.

[4] A. Biś, *An analogue of the Variational Principle for group and pseudogroup actions*, Ann. Institut Fourier, to appear.

[5] Ž. Bižaca, *Smooth structures on collarable ends of 4-manifolds*, Topology, **37** (1998), 461–467.

[6] F. Brito, R. Langevin and H. Rosenberg, *Intégrales de courbure sur des variétés feuilletées*, J. Diff. Geom., **16** (1981), 19–50.

[7] C. Bonatti and H. Eynard, *Connectedness of the space of smooth actions of $\mathbb{Z}^n$ on the interval*, arXiv:1209.1601, (2012).

[8] M. Brin and A. Katok, *On local entropy*, in Geometric dynamics (Rio de Janeiro, 1981), Lecture Notes in Math., **1007**, Springer, Berlin, 1983, 30–38.

[9] D. Calegari, *Problems in foliations and laminations of 3-manifolds*, in Topology and geometry of manifolds (Athens, GA, 2001), Proc. Sympos. Pure Math. Vol. 71, Amer. Math. Soc., Providence, R.I., 2003, 297–335.

[10] J. Cantwell and L. Conlon, *Foliations and subshifts*, Tohoku Math. J., **40** (1988), 165–187.

[11] J. Cantwell and L. Conlon, *Endsets of exceptional leaves; a theorem of G. Duminy*, in Foliations: Geometry and Dynamics (Warsaw, 2000), World Sci. Publ. Singapore, 2002, 225–261.

[12] Y. Carrière and É. Ghys, *Feuilletages totalement géodésiques*, An. Acad. Brasil. Ciênc., **53** (1981), 427–432.

[13] J. Cerf, *Sur les difféomorphismes de la sphère de dimension trois ($\Gamma_4 = 0$)*, Lecture Notes in Math., No. 53, Springer-Verlag, Berlin, 1968.

[14] A. Clark, S. Hurder and O. Lukina, *Shape of matchbox manifolds*, preprint, (2013).

[15] A. Clark and S. Hurder, *Homogeneous matchbox manifolds*, Trans. Amer. Math. Soc., to appear, arXiv:1006.5482v2.

[16] A. Clark and S. Hurder, *Embedding solenoids in foliations*, Topology Appl., **158** (2011), 1249–1270.

[17] L. Conlon, *Transversally parallelizable foliations of codimension two*, Trans. Amer. Math. Soc., **194** (1974), 79–102, erratum, ibid. 207.

[18] G. Deschamps, *Feuilletage lisse de $\mathbb{S}^5$ par surfaces complexes*, C. R. Math. Acad. Sci. Paris, **348** (2010), 1303–1306.

[19] F. Ding, *Smooth structures on some open 4-manifolds*, Topology, **36** (1997), 203–207.

[20] H. Eynard, *A connectedness result for commuting diffeomorphisms of the interval*, Ergodic Theory Dynam. Systems, **31** (2011), 1183–1191.

[21] E. Fedida, *Sur les feuilletages de Lie*, C. R. Acad. Sci. Paris Sér. A-B, **272** (1971), A999–A1001.

[22] S. Friedl and S. Vidussi, *Twisted Alexander polynomials detect fibered 3-manifolds*, Ann. of Math. (2), **173** (2011), 1587-1643.

[23] H. Furstenberg, *The structure of distal flows*, Amer. J. of Math., **85** (1963), 477–515.

[24] M. Furuta and H. Ohta, *Differentiable structures on punctured 4-manifolds* Topology Appl., **51** (1993), 291–301.

[25] É. Ghys and V. Sergiescu, *Sur un groupe remarquable de difféomorphismes du cercle*, Comment. Math. Helv., **62** (1987), 185–239.

[26] É. Ghys, R. Langevin, and P. Walczak, *Entropie géométrique des feuilletages*, Acta Math., **160** (1988), 105–142.

[27] É. Ghys, *Classification des feuilletages totalement géodésiques de codimension un*, Comment. Math. Helv., **58** (1983), 543–572.

[28] E. Glasner and X. Ye, *Local entropy theory*, Ergodic Theory Dynam. Systems, **29** (2009), 321–356.

[29] V. Guillemin, E. Miranda and A.R. Pires, *Codimension one symplectic foliations and regular Poisson structures*, Bull. Braz. Math. Soc. (N.S.), **42** (2011), 607–623.

[30] W. Goldman, *The symplectic nature of fundamental groups of surfaces*, Adv. in Math., **54** (1984), 200–225.

[31] R. Gompf, *An exotic menagerie*, J. Diff. Geom., **37** (1993), 199–223.

[32] A. Haefliger and É. Salem, *Riemannian foliations on simply connected manifolds and actions of tori on orbifolds*, Illinois J. Math., **34** (1990), 706–730.

[33] A. Haefliger, *Foliations and compactly generated pseudogroups* in Foliations: Geometry and Dynamics (Warsaw, 2000), World Scientific Publishing Co. Inc., River Edge, N.J., 2002, 275–295.

[34] B. Hasselblatt, *Regularity of the Anosov splitting and of horospheric foliations*, Ergodic Theory Dynam. Systems, **14** (1994), 645–666.

[35] A. Hatcher, *A proof of the Smale conjecture*, $\mathrm{Diff}(S^3) \simeq O(4)$, Ann. of Math. (2) , **117** (1983), 553–607.

[36] J. Heitsch and S. Hurder, *Secondary classes, Weil measures and the geometry of foliations*, J. Diff. Geom., **20** (1984), 291–309.

[37] S. Hurder and A. Katok, *Differentiability, rigidity and Godbillon-Vey classes for Anosov flows*, Inst. Hautes Études Sci. Publ. Math., **72** (1990), 5–61.

[38] S. Hurder, *Exceptional minimal sets of $C^{1+\alpha}$ actions on the circle*, Ergodic Theory Dynam. Systems, **11** (1991), 455-467.

[39] S. Hurder, *Dynamics and the Godbillon-Vey class: a History and Survey*, in Foliations: Geometry and Dynamics (Warsaw, 2000), World Sci. Publ., Singapore, 2002, 29–60.

[40] S. Hurder, *Foliation Geometry/Topology Problem set*, Geometry and Foliations 2003, Ryukoku University, Fukakusa, Kyoto, Japan. Available at http://www.foliations.org/surveys/FoliationProblems2003.pdf

[41] S. Hurder, *Exceptional minimal sets and the Godbillon-Vey class*, preprint, (2005).

[42] S. Hurder, *Problem set*, in Foliations 2005, World Sci. Publ., Singapore, 2006, 441–475.

[43] S. Hurder, *Classifying foliations*, in Foliations, Geometry and Topology. Paul Schweitzer Festschrift, (eds. Nicolau Saldanha et al), Contemp Math. Vol. 498, American Math. Soc., Providence, RI, 2009, 1–61.

[44] S. Hurder, *Lectures on Foliation Dynamics: Barcelona 2010*, Proceedings of Conference on Geometry and Topology of Foliations (C.R.M. 2010),

arXiv:1104.4852.

[45] T. Inaba, *Examples of exceptional minimal sets*, in A Fête of Topology, Academic Press, Boston, MA, 1988, 95–100.

[46] T. Inaba and S. Matsumoto, *Resilient leaves in transversely projective foliations*, Journal of Faculty of Science, University of Tokyo, **37** (1990), 89–101.

[47] R. Langevin, *A list of questions about foliations*, in Differential Topology, Foliations and Group Actions. Rio de Janeiro 1992, Contemp. Math. Vol. 161, Amer. Math. Soc., Providence, R.I., 1991, 59–80.

[48] R. Langevin, *Introduction to a few metric aspects of foliation theory*, TWMS J. Pure Appl. Math., **2** (2011), 74–96.

[49] R. Langevin and G. Levitt, *Courbure totale des feuilletages des surfaces*, Comment. Math. Helv., **57** (1982), 175–195.

[50] R. Langevin and C. Possani, *Total curvature of foliations*, Illinois J. Math., **37** (1993), 508–524.

[51] R. Langevin and P. Walczak, *Conformal geometry of foliations*, Geom. Dedicata, **132** (2008), 135–178.

[52] R. Langevin and P. Walczak, *Canal foliations of S^3*, J. Math. Soc. Japan, **64** (2012), 659–682.

[53] H.B. Lawson, *Codimension-one foliations of spheres*, Ann. of Math. (2) , **94** (1971), 494–503.

[54] D. Lehmann, *Sur l'approximation de certains feuilletages nilpotents par des fibrations*, C. R. Acad. Sci. Paris Sér. A-B, **286** (1978), A251–A254.

[55] E. Lindenstrauss, *Measurable distal and topological distal systems*, Ergodic Theory Dynam. Systems, **19** (1999), 1063–1076.

[56] X. Masa, E. Macias-Virgós and J. Álvarez López, *Open Problems*, in Analysis and geometry in foliated manifolds (Santiago de Compostela, 1994), World Sci. Publ., River Edge, NJ, 1995, 239–243.

[57] Y. Matsuda, *Groups of real analytic diffeomorphisms of the circle with a finite image under the rotation number function*, Ann. Inst. Fourier (Grenoble), **59** (2009), 1819–1845.

[58] S. Matsumoto, *Measure of exceptional minimal sets of codimension one foliations*, in A Fête of Topology, Academic Press, Boston, 1988, 81–94.

[59] L. Meersseman, *Feuilletages par variétés complexes et problèmes d'uniformisation*, preprint, (2011).

[60] G. Meigniez, *Regularization and minimization of Γ_1-structures*, arXiv:0904.2912v4, (2011).

[61] J. Milnor, *On the existence of a connection with curvature zero*, Comment. Math. Helv., **32** (1958), 215–223.

[62] Y. Mitsumatsu, *A relation between the topological invariance of the Godbillon-Vey invariant and the differentiability of Anosov foliations*, In Foliations (Tokyo, 1983), Advanced Studies in Pure Math., vol. 5, North-Holland, 1985, 159–167.

[63] Y. Mitsumatsu, *Anosov flows and non-Stein symplectic manifolds*, Ann. Inst. Fourier, Grenoble, **45** (1995), 1407–1421.

[64] Y. Mitsumatsu, *Leafwise symplectic structures on Lawson's foliations*, arXiv:1101.2319v3, (2011).

[65] P. Massot, K. Niederkrüger and C. Wendl, *Weak and strong fillability of higher dimensional contact manifolds*, arXiv:1111.6008, (2011).

[66] L. Meersseman and A. Verjovsky, *A smooth foliation of the 5-sphere by complex surfaces*, Ann. of Math. (2) , **156** (2002), 915–930. Corrigendum: Ann. of Math. (2), **174** (2011), 1951–1952.

[67] L. Meersseman and A. Verjovsky, *On the moduli space of certain smooth codimension-one foliations of the 5-sphere by complex surfaces*, J. Reine Angew. Math., **632** (2009), 143–202.

[68] P. Molino, *Étude des feuilletages transversalement complets et applications*, Ann. Sci. École Norm. Sup. (4), **10** (1977), 289–307.

[69] P. Molino, Riemannian foliations, Birkhäuser Boston Inc., Boston, MA, 1988.

[70] S. Morita, *Characteristic classes of surface bundles*, Invent. Math., **90** (1987), 551–577.

[71] M. Mostow and P.A. Schweitzer, *Foliation problem session*, in Algebraic and geometric topology (Proc. Sympos. Pure Math., Stanford Univ., Stanford, Calif., 1976), Part 2, Proc. Sympos. Pure Math., XXXII, Amer. Math. Soc., Providence, R.I., 1978, 269–271.

[72] Ya.B Pesin, Dimension theory in dynamical systems, Chicago Lectures in Mathematics, Contemporary views and applications, University of Chicago Press, Chicago, IL, 1997.

[73] B. Reinhart and J. Wood, *A metric formula for the Godbillon-Vay invariant for foliations*, Proc. Amer. Math. Soc., **38** (1973), 427–430.

[74] H. Rosenberg and R. Roussarie, *Les feuilles exceptionnelles ne sont pas exceptionnelles*, Comment. Math. Helv., **45** (1970), 517–523.

[75] V. Rovenski and L. Zelenko, *The mixed scalar curvature flow on a fiber bundle*, arXiv:1203.6361, (2012).

[76] V. Rovenski and P. Walczak, *Integral formulae on foliated symmetric spaces*, Math. Ann., **352** (2012), 223–237.

[77] H. Rummler, *Quelques notions simples en géométrie riemannienne et leurs applications aux feuilletages compacts*, Comment. Math. Helv., **54** (1979), 224–239.

[78] R. Sacksteder, *On the existence of exceptional leaves in foliations of co-dimension one*, Ann. Inst. Fourier, **14** (1964), 221–225.

[79] R. Sacksteder, *Foliations and pseudogroups*, Amer. J. Math., **87** (1965), 79–102.

[80] É. Salem, *Une généralisation du théorème de Myers-Steenrod aux pseudogroupes d'isométries*, Ann. Inst. Fourier, **38** (1988), 185–200.

[81] P.A. Schweitzer, *Some problems in foliation theory and related areas*, in Differential topology, foliations and Gelfand-Fuks cohomology (Proc. Sympos., Pontifícia Univ. Católica, Rio de Janeiro, 1976), Lect. Notes in Math., **652**, Springer Verlag, Berlin, 1978, 240–252.

[82] V. Sergiescu and T. Tsuboi, *Acyclicity of the groups of homeomorphisms of the Menger compact spaces*, Amer. J. Math., **118** (1996), 1299–1312.

[83] F. Takens and E. Verbitski, *Multifractal analysis of dimensions and entropies*, Regul. Chaotic Dyn., **5** (2000), 361–382.

[84] W.P. Thurston, Foliations of Three-Manifolds Which Are Circle Bundles,

Thesis (Ph.D.) – University of California, Berkeley, 1972.

[85] W.P. Thurston, *Noncobordant foliations of S^3*, Bull. Amer. Math. Soc., **78** (1972), 511–514.

[86] D. Tischler, *On fibering certain foliated manifolds over S^1*, Topology, **9** (1970), 153–154.

[87] T. Tsuboi, *On the uniform simplicity of diffeomorphism groups*, in Differential geometry, World Sci. Publ., Singapore, 2009, 43–55.

[88] T. Tsuboi, *On the uniform perfectness of the groups of diffeomorphisms of even-dimensional manifolds*, Comment. Math. Helv., **87** (2012), 141–185.

[89] T. Tsuboi, *Homeomorphism groups of commutator width one*, Proc. Amer. Math. Soc., to appear.

[90] P. Walczak, *Dynamics of the geodesic flow of a foliation*, Ergodic Theory Dynam. Systems, **8** (1988), 637–650.

[91] P. Walczak, *An integral formula for a Riemannian manifold with two orthogonal complementary distributions*, Colloq. Math., **58** (1990), 243–252.

[92] P. Walczak, *Foliations invariant under the mean curvature flow*, Illinois J. Math., **37** (1993), 609–623.

[93] P. Walczak, Dynamics of foliations, groups and pseudogroups, Monografie Matematyczne, **64**, Birkhäuser Verlag, 2004.

Received January 22, 2013.

List of Participants

1. ABE KŌJUN, Shinshu University, Japan,
   e-mail: kojnabe@shinshu-u.ac.jp
2. ÁLVAREZ LÓPEZ JESÚS A., Universidade de Santiago de Compostela,
   Spain, e-mail: jesus.alvarez@usc.es
3. ANDRZEJEWSKI KRZYSZTOF, Uniwersytet Łódzki, Poland,
   e-mail: kandrzejwski@tlen.pl
4. ASAOKA MASAYUKI, Kyoto University, Japan,
   e-mail: asaoka@math.kyoto-u.ac.jp
5. ASUKE TARO, University of Tokyo, Japan,
   e-mail: asuke@ms.u-tokyo.ac.jp
6. BADURA MAREK, Uniwersytet Łódzki, Poland,
   e-mail: marekbad@math.uni.lodz.pl
7. BARTOSZEK ADAM, Uniwersytet Łódzki, Poland,
   e-mail: mak@math.uni.lodz.pl
8. BIŚ ANDRZEJ, Uniwersytet Łódzki, Poland,
   e-mail: andbis@math.uni.lodz.pl
9. BOHNET DORIS, Department Mathematik, Universität Hamburg, Germany, e-mail: bohnet@math.uni-hamburg.de
10. BOWDEN JONATHAN, Max-Planck-Institut für Mathematik, Germany,
    e-mail: jonathan.bowden@math.uni-augsburg.de
11. CZARNECKI ANDRZEJ, Uniwersytet Jagielloński, Poland,
    e-mail: Andrzej.Czarnecki@im.uj.edu.pl
12. CZARNECKI MACIEJ, Uniwersytet Łódzki, Poland,
    e-mail: maczar@math.uni.lodz.pl
13. DOLGONOSOVA ANNA, Nizhny Novgorod State University, Russia,
    e-mail: dolgonosova@rambler.ru
14. EYNARD-BONTEMPS HÉLÈNE, Université Pierre et Marie Curie,
    France, e-mail: heynardb@math.jussieu.fr
15. FUKUI KAZUHIKO, Kyoto Sangyo University, Japan,
    e-mail: fukui@cc.kyoto-su.ac.jp
16. HECTOR GILBERT, Université Claude Bernard Lyon 1, France,
    e-mail: gilberthector@orange.fr

17. HURDER STEVEN, University of Illinois at Chicago, United States, e-mail: hurder@uic.edu

18. ISHIDA TOMOHIKO, University of Tokyo, Japan, e-mail: ishidat@ms.u-tokyo.ac.jp

19. KASUYA NAOHIKO, University of Tokyo, Japan, e-mail: nkasuya@ms.u-tokyo.ac.jp

20. KLEPTSYN VICTOR, Institut de Recherche Mathématique de Rennes, France, e-mail: victor.kleptsyn@univ-rennes1.fr

21. KODAMA HIROKI, University of Tokyo, Japan, e-mail: kodama@ms.u-tokyo.ac.jp

22. KORDYUKOV YURI, Russian Academy of Science, Russia, e-mail: ykordyukov@yahoo.com

23. KOWALIK AGNIESZKA, AGH University of Science and Technology, Poland, e-mail: kowalik@wms.mat.agh.edu.pl

24. KOZŁOWSKI WOJCIECH, Uniwersytet Łódzki, Poland, e-mail: wojciech@math.uni.lodz.pl

25. KUBARSKI JAN , Politechnika Łódzka, Poland

26. KUREŠ MIROSLAV, Brno University of Technology, Czech Republic, e-mail: kures@fme.vutbr.cz

27. LANGEVIN RÉMI, Université de Bourgogne, France, e-mail: langevin@u-bourgogne.fr

28. LUŻYŃCZYK MAGDALENA, Uniwersytet Łódzki, Poland, e-mail: luzynczyk@math.uni.lodz.pl

29. MATSUDA YOSHIFUMI, University of Tokyo, Japan, e-mail: ymatsuda@ms.u-tokyo.ac.jp

30. MATSUMOTO SHIGENORI, Nihon University, Japan, e-mail: matsumo@math.cst.nihon-u.ac.jp

31. MEIGNIEZ GAËL, Université de Bretagne Sud, France, e-mail: Gael.Meigniez@univ-ubs.fr

32. MENIÑO CARLOS, Universidade Santiago de Compostela, Spain, e-mail: carlos.meninho@gmail.com

33. MICHALIK ILONA, AGH University of Science and Technology, Poland, e-mail: imichali@wms.mat.agh.edu.pl

34. MIRANDA EVA, Universitat Politècnica de Catalunya, Spain, e-mail: eva.miranda@upc.edu

35. MITSUMATSU YOSHIHIKO, Chuo University, Japan, e-mail: yoshi@math.chuo-u.ac.jp

36. MOREIRA GALICIA MANUEL F., Universidade de Santiago de Compostela, Spain, e-mail: morgal2002@gmail.com

37. MOZGAWA WITOLD, Uniwersytet Marii Curie-Skłodowskiej, Poland, e-mail: mozgawa@hektor.umcs.lublin.pl
38. NAKAE YASUHARU, Akita University, Japan, e-mail: nakae@math.akita-u.ac.jp
39. NAKAYAMA HIROMICHI, Aoyama Gakuin University, Japan, e-mail: nakayama@gem.aoyama.ac.jp
40. NGUIFO BOYOM MICHEL, Universite Montpellier, France, e-mail: boyom@math.univ-montp2.fr
41. NIEDZIAŁOMSKI KAMIL, Uniwersytet Łódzki, Poland, e-mail: kamiln@math.uni.lodz.pl
42. NOZAWA HIRAKU, European Post-Doctoral Institute for Mathematical Sciences, France, e-mail: nozawahiraku@06.alumni.u-tokyo.ac.jp
43. PABINIAK MILENA, Cornell University, USA, e-mail: milenapabiniak@gmail.com
44. REBELO JULIO, Institut de Mathematiques de Toulouse, France, e-mail: rebelo@math.univ-toulouse.fr
45. RECHTMAN ANA, Université de Strasbourg, France, e-mail: rechtman@math.unistra.fr
46. ROGOWSKI JACEK, Politechnika Łódzka, Poland, e-mail: jacekrog@p.lodz.pl
47. ROVENSKI VLADIMIR, University of Haifa, Israel, e-mail: rovenski@math.haifa.ac.il
48. ROY INDRAVA, Mathematisches Forschungsinstitut Oberwolfach, Germany, e-mail: indrava@gmail.com
49. RYBICKI TOMASZ, AGH University of Science and Technology, Poland, e-mail: tomasz@agh.edu.pl
50. SCHWEITZER S.J. PAUL, PUC-Rio de Janeiro, Brazil, e-mail: paul37sj@gmail.com
51. SKRZYPIEC MAGDALENA, Uniwersytet Marii Curie-Skłodowskiej, Poland, e-mail: mskrzypiec@hektor.umcs.lublin.pl
52. SLESAR VLADIMIR, University of Craiova, Romania, e-mail: vlslesar@yahoo.com
53. TSUBOI TAKASHI, University of Tokyo, Japan, e-mail: tsuboi@ms.u-tokyo.ac.jp
54. VOGEL THOMAS, Max-Planck-Institut für Mathematik, Germany, e-mail: tvogel@mpim-bonn.mpg.de
55. WALCZAK PAWEŁ, Uniwersytet Łódzki, Poland, e-mail: pawelwal@math.uni.lodz.pl
56. WALCZAK SZYMON, Uniwersytet Łódzki, Poland,

e-mail: szymon.walczak@math.uni.lodz.pl

57. WALCZAK ZOFIA, Uniwersytet Łódzki, Poland,
    e-mail: zofia.walczak@math.uni.lodz.pl
58. WOLAK ROBERT, Uniwersytet Jagielloński, Poland,
    e-mail: Robert.Wolak@im.uj.edu.pl
59. ZAWADZKI TOMASZ, Uniwersytet Łódzki, Poland,
    e-mail: zawadzki@math.uni.lodz.pl
60. ZHUKOVA NINA, Nizhny Novgorod State University, Russia,
    e-mail: nina.i.zhukova@yandex.ru

Program

Y. MATSUDA, *Limits of geometrically finite convergence actions of a group*

M. LUŻYŃCZYK, *New integral formula on Riemannian manifolds with a pair of distributions*

A. ZHIROV, *How many different pseudo-Anosov homeomorphims may have the same invariant foliations*

WEDNESDAY, JUNE 27

A. RECHTMAN, *The minimal set of Kuperberg's plug*

T. ASUKE, *On independent rigid classes in $H^*(WU_q)$, III*

R. LANGEVIN, *Foliations of $\mathbb{S}^3$ by Dupin cyclides singular along a curve*

K. ABE, *Uniform perfectness of diffeomorphism groups and its application*

THURSDAY, JUNE 28

M. ASAOKA, *Rigidity of certain solvable actions on the sphere*

G. HECTOR, *Topological Canal Foliations*

V. ROVENSKI, *Deforming metrics of foliations*

A. BIŚ, *An analogue of the Variational Principle for group and pseudogroup actions*

V. SLESAR, *Spectral properties of Dirac-type operators defined on foliated manifolds*

H. NOZAWA, *Finiteness of characteristic classes of transversely homogeneous foliations*

I. ROY, *Rho-invariants for foliations and their stability properties*

H. KODAMA, *Minimal C^1-diffeomorphisms of the circle which admit measurable fundamental domains*

C. MENIÑO COTÓN, *Dynamical Lyusternik-Schnirelmann category of foliations*

FRIDAY, JUNE 29

H. NAKAYAMA, *Smooth embedding of the minimal homeomorhism of Gottschalk and Hedlund*

Y. MITSUMATSU, *Leafwise symplectic structures on codimension one foliations on the 5-sphere*

H. EYNARD-BONTEMPS, *Connectedness of the space of smooth Z^2 actions on the interval*

M. PABINIAK, *Gelfand-Tsetlin system of action coordinates and Gromov width of coadjoint orbits*

Y.A. KORDYUKOV, *Transverse Hamiltonian dynamics on foliated manifolds*

E. MIRANDA, *Rigidity for Hamiltonian actions on Poisson manifolds and a Nash-Moser abstract normal form theorem*

M. KUREŠ, *On natural affinor fields induced by minimal ideals of a Weil algebra*

A. CZARNECKI, *Homotopy approach to basic cohomology*

K. NIEDZIAŁOMSKI, *Harmonic distributions and conformal deformations*

SATURDAY, JUNE 30

Problem session
Closing

Author Index